Plateau Lake Water Quality and Eutrophication: Status and Challenges

Plateau Lake Water Quality and Eutrophication: Status and Challenges

Editors

Hucai Zhang
Jingan Chen
G.D. Haffner

MDPI • Basel • Beijing • Wuhan • Barcelona • Belgrade • Manchester • Tokyo • Cluj • Tianjin

Editors

Hucai Zhang
Institute for Ecological
Research and Pollution
Control of Plateau Lakes
Yunnan University
Kunming
China

Jingan Chen
Institute of Geochemistry
Chinese Academy of Sciences
Guiyang
China

G.D. Haffner
Great Lakes Institute for
Environmental Research
(GLIER)
University Windsor
Windsor Ontario
Canada

Editorial Office
MDPI
St. Alban-Anlage 66
4052 Basel, Switzerland

This is a reprint of articles from the Special Issue published online in the open access journal *Water* (ISSN 2073-4441) (available at: www.mdpi.com/journal/water/special_issues/Plateau_Lake_Eutrophication).

For citation purposes, cite each article independently as indicated on the article page online and as indicated below:

LastName, A.A.; LastName, B.B.; LastName, C.C. Article Title. *Journal Name* **Year**, *Volume Number*, Page Range.

ISBN 978-3-0365-6465-4 (Hbk)
ISBN 978-3-0365-6464-7 (PDF)

Cover image courtesy of Hucai Zhang

Contents

About the Editors

Hucai Zhang

Dr. Hucai Zhang is a Distinguished Professor at Yunnan University in Kunming, China and the Academician of the Russian Academy of Natural Sciences (RAEN). Professor Zhang has worked on Late Pleistocene paleolake evolution and environmental changes in arid western China and eastern Sahara Desert, the relationship between environmental changes and faunas (human) in the Middle-Late Pleistocene in Northeast China, limnology and its implications for changes in the Asian monsoon and regional ecosystems on the Qinghai-Tibet and Yunnan-Guizhou Plateaus, changes in the southern African monsoon regime and its relationship to human evolution, and heavy metal pollution and eutrophication in plateau lakes.

Professor Zhang participated in the DFG Africa Sahara Special Research Project SFB69 (1992–1993) and the International Ocean Discovery Project (IODP) Expedition 361—Southern African Climates. He has undertaken Key Programs, Key International Cooperation Projects, General Programs of the National Natural Science Foundation of China, the Leading Scientist Plan of Yunnan Province and Senior Talent Projects. He has published more than 300 research papers in internationally recognized journals including Cell, Nature Communication, Science, Radiocarbon, Palaeogeography Palaeoclimatology Palaeoecology, Quaternary Research, the Journal of Geophysical Research, Science China, Science Bulletin and Quaternary Sciences among others. His books include Superficial Elemental Geochemistry and Theoretical Principles (Lanzhou University Press, 1997).

Jingan Chen

Prof. Dr. Jingan Chen is a professor of the Institute of Geochemistry, Chinese Academy of Sciences. He has engaged in water environment and security, lake eutrophication, sediment pollution and control, biogeochemical cycles of nutrients and isotope geochemistry. In recent years, the main work has focused on lake eutrophication, sediment pollution and control, biogeochemical cycles of nutrients and construction of dynamic models of pollutants in deep and shallow waters, including lakes and reservoirs in order to establish a comprehensive understanding of the migration dynamics of pollutants in water bodies. He has published more than 200 scientific research articles.

G.D. Haffner

Prof. Douglas G. Haffner is a professor of the University of Windsor, Canada Research Chair, founder of the Great Lakes Institute for Environmental Research (GLIER). He has engaged in water pollutant detection, chemical impact research. In recent years, the main work has focused on the construction of dynamic models of pollutants in deep and shallow waters in order to establish a comprehensive understanding of the migration dynamics of pollutants in water bodies. He has published more than 150 scientific research articles. He has worked with the scientific researchers from the United States, New Zealand, Indonesia, Caribbean island countries, and China for many years.

Editorial

Plateau Lake Water Quality and Eutrophication: Status and Challenges

Hucai Zhang [1,*] , Jingan Chen [2] and Douglas G. Haffner [3]

[1] Institute for Ecological Research and Pollution Control of Plateau Lakes, Yunnan University, Kunming 650500, China
[2] Institute of Geochemistry, Chinese Academy of Sciences, Guiyang 550081, China
[3] Great Lakes Institute for Environmental Research (GLIER), University of Windsor, Windsor, ON N9B 3P4, Canada
* Correspondence: zhanghc@ynu.edu.cn

1. Introduction

The continuous and widespread deterioration of lake water quality and eutrophication is not only a local problem, but also a global phenomenon. It is not only destroying, or at least limiting the valuable water resources for daily life, but threatening the water security for sustainable social development. More importantly, along with the rapid accumulation of the alga, including novel hypertoxic viruses and new toxic chemicals as well as other organic compounds, these stressors threaten aquatic life, biodiversity and endanger our health [1]. The scientific community, especially environmental scientists and ecologists, needs to pay special attention to this issue and alert the public of the potential ecological and human health effects. The unknown consequences of utilizing trans-watershed or long-distance water diversion to dilute highly polluted lake water as a solution to eutrophication and contamination, an approach often preferred by engineers and hydrologists, must be avoided to provide sustainable water resources.

The costs necessary to control lake pollution and eutrophication are high, not to mention the invisible influences on ecosystem productivity and potential persistent threat to public health. There is a need for watershed management approaches where remedial build on each other to restore and protect water quality in an efficient manner.

The so-called Nine Large Lakes (>30 km^2) in Yunnan Province of southwestern China have experienced dramatic changes, seven of the nine lakes have become heavily polluted during last few decades [1]. These closed and/or semi-closed inland lakes are strongly influenced by the monsoonal climate, which results in a rainy season from mid-May to October and a dry season from November to early May, with a south-southwestern-dominated wind. At the same time, the elevated lake surfaces are exposed to strong ultraviolet radiation. The distinct seasonal contrast leads to large seasonal lake-level fluctuations, influencing the function and structure of the ecological system. These climato-environmental scenarios are what make the lakes in the Yunnan Plateau different to and more sensitive than lakes in other locations. This situation implies that we need to adopt different approaches to manage water pollution and eutrophication, instead of replicating management paradigms from other systems. Understanding the status, evolutionary processes, and mechanisms of lake water systems, as well as predicting future trends, are critical for us to address current and future problems and challenges.

2. Articles

In total, fifteen papers were published in this Special Issue. The article titles, authors and keywords are summarized in Table 1.

Citation: Zhang, H.; Chen, J.; Haffner, D.G. Plateau Lake Water Quality and Eutrophication: Status and Challenges. *Water* **2023**, *15*, 337. https://doi.org/10.3390/w15020337

Received: 28 December 2022
Revised: 4 January 2023
Accepted: 4 January 2023
Published: 13 January 2023

Table 1. Summary of the papers published in the Special Issue entitled "Plateau Lake Water Quality and Eutrophication: Status and Challenges" for the journal *Water*.

Title	Authors	Keywords
Lake Management and Eutrophication Mitigation: Coming down to Earth-In Situ Monitoring, Scientific Management and Well-Organized Collaboration Are Still Crucial	Hucai Zhang	None
Seasonal Stratification Characteristics of Vertical Profiles and Water Quality of Lake Lugu in Southwest China	Fengqin Chang Pengfei Hou Xinyu Wen Lizeng Duan Yang Zhang Hucai Zhang	Lake Lugu Yunnan thermal stratification seasonal changes water quality
Seasonal Variation and Spatial Heterogeneity of Water Quality Parameters in Lake Chenghai in Southwestern China	Pengfei Hou Fengqin Chang Lizeng Duan Yang Zhang Hucai Zhang	Lake Chenghai Yunnan water quality parameters seasonality spatial heterogeneity
Spatiotemporal Changes in Water Quality Parameters and the Eutrophication in Lake Erhai of Southwest China	Kun Chen Lizeng Duan Qi Liu Yang Zhang Xiaonan Zhang Fengwen Liu Hucai Zhang	Lake Erhai Temperature Chl-a seasonal changes eutrophication
Effects of Seasonal Variation on Water Quality Parameters and Eutrophication in Lake Yangzong	Weidong Xu Lizeng Duan Xinyu Wen Huayong Li Donglin Li Yang Zhang Hucai Zhang	Lake Yangzong water quality parameters temporal and spatial variations cyanophyte relative quantity index nutrient reduction
Seasonal Variation in the Water Quality and Eutrophication of Lake Xingyun in Southwestern China	Yanbo Zeng Fengqin Chang Xinyu Wen Lizeng Duan Yang Zhang Qi Liu Hucai Zhang	Lake Xingyun water quality spatial variation temporal variation
Seasonal Variations in Water Quality and Algal Blooming in Hypereutrophic Lake Qilu of Southwestern China	Donglin Li Fengqin Chang Xinyu Wen Lizeng Duan Hucai Zhang	Lake Qilu seasonal variation water temperature dissolved oxygen chlorophyll-a pH turbidity
Seasonal Water Quality Changes and the Eutrophication of Lake Yilong in Southwest China	Qingyu Sui Lizeng Duan Yang Zhang Xiaonan Zhang Qi Liu Hucai Zhang	water quality Lake Yilong spatial-temporal variations anthropogenic activities

Table 1. *Cont.*

Title	Authors	Keywords
Release of Endogenous Nutrients Drives the Transformation of Nitrogen and Phosphorous in the Shallow Plateau of Lake Jian in Southwestern China	Yang Zhang Fengqin Chang Xiaonan Zhang Donglin Li Qi Liu Fengwen Liu Hucai Zhang	plateau lake eutrophication nutrient limitation transformation
Nutrient Thresholds Required to Control Eutrophication: Does It Work for Natural Alkaline Lakes?	Jing Qi Le Deng Yongjun Song Weixiao Qi Chengzhi Hu	nutrient threshold alkaline lake pH phytoplankton blooms
Synergistic Effects and Ecological Responses of Combined In Situ Passivation and Macrophytes toward the Water Quality of a Macrophytes-Dominated Eutrophic Lake	Wei Yu Haiquan Yang Yongqiong Yang Jingan Chen Peng Liao Jingfu Wang Jiaxi Wu Yun He Dan Xu	La-modified material macrophyte Sediments Phosphorus eutrophication
Effect of Ecosystem Degradation on the Source of Particulate Organic Matter in a Karst Lake: A Case Study of the Caohai Lake, China	Jiaxi Wu Haiquan Yang Wei Yu Chao Yin Yun He Zheng Zhang Dan Xu Qingguang Li Jingan Chen	particulate organic matter (POM) carbon and nitrogen stable isotopes source tracing ecosystem degradation Caohai Lake
Influence of Cascade Hydropower Development on Water Quality in the Middle Jinsha River on the Upper Reach of the Yangtze River	Tianbao Xu Fengqin Chang Xiaorong He Qingrui Yang Wei Ma	middle reach of the Jinsha River cascade hydropower development water quality regression discontinuity analysis
Spatial and Temporal Distribution Characteristics of Nutrient Elements and Heavy Metals in Surface Water of Tibet, China and Their Pollution Assessment	Jiarui Hong Jing Zhang Yongyu Song Xin Cao	nutrient elements heavy metal elements spatiotemporal characteristics entropy method-fuzzy evaluation method principal component analysis
Tributary Loadings and Their Impacts on Water Quality of Lake Xingyun, a Plateau Lake in Southwest China	Liancong Luo Hucai Zhang Chunliang Luo Chrisopher McBridge Kohji Muraoka Hong Zhou Changding Hou Fenglong Liu Huiyun Li	external loading internal loading water quality tributary Lake Xingyun

In these papers, the seasonal variation in the water quality and eutrophication of Lake Lugu [2], Lake Chenghai [3], Lake Erhai [4], Lake Yangzong [5], Lake Xingyun [6], Lake Qilu [7], Lake Yilong [8], and a small water body (Lake Jian) [9] in Yunnan Plateau of Southwestern China was examined using current monitoring and research data. As all these lakes are heavily polluted, the situation is such that even if there might exist a nutrient "threshold" to control the eutrophication in one lake [10], it is difficult to identify a common management approach for all lakes. Fundamentally, many lake processes are almost "dead". Restoring the ecological functioning of the lake should be the priority before, or at least in parallel with, other engineering management approaches.

The quantifying the dynamics of phosphates in lake systems is a crucial first step in controlling eutrophication. La-modified materials and La/Al-co-modified attapulgite along with macrophytes, *Hydrilla verticillata royle* and *Ceratophyllum demersum* L., were investigated and tested as an approach to modify P dynamics. The results indicate that mineralization of organophosphates is an important factor for regulating high internal P loadings and P concentrations. The combination of LMM and macrophytes led to synergistic effects in the efficiency of aquatic ecological restoration compared with individual treatments. It was also concluded that LMM enhanced the conversion rates of redox-sensitive P forms in surface sediments [11].

Organic matter is is a keey component of lakes, and the origins or organic matter vary from one lake to another. The temporal and spatial distributions of particulate organic matter (POM) prior to and after ecosystem degradation in the karst lake (Caohai Lake) were analyzed using a combination of carbon and nitrogen stable isotopes (δ^{13}C–δ^{15}N). This study revealed that environmental factors, including DO, turbidity, water depth, and water temperature, that regulate both photosynthesis and sediment resuspension, are key factors determining the spatiotemporal distribution of POM. Meanwhile, the POM in water is closely related to the dissolved oxygen concentrations and pH, such that decreased dissolved oxygen (DO) concentrations and pH values resulted in an increase of POM [12].

The impact of cascade hydro-power development on water quality was investigated at six hydropower stations that have been in joint operation for seven years along the main course of the middle reach of the Jinsha River in Yunnan and Sichuan Provinces. This study reveals that cascade hydropower development resulted in a decrease in TP concentrations but an increase in the concentration of CODMn and NH_3-N along that section of the river. The concentrations of CODMn and TP are higher during the rainy season and lower in the dry season, which is directly related to the input of non-point-source pollutants in the basin during the period of high surface runoff [13].

To understand the pollution status of the surface water of the Tibet Plateau, the spatial and temporal variation of nutrients, heavy metals with respect to, water quality conditions and pollutant sources were studied in surface water from 41 cross-sectional monitoring sites in 2021. The results revealed that 12 polluting elements, except lead (Pb), had significant seasonal variation. In general, the water quality in most parts of Tibet was observed to be good. The water quality of the 41 monitoring sections met the Class I water standard as per the entropy –fuzzy evaluation method [14].

Models have also been applied to better understand the effects of external nutrient loading impacts on the water quality of the lakes. It was found that the annual inputs of total nitrogen (TN) had higher variability than total phosphorus (TP) in Lake Xingyun, and the highest loadings were during the wet season and the lowest during the dry season. The poor correlation between in-lake nutrient concentrations and tributary nutrient inputs at monthly and annual time scales suggests that both external and internal loadings were regulating lake eutrophication [15].

3. Conclusions

This Special Issue highlights and discusses major threats to Plateau Lakes water quality, and provides an update on both lake current status as well as future challenges. Lake problems, such as pollution and eutrophication, require that we quantify how serious the situation is, identify the probable causes, and recommend how to control the pollution in order to restore and protect water quality. More importantly, this issues stresses the need to manage the lakes and the watershed under a unified approach.

It is important to point out that there are still many other aspects of plateau lake water quality and eutrophication that require further research and monitoring. Although water pollution and lake eutrophication conditions have recently been protected from further deterioration, this is not due to actions we have taken; instead, it is the result of reduced human disturbance of the plateau lake water and the discharge of nutrients into the lake, which is attributable to the COVID-19 pandemic as well as the abnormal climatic conditions during the last three years. According to our monitoring data, the nutrient concentrations in the plateau lake water are still higher than acceptable values in other lakes. Nevertheless, the fundamental problems still need to be addressed, and therefore, it is not the time to take pride and become complacent. The war continues, even though the battle fields are almost empty. We still have a long way to go to regain the beauty of harmony in mountain–river–lake–water–plants scenery.

Funding: This research received no external funding.

Institutional Review Board Statement: Not applicable.

Informed Consent Statement: Not applicable.

Data Availability Statement: Not applicable.

Acknowledgments: We express our sincere thanks for all contributions to this Special Issue, the time invested by each author, as well as the anonymous reviewers who contributed to the development and improvement of the articles. We greatly appreciate the efficient and accountable review processes and management of this Special Issue.

Conflicts of Interest: The author declares no conflict of interest.

References

1. Zhang, H. Lake Management and Eutrophication Mitigation: Coming down to Earth-In Situ Monitoring, Scientific Management and Well-Organized Collaboration Are Still Crucial. *Water* **2022**, *14*, 2878. [CrossRef]
2. Chang, F.; Hou, P.; Wen, X.; Duan, L.; Zhang, Y.; Zhang, H. Seasonal Stratification Characteristics of Vertical Profiles and Water Quality of Lake Lugu in Southwest China. *Water* **2022**, *14*, 2554. [CrossRef]
3. Hou, P.; Chang, F.; Duan, L.; Zhang, Y.; Zhang, H. Seasonal Variation and Spatial Heterogeneity of Water Quality Parameters in Lake Chenghai in Southwestern China. *Water* **2022**, *14*, 1640. [CrossRef]
4. Chen, K.; Duan, L.; Liu, Q.; Zhang, Y.; Zhang, X.; Liu, F.; Zhang, H. Spatiotemporal Changes in Water Quality Parameters and the Eutrophication in Lake Erhai of Southwest China. *Water* **2022**, *14*, 3398. [CrossRef]
5. Xu, W.; Duan, L.; Wen, X.; Li, H.; Li, D.; Zhang, Y.; Zhang, H. Effects of Seasonal Variation on Water Quality Parameters and Eutrophication in Lake Yangzong. *Water* **2022**, *14*, 2732. [CrossRef]
6. Zeng, Y.; Chang, F.; Wen, X.; Duan, L.; Zhang, Y.; Liu, Q.; Zhang, H. Seasonal Variation in the Water Quality and Eutrophication of Lake Xingyun in Southwestern China. *Water* **2022**, *14*, 3677. [CrossRef]
7. Li, D.; Chang, F.; Wen, X.; Duan, L.; Zhang, H. Seasonal Variations in Water Quality and Algal Blooming in Hypereutrophic Lake Qilu of Southwestern China. *Water* **2022**, *14*, 2611. [CrossRef]
8. Sui, Q.; Duan, L.; Zhang, Y.; Zhang, X.; Liu, Q.; Zhang, H. Seasonal Water Quality Changes and the Eutrophication of Lake Yilong in Southwest China. *Water* **2022**, *14*, 3385. [CrossRef]
9. Zhang, Y.; Chang, F.; Zhang, X.; Li, D.; Liu, Q.; Liu, F.; Zhang, H. Release of Endogenous Nutrients Drives the Transformation of Nitrogen and Phosphorous in the Shallow Plateau of Lake Jian in Southwestern China. *Water* **2022**, *14*, 2624. [CrossRef]
10. Qi, J.; Deng, L.; Song, Y.; Qi, W.; Hu, C. Nutrient Thresholds Required to Control Eutrophication: Does It Work for Natural Alkaline Lakes? *Water* **2022**, *14*, 2674. [CrossRef]
11. Yu, W.; Yang, H.; Yang, Y.; Chen, J.; Liao, P.; Wang, J.; Wu, J.; He, Y.; Xu, D. Synergistic Effects and Ecological Responses of Combined In Situ Passivation and Macrophytes toward the Water Quality of a Macrophytes-Dominated Eutrophic Lake. *Water* **2022**, *14*, 1847. [CrossRef]
12. Wu, J.; Yang, H.; Yu, W.; Yin, C.; He, Y.; Zhang, Z.; Xu, D.; Li, Q.; Chen, J. Effect of Ecosystem Degradation on the Source of Particulate Organic Matter in a Karst Lake: A Case Study of the Caohai Lake, China. *Water* **2022**, *14*, 1867. [CrossRef]
13. Xu, T.; Chang, F.; He, X.; Yang, Q.; Ma, W. Influence of Cascade Hydropower Development on Water Quality in the Middle Jinsha River on the Upper Reach of the Yangtze River. *Water* **2022**, *14*, 1943. [CrossRef]

14.   Hong, J.; Zhang, J.; Song, Y.; Cao, X. Spatial and Temporal Distribution Characteristics of Nutrient Elements and Heavy Metals in Surface Water of Tibet, China and Their Pollution Assessment. *Water* **2022**, *14*, 3664. [CrossRef]
15.   Luo, L.; Zhang, H.; Luo, C.; McBridge, C.; Muraoka, K.; Zhou, H.; Hou, C.; Liu, F.; Li, H. Tributary Loadings and Their Impacts on Water Quality of Lake Xingyun, a Plateau Lake in Southwest China. *Water* **2022**, *14*, 1281. [CrossRef]

Editorial

Lake Management and Eutrophication Mitigation: Coming down to Earth—In Situ Monitoring, Scientific Management and Well-Organized Collaboration Are Still Crucial

Hucai Zhang 

Institute for Ecological Research and Pollution Control of Plateau Lakes, School of Ecology and Environmental Science, Yunnan University, Kunming 650500, China; zhanghc@ynu.edu.cn

Citation: Zhang, H. Lake Management and Eutrophication Mitigation: Coming down to Earth—In Situ Monitoring, Scientific Management and Well-Organized Collaboration Are Still Crucial. *Water* **2022**, *14*, 2878. https://doi.org/10.3390/w14182878

Received: 18 July 2022
Accepted: 13 September 2022
Published: 15 September 2022

Publisher's Note: MDPI stays neutral with regard to jurisdictional claims in published maps and institutional affiliations.

Lakes, together with rivers and subterranean aquifers, are indispensable natural resources for humans and other organisms. Globally, there are more than 100 million lakes [1], holding 87% of Earth's liquid surface freshwater [2] and covering an area of 4.2×10^6 km^2, including water bodies smaller than 1 km^2 [3]. Lakes not only play a crucial role in water supply, food production, and climate regulation [4] but also function as a cornerstone for socio-economic development.

During the last century, anthropogenic climate changes, especially seasonal climate alternations, intensified widespread use of agricultural chemicals (e.g., fertilizers and pesticides), and rapidly increasing urbanization, have dramatically changed regional watershed and hydrological patterns, exerting excessive pressure on lacustrine ecosystems [5]. As both air and water temperatures are key controlling factors of lake thermal regimes [6] and ecosystem metabolism [7], rising air temperatures and persistent nutrient input have direct effects on the physical and ecological properties of lakes [8], often resulting in nuisance algal blooms worldwide.

Harmful algal blooms affect ecosystem productivity and public health globally [9], and the costs are high. For example, primarily as a result of harm to drinking water supplies, aquatic food production, and diminished tourism, economic losses of more than a billion dollars occur annually in the United States alone [10]. During the last few years, the equivalent of tens of billions of US dollars have been allocated by the Chinese government to mitigate eutrophication of lakes. In Yunnan Province in southwestern China, conservation and pollution control of the so-called Nine Large Lakes (>30 km^2) alone has cost more than RMB 1.16 billion (~USD 180 million) during the last decade, but the situation is still serious.

For many years, scientists have spared no effort to understand algal blooms and have struggled to find effective measures to mitigate their harmful effects. Two of the main foci of the United Nations' Sustainable Development Goals are a commitment to water resources (Goal #6) and the impacts of climate change (Goal #13). These concerns are also essential components of the United Nations Framework Convention on Climate Change (UNFCCC) and the Intergovernmental Panel on Climate Change (IPCC).

People have long realized that satellites might play an important role in the scientific study and operational management of hydrology and water systems [11]. Space-based remote sensing was expected to revolutionize the monitoring of algal blooms and the water quality of large lakes [12], but it has proven difficult to draw statistically accurate pictures from such data [13]. Before the development of advanced technical equipment and practical theories, we must first focus on understanding lacustrine eutrophication and algal blooming [14]. In situ monitoring and sustained analyses of various samples are crucial, not only with respect to adequately understanding lacustrine systems themselves, but also to provide valuable background and crosschecks to ensure reliable application of advanced technologies in the future.

Lakes themselves and their drainages involve many important systems and dynamic processes (Figure 1). Humans directly change both global and regional climate dynamics, catchment hydrological and transportation patterns and processes, lake eco-dynamics and deposition–evolution processes. Most importantly, serious disturbance of all these processes results in the shutoff of three critical interactions: depositional processes and geochemical and bio-geochemical interactions, which can lead to the deterioration or collapse of lakes' self-clarification ability and self-restoring capacity, effectively leaving them "dead."

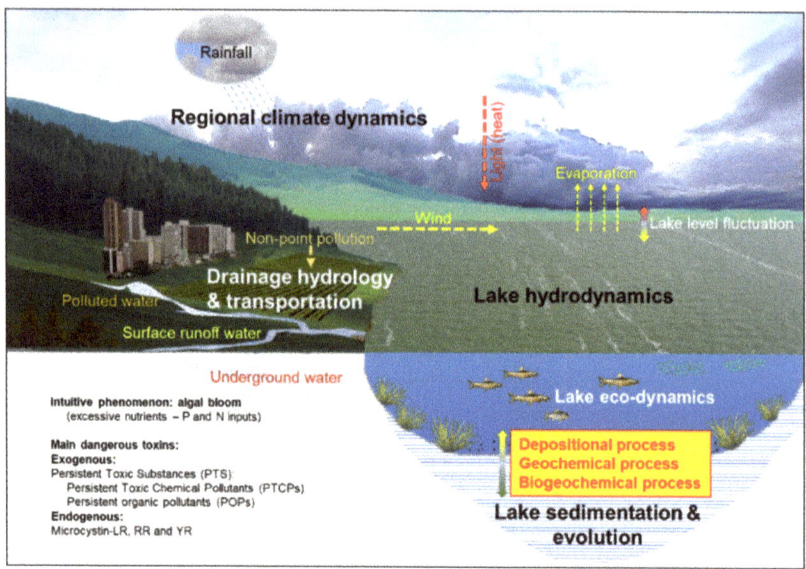

Figure 1. Watershed lake systems and main processes.

Strengthening anthropogenic environmental changes driving biodiversity loss decreases ecosystem stability [15]. Maintaining healthy biodiversity is crucial to stabilize ecosystem productivity [16–19], as greater biodiversity generally provides greater resistance to the extreme climate events [20].

Ongoing climate change is expected to accelerate hydrological cycles and thereby increase available renewable freshwater resources. However, changes in seasonal patterns and the increasing probability of extreme events may offset this effect [21]. This will inevitably induce fundamental variations in lake systems and their functions. In this expectation, we face the brutal reality that much more time and effort than expected are needed to restore polluted lakes to their health condition. In particular, (1) we must pay special attention and alert that the potential harmful effects and unrealized consequences of highly eutrophicated lake waters, e.g., novel hypertoxic viruses and new toxic chemical and organic compounds are overwhelming; (2) we should pay attention to the large long-distance trans-regional water drainage claimed to mitigate lake water pollution, as this process might result in abrupt changes in the established watershed ecosystems.

Lakes support a global heritage of biodiversity and supply key ecosystem resources. Securing a sustainable future for lakes ultimately lies in the scientific management of these treasured natural resources, and concerted efforts at the local governance through national and international levels. It is necessary to work from individual to regional clusters of lakes because the lake status varies depending upon the location, depth, area, agricultural and industrial intensity, and trophic status. "One alone is good," but only through close and coherent collaboration can we successfully address global challenges, pursuing common goals to maintain and protect lake health synergistically.

Funding: This research was funded by the Scientist workshop and the Scientist workshop and the Special Project for Social Development of Yunnan Province (Grant No. 202103AC100001).

Data Availability Statement: Data inquiries can be directed to the author.

Conflicts of Interest: The author declares no conflict of interest.

References

1. Verpoorter, C.; Kutser, T.; Seekell, D.A.; Tranvik, L.J. A global inventory of lakes based on high-resolution satellite imagery. *Geophys. Res. Lett.* **2014**, *41*, 6396–6402. [CrossRef]
2. Gleick, P.H. Water and conflict: Fresh water resources and international security. *Int. Secur.* **1993**, *18*, 79–112. [CrossRef]
3. Downing, J.A.; Prairie, Y.T.; Cole, J.J.; Duarte, C.M.; Tranvik, L.J.; Striegl, R.G.; Middelburg, J.J. The global abundance and size distribution of lakes, ponds, and impoundments. *Limnol. Oceanogr.* **2006**, *51*, 2388–2397. [CrossRef]
4. Tao, S.; Fang, J.; Ma, S.; Cai, Q.; Xiong, X.; Tian, D.; Zhao, S. Changes in China's lakes: Climate and human impacts. *Natl. Sci. Rev.* **2020**, *7*, 132–140. [CrossRef] [PubMed]
5. Carvalho, L.; McDonald, C.; de Hoyos, C.; Mischke, U.; Phillips, G.; Borics, G.; Cardoso, A.C. Sustaining recreational quality of European lakes: Minimizing the health risks from algal blooms through phosphorus control. *J. Appl. Ecol.* **2013**, *50*, 315–323. [CrossRef]
6. Livingstone, D.M. A change of climate provokes a change of paradigm: Taking leave of two tacit assumptions about physical lake forcing. *Int. Rev. Hydrobiol.* **2008**, *93*, 404–414. [CrossRef]
7. Yvon-Durocher, G.; Caffrey, J.M.; Cescatti, A.; Dossena, M.; Giorgio, P.D.; Gasol, J.M.; Allen, A.P. Reconciling the temperature dependence of respiration across timescales and ecosystem types. *Nature* **2012**, *487*, 472–476. [CrossRef] [PubMed]
8. Woolway, R.I.; Merchant, C.J. Worldwide alteration of lake mixing regimes in response to climate change. *Nat. Geosci.* **2019**, *12*, 271–276. [CrossRef]
9. Ndlela, L.L.; Oberholster, P.J.; Van Wyk, J.H.; Cheng, P.H. An overview of cyanobacterial bloom occurrences and research in Africa over the last decade. *Harmful Algae* **2016**, *60*, 11–26. [CrossRef] [PubMed]
10. Kudela, R.M.; Berdalet, E.; Bernard, S.; Burford, M.; Fernand, L.; Lu, S.; Urban, E. Harmful Algal Blooms. A Scientific Summary for Policy Makers (IOC/UNESCO, 2015). Available online: www.globalhab.info (accessed on 12 September 2022).
11. Famiglietti, J.S.; Cazenave, A.; Eicker, A.; Reager, J.T.; Rodell, M.; Velicogna, I. Satellites provide the big picture. *Science* **2015**, *349*, 684–685. [CrossRef] [PubMed]
12. Ho, J.C.; Michalak, A.M.; Pahlevan, N. Widespread global increase in intense lake phytoplankton blooms since the 1980s. *Nature* **2019**, *574*, 667–670. [CrossRef] [PubMed]
13. Feng, L.; Dai, Y.; Hou, X.; Xu, Y.; Liu, J.; Zheng, C. Concerns about phytoplankton bloom trends in global lakes. *Nature* **2021**, *590*, E35–E38. [CrossRef] [PubMed]
14. Taranu, Z.E.; Gregory-Eaves, I.; Leavitt, P.R.; Bunting, L.; Buchaca, T.; Catalan, J.; Vinebrooke, R.D. Acceleration of cyanobacterial dominance in north temperate–subarctic lakes during the Anthropocene. *Ecol. Lett.* **2015**, *18*, 375–384. [CrossRef] [PubMed]
15. Hautier, Y.; Tilman, D.; Isbell, F.; Seabloom, E.W.; Borer, E.T.; Reich, P.B. Plant ecology. Anthropogenic environmental changes affect ecosystem stability via biodiversity. *Science* **2015**, *348*, 336–340. [CrossRef] [PubMed]
16. Naeem, S.; Bunker, D.E.; Hector, A.; Loreau, M.; Perrings, C. *Biodiversity, Ecosystem Functioning and Human Wellbeing: An Ecological and Economic Perspective*; Naeem, S., Ed.; Oxford University Press: Oxford, UK, 2009; pp. 78–93.
17. Naeem, S.; Li, S. Biodiversity enhances ecosystem reliability. *Nature* **1997**, *390*, 507–509. [CrossRef]
18. Bai, Y.; Han, X.; Wu, J.; Chen, Z.; Li, L. Ecosystem stability and compensatory effects in the Inner Mongolia grassland. *Nature* **2004**, *431*, 181–184. [CrossRef] [PubMed]
19. Tilman, D.; Reich, P.B.; Knops, J.M.H. Biodiversity and ecosystem stability in a decade-long grassland experiment. *Nature* **2006**, *441*, 629–632. [CrossRef] [PubMed]
20. Isbell, F.; Craven, D.; Connolly, J.; Loreau, M.; Schmid, B.; Beierkuhnlein, C.; Eisenhauer, N. Biodiversity increases the resistance of ecosystem productivity to climate extremes. *Nature* **2015**, *526*, 574–577. [CrossRef] [PubMed]
21. Oki, T.; Kanae, S. Global Hydrological Cycles and World Water Resources. *Science* **2006**, *313*, 1068–1072. [PubMed]

Article

Seasonal Stratification Characteristics of Vertical Profiles and Water Quality of Lake Lugu in Southwest China

Fengqin Chang [1], Pengfei Hou [1,*], Xinyu Wen [2], Lizeng Duan [1], Yang Zhang [1] and Hucai Zhang [1,*]

[1] Institute for Ecological Research and Pollution Control of Plateau Lakes, School of Ecology and Environmental Science, Yunnan University, Kunming 650500, China
[2] School of Geography and Engineering of Land Resources, Yuxi Normal University, Yuxi 653100, China
* Correspondence: houyu8806@163.com (P.H.); zhanghc@ynu.edu.cn (H.Z.)

Abstract: According to the vertical section monitoring data of Lake Lugu water temperature (WT), electrical conductivity (EC), dissolved oxygen (DO), pH and chlorophyll-*a* (Chl-*a*) parameters in January (winter), April (spring), July (summer), and October (autumn) in 2015, the vertical stratification structure of WT and the null seasonality of water chemistry were analyzed. The relationship between the seasonal variation of WT stratification and the spatial and temporal distribution of EC, pH, DO and Chl-*a* was explored. The relationship between EC and WT was found for the epilimnion, thermocline and hypolimnion. The results of the study showed that: (1) The Lake Lugu water body shows obvious thermal stratification in spring, summer and autumn. In winter, the WT is close to isothermal condition in the vertical direction; in summer, the thermocline is located at 10–25 m water depth; while in autumn, the thermocline moves down to 20–30 m. (2) The Hypolimnion WT was maintained at 9.5 °C~10 °C, which is consistent with the annual mean temperature of Lake Lugu, indicating that the hypolimnion water column is stable and relatively constant, and reflects the annual mean temperature of the lake. The thermally stratified structure has some influence on the changes of EC, DO, pH and Chl-*a*, resulting in the obvious stratification of EC, DO and pH in the water body. (3) Especially in summer, when the temperature increased, the thermal stratification phenomenon was significant, and DO and pH peaked in thermocline, with a decreasing trend from the peak upward and downward, and the hypolimnion was in an anoxic state and the pH value was small. Although chlorophyll a remained low below thermocline and was not high overall, there was a sudden increase in the surface layer, which should be highly warned to prevent a large algal bloom or even a localized outbreak in Lake Lugu. (4) There is a simple linear function between EC and WT in both vertical section and Epilimnion, thermocline and hypolimnion, which proves that Lake Lugu is still influenced by natural climate and maintains natural water state, and is a typical warm single mixed type of lake. (5) It is suggested to strengthen water quality monitoring, grasp its change pattern and influence factors, and take scientific measures to prevent huge pressure on the closed ecological environment of Lake Lugu, and provide scientific basis for the protection of high-quality freshwater lakes in the plateau.

Keywords: Lake Lugu; Yunnan; thermal stratification; seasonal changes; water quality

Citation: Chang, F.; Hou, P.; Wen, X.; Duan, L.; Zhang, Y.; Zhang, H. Seasonal Stratification Characteristics of Vertical Profiles and Water Quality of Lake Lugu in Southwest China. *Water* **2022**, *14*, 2554. https://doi.org/10.3390/w14162554

Academic Editor: Danny D. Reible

Received: 18 July 2022
Accepted: 17 August 2022
Published: 19 August 2022

Publisher's Note: MDPI stays neutral with regard to jurisdictional claims in published maps and institutional affiliations.

1. Introduction

In the context of global change, the study of the mechanism of the effect of temperature increases and eutrophication on the thermal stratification of lakes and reservoirs and its ecological and environmental effects has become one of the most common problems in current international research [1,2]. Lake thermal stratification and thermal cycling are important factors governing various physicochemical processes (e.g., dissolved oxygen (DO) distribution, nutrient exchange, microbial activity [3], bottom sediment nutrient release, etc.) and kinetic phenomena such as upstream and downstream water mixing and convection, and are important indicators affecting lake biological production and ecosystem evolution [4,5]. In general, the vertical distribution of water temperature (WT) in lakes varies with the difference in depth of the lake. Compared with shallow lakes, deep-water lakes have large and long-lasting temperature gradients [6], and deep-water lakes are less affected by wind, have strong heat storage capacity, and have large vertical temperature differences, which make it easy to form stable stratification [7], so thermal stratification is a natural phenomenon that exists in deeper lakes (water depth > 10 m) [8]. The thermocline of deep-water lakes is like a blocking layer in which the vertical gradient of the physicochemical properties of the lake water is large, while the physicochemical properties of the lake water in the epilimnion and hypolimnion are more uniform, which is due to the existence of the thermocline that can effectively impede convection, turbulence and molecular exchange in the upper and lower water bodies, affecting the distribution of light and nutrients in the lake water column, thus influencing the vertical distribution of water chemistry parameters [9,10]. For deep-water lakes (including reservoirs) with great and continuous temperature differences, the vertical distribution and variation patterns of WT determine the vertical stratification and mixed exchange of chemical factors as well as biological factors (phytoplankton, animals, etc.), which in turn profoundly affect the lake ecosystem [11,12]. Therefore, to understand the significance of water chemistry parameters in deep-water lakes, it is necessary to conduct an in-depth study of seasonal thermal stratification in lakes [13].

With the frequent occurrence and increasing intensity of extreme climatic and meteorological events [14], a series of changes and responses will also occur in the climatic environment of the monsoon region, especially in the highly variable southwest (Indian) monsoon region, causing corresponding changes in lakes [15]. All these changes will lead to changes in the hydrological cycle and water resources, and will have a significant impact on our water resources not only in terms of quantity but also in terms of regional distribution [16–18]. This requires an understanding and knowledge of the basic characteristics of lakes. Several scholars have conducted in-depth and systematic studies on the seasonal stratification and water chemistry characteristics of many natural deep-water lakes, shallow lakes, and large artificial reservoirs [19,20]. These results indicate that parameters such as WT, DO, chlorophyll-*a* (Chl-*a*), pH, electrical conductivity (EC), cell density of cyanobacteria, and turbidity are prone to vertical seasonal stratification in summer, especially in deep-water lakes and reservoirs [21,22]; vertical changes in lake thermal stratification affect the upward and downward distribution of water chemistry parameters such as DO, pH, and Chl-*a*, and seasonal stratification of water chemistry parameters caused by changes in WT in a deep-water lake [23–25]. Lugu is a warm temperate semi-enclosed plateau deep-water lake, the lake water does not freeze throughout the year, and it is easy to form a stable thermal stratification phenomenon; and the influence of inflow and outflow on the heat balance of the lake is minimal, so it is the best choice to study the seasonal variation of lake water bodies. Zhao et al. [26] studied the stability, mixing depth, thermocline depth, and buoyancy frequency to determine the onset, development, and termination of seasonal temperature stratification. However, this study only sampled the northern part of the lake and did not discuss the relationship between conductivity and temperature difference. Wang et al. [27] analyzed the variation of DO concentration at the surface WT and bottom of Lake Lugu and concluded that Lake Lugu is a warm single mixed lake with higher DO in autumn and winter and the highest DO in the hyaline zone below the thermocline in summer and autumn. Chen et al. [28] used the DYRESM model to investigate the stratification and other thermodynamic conditions in Lake Lugu in southwest China. the understanding of the changes of water chemistry parameters such as lake temperature, Chl-*a*, and DO concentrations, and pH is not only of practical significance for lake eutrophication control and water quality protection, but also very important for local and even global climate change studies.

Lake Lugu, as an important deep-water lake in the southwest monsoon region, is relatively little understood by us, lacking detailed studies on its seasonal stratification of temperature and vertical variation of water chemistry parameters. Especially in recent years, more and more human traffic has not only changed the original production and living style of local residents, but also caused great pressure on the relatively closed ecological environment of Lake Lugu, which made the garbage around the lake piled up and sewage directly into the lake, polluting the local natural ecological environment. Based on this, the vertical stratification structure of WT and the spatial stratification of water chemistry were analyzed based on the vertical section monitoring data of Lake Lugu WT, EC, pH, DO, and Chl-*a* parameters in January (winter), April (spring), July (summer), and October (autumn) in 2015. The seasonal stratification characteristics and patterns of Lake Lugu WT were revealed. Epilimnion, thermocline, and hypolimnion as a function of EC and WT. The results of the study can help improve the health of lake ecosystems and provide a scientific basis for the conservation of high-quality freshwater lakes in the plateau.

2. Data and Methods

2.1. Background of Lake Lugu

Lake Lugu is located at the junction of two provinces, northwestern Yunnan Province and southwestern Sichuan Province, and is a plateau faulted solution trap lake (Figure 1). Its main fault structure system consists of one northwest–southeast and two east–west faults together [29]. The geographic coordinates of Lake Lugu are 27°41′ to 27°45′ N, 100°45′ to 100°51′ E. Lake Lugu is a natural freshwater lake belonging to the Jinsha River. It slightly trends from northwest to southeast, with a length of 9.5 km from north to south, a width of 5.2 km from east to west, and a coastline of about 44 km. According to data measured in 2005 [30], the lake is at an altitude of 2692.2 m, consists of an area of 57.7 km^2, has a maximum depth of 105.3 m, and an average depth of 38.4 m. The water storage capacity is 1.953 billion m^3. Its annual amount of water entering is 110 million m^3, maximum water transparency is 12–14 m, and the nutritional level of the water body is steadily maintained at a Class I water quality.

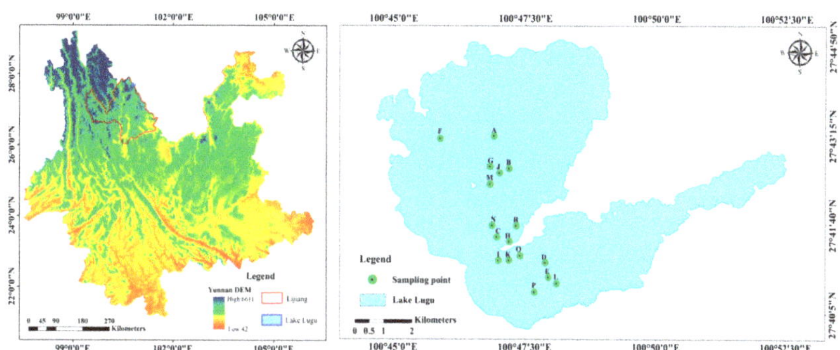

Figure 1. Distribution of monitoring and sampling sites in Lake Lugu.

Lake Lugu is located in the southwest monsoon climate zone, belongs to the subtropical highland monsoon climate zone, and is characterized by a warm, humid mountainous monsoon climate. It is controlled by dry continental winds in winter and humid Indian Ocean monsoon in summer, with distinct dry and wet seasons. The average annual rainfall is 730 to 830 mm, with 89% of the annual rainfall concentrated in the rainy season (June to October). Lake Lugu obtains its water supply primarily from precipitation, surface, and groundwater. The lake outlet is located on the eastern shore of the southern lake area, with the Caohai wetland being the only outlet. In the annual dry season (January to May), the lake has almost no outflow. The lake's water replacement duration is up to 18.9 years and it is a semi-closed lake [31,32].

2.2. Data Collection and Index Determination

Due to the large area of Lake Lugu and the influence of geological structure, the lake was divided into two parts: south and north. To improve the monitoring efficiency of the lake and assess the entire lake, monitoring sections were established in January, April, July, and October 2015 in the north and south regions of Lake Lugu (Figure 1). In January, the monitoring sections were marked as points A (100°46′28″ E~27°43′15″ N), B (100°46′33″ E~27°41′29″ N), C (100°46′33″ E~27°40′51″ N), D (100°46′47″ E~27°42′38″ N) and E (100°47′18″ E~27°41′00″ N); April as F (100°45′27.15″ E~27°43′13.34″ N), G (100°46′25.08″ E~27°42′42.39″ N), H (100°46′41.14″ E~27°40′57.72″ N) and I (100°46′40.71″ E ~27°41′20.48″ N); July as J (100°45′40.9″ E~27°43′7.2″ N), K (100°46′8.5″ E~26°42′43.4″ N), L (100°47′7.3″ E~27°41′0.8″ N) and R (100°48′6.3″ E~27°40′1.8″ N), and; M (100°46′24″ E~ 27°42′24″ N), N (100°46′28″ E~27°41′40″ N), O (100°46′57″ E~27°41′08″ N), P (100°47′13″ E~ 27°40′37″ N) in October. Sampling locations were established using the satellite-based Global Positioning System (GPS). Water quality parameters, including WT, DO, Chl-*a* concentration, pH value, and phycocyanin concentration, among others, were measured with a Xylem Analytics YSI-6600 multi-parameter sonde (Xylem, Milford, OH, USA). A vertical line was established at each site to monitor water quality at different depths. Data were first collected at 0.1–1 m below the water surface, and the last data were monitored 0.5 m above the lake bottom, with additional data collected at 1 m intervals. To ensure data accuracy, each depth was measured six times.

2.3. Analysizing Methods

2.3.1. Lake Quality Level

River water quality classification was based on the national quality standards (GB 3838-2002) [33]. According to the environmental functions and protection objectives of surface waters, the lake quality level was divided into the following five functional level categories (Table 1):

Table 1. Water function and standard classification.

Water Quality Classification	Scope of Application
Class I	Mainly applicable to source waters and national nature reserves.
Class II	Mainly applicable to centralized drinking water, surface water sources, first-class protected areas, etc.
Class III	Mainly applicable to secondary protection zones, fisheries, and swimming areas of centralized drinking water surface water sources.
Class IV	Mainly applicable to general industrial water use areas and recreational areas where the human body is not in direct contact with water.
Class V	Mainly applicable to agricultural water use areas and general landscape requirements.

2.3.2. Correlation Analysis Method

Pearson's correlation coefficient is a metric used to describe relationships among variables. This method uses the covariance matrix of data to evaluate the strength of the relationship between two vectors. Normally, the Pearson's correlation coefficient between two variables, β_i and β_j, can be calculated as shown in Equation, where $cov(\beta_i, \beta_j)$ is the covariance, $var(\beta_i)$ is the variance of β_i, and $var(\beta_j)$ is the variance of β_j [34].

$$R(\beta_i, \beta_j) = \frac{cov(\beta_i, \beta_j)}{\sqrt{var(\beta_i) \times var(\beta_j)}} \tag{1}$$

3. Results and Discussion

3.1. Subsection Vertical Stratification and Seasonal Temperature Fluctuations

Under normal circumstances, the WT of deep-water plateau lakes reacts sensitively to changes in seasonal temperature [35]. Thus, in summer, thermal stratification of the body of water results in changes in the lake's WT. The isothermal layer gradually decreases with increasing depth, resulting in a sharp drop in the thermocline. Similar to other deep-water lakes on the plateau (alpine), the WT of Lake Lugu has distinct stratification and mixing features in the vertical section during spring, summer, autumn and winter. The vertical distribution of WT in January (winter), April (spring), July (summer), and October (autumn) at each sampling site in Lugu Lake (Figure 2) clearly shows that the WT in Lake Lugu varies seasonally, with a significant temperature gradient of the water column in the vertical section in April, July, and October, and a gradual change in the depth in the thermocline. In April, July, and October, the temperature gradient in the vertical section of the water column were obvious, and the depth of the thermocline changed gradually with time, but in January, as the temperature decreased, the overall temperature of the water column also decreased, showing no temperature gradient in the vertical section.

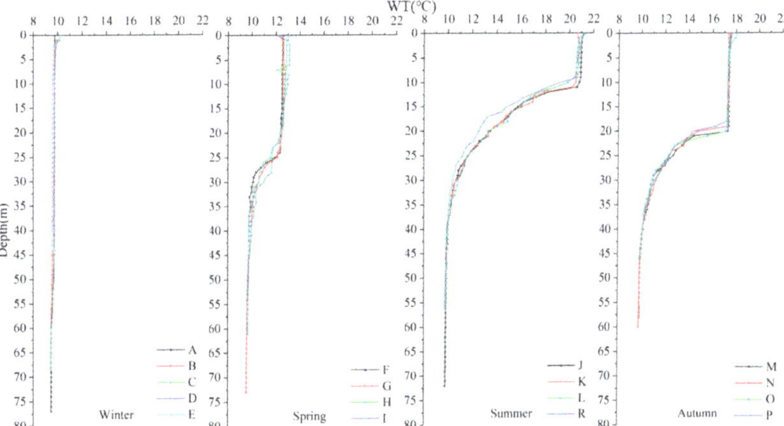

Figure 2. Seasonal variations of vertical water temperature (WT) section in Lake Lugu.

In April, with an increase in solar radiation and temperature, the surface water body rapidly warmed up, causing a gradual increase in temperature difference between the deep-water body and the appearance of the stratification phenomenon. The temperature of Lake Lugu reached its maximum in July, which made the external heat transfer from the surface layer to be continuously downward, decreased the maximum temperature difference, and shifted the depth of the thermocline upward, leading to an obvious temperature stratification phenomenon. According to the thermocline definition, which suggests a water layer having a WT gradient > 0.2 °C/m [36,37], Lake Lugu's WT in July could be divided into three layers in a vertical section: epilimnion (0–10 m), thermocline (11–25 m), hypolimnion (<25 m). As the temperature dropped in October, the depth of the thermocline decreased to a 20 m water level. In general, reservoirs with a depth of ≥7 m have a thermocline, which causes lakes' WT to drop by more than 1 °C with every 1 m drop in water depth. The distribution of WT in a layered lake reservoir, formation of thermocline or the up-and-down exchange of water has significant effects on the vertical distribution of DO, distribution of nutrients, and distribution of aquatic organisms, especially phytoplankton. With weak solar radiation and low temperature in January, the lake water bodies release latent heat, eventually forming a uniform temperature distribution above and below the water body without obvious stratification. During seasonal scale change, the position of the thermocline constantly shifted downward due to turbulence, convection, and the molecular diffusion of the lake water after its formation. Therefore, the position of the thermocline in Lake Lugu was about 15 m and 10 m lower in spring and autumn than in summer and disappeared completely in winter.

The WT in the surface layer of Lake Lugu was influenced by external climatic factors and varied significantly in July and October, ranging from 17.1 °C to 22 °C. However, the deep-water layer (less than 40 m) ranged between 9.5 °C to 10 °C, which was constant throughout the year. According to three weather stations near Lake Lugu, Yanyuan (27.27° N, 101.37° E; 2439.4 m), Zhongdian (27.50° N, 99.42° E; 3276.1 m), and Muli (28.08° N, 100.50° E; 2666.6 m), the annual average temperature of Lake Lugu is 10.3 °C. The mean temperature of the water from 1951 to 1980 was 12.6 °C, 5.4 °C, and 11.5 °C, respectively, and after correction, the annual average temperature layer of Lake Lugu and the surrounding lake area was basically consistent, indicating that the water in the mean temperature layer of Lake Lugu has been at a constant temperature for many years, reflecting the mean annual temperature of the lake area.

From a spatial point of view, the surface WT in the northern part of Lake Lugu was lower than in the southern part of the investigated year. In spring, summer, and autumn, there was thermal stratification in the water body north of Lake Lugu, especially in summer and autumn, and the WT had a variable temperature layer, thermocline layer, uniform temperature layer in the vertical section. In April, the thermocline appeared in the northern water body when the water layer was 25 m and re-appeared in the southern water body when the water layer was 21 m. In July, the thermocline of the north and south water bodies appeared at the water layer at 10 m, and in late October, the thermocline appeared in both the northern and southern water bodies at the 20 m water layer. The warming and cooling of the surface WT in the northern water column were slower than those in the south due to the difference in water depth or wind effects in the same lake, resulting in differences in the horizontal distribution of WT.

3.2. Vertical Variation Characteristics and Seasonal Dynamics of Hydrochemical Parameters
3.2.1. Electrical Conductivity (EC)

Electrical conductivity is a measure of the ability of a substance to carry an electrical current and is influenced by factors such as salinity, dissolved solids in water, temperature, and water supply. Since the EC of water is affected by WT, nutrients, and water supply, seasonal stratification led to evident EC changes (Figure 3), with the vertical distribution trend of water EC in Lake Lugu consistent with WT. Figure 3 shows a small variation in EC in January, maximum EC in July, and minimum EC in April. The EC of Lake Lugu changed significantly on the vertical section, first showing a decreasing trend and then remaining constant. The EC changed slightly in the thermotropic and hypolimnion layers but significantly changed in different months.

The trend was consistent with the vertical variation of WT in January, and there was no significant difference in conductivity except for a very slight increase in the water layer below 49 m. However, EC varied vertically in April, July, and October, resulting in obvious stratification. Among them, EC changed abruptly at a water layer of 25 m in April and stabilized below 31 m. In April, July, and October, EC varied vertically and was stratified. In April, EC changed abruptly at 25 m and stabilized below 31 m. In July, EC fluctuated from 10 to 25 m and stabilized below 25 m. In October, EC changed abruptly at 20 m and stabilized above this depth, decreased below 20 m to 30 m, and stabilized below 30 m. In October, the EC changed abruptly at 20 m and stabilized above this depth. According to the measurement results in 2015, the spatial difference in water conductivity in the northern and southern parts of Lake Lugu was not large, with a very similar vertical change trend. From a seasonal point of view, with the same trend of change in spring, summer and autumn, EC changed significantly, peaked in summer and autumn, and was lowest in winter.

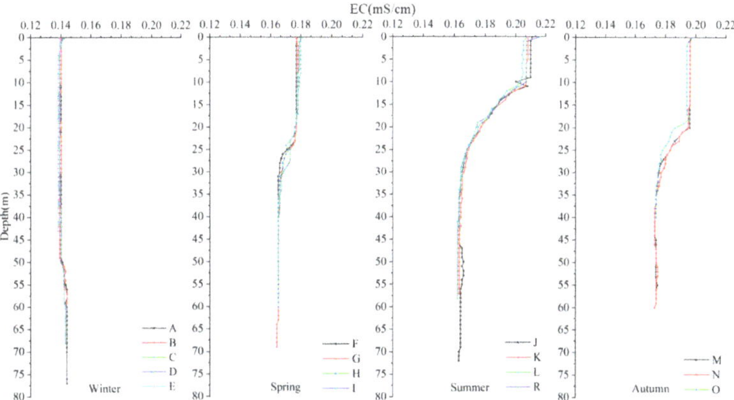

Figure 3. Seasonal variations of vertical electrical conductivity (EC) in Lake Lugu.

3.2.2. Dissolved Oxygen (DO)

With the trend of changes in WT, the vertical distribution of DO in the water body of Lake Lugu demonstrated a clear seasonal stratification (Figure 4). Based on the measured data, the DO concentration in the surface layer of Lake Lugu in January, April, July, and October was almost unchanged from 6.5 to 8.5 mg/L, but the vertical changes were significantly different. In January, the DO of the water body did not change significantly within 48 m of the vertical section, with a mass concentration of 7.0 to 8.5 mg/L. The DO of the water body with a large water depth in the north changed suddenly at 50 m, and the DO of the water layer below 55 m was <4.0 mg/L. However, when the water depth exceeded 60 m, the DO was <2.0 mg/L, forming an anaerobic environment. As of April, the mass concentration of DO increase was due to the formation of the recessive layer in the lake. We observed that DO was stabilized at about 8.0 mg/L above the 25 m depth. When the water depth increased, DO gradually decreased to 5.5 mg/L. The change in DO in July was more complex and diverse, and the vertical distribution of DO in the changed temperature layer corresponding to 0~10 m was relatively stable. DO gradually increased downward from the depth of 11 m and reached its maximum value at the 20 m depth, forming a high DO layer in the 11~33 m section. In October, DO was evenly distributed vertically in the upper layer of the water at 0~21 m, with a value slightly higher than 7.0 mg/L. At 21 to 45 m depths, the DO concentration dropped sharply and was 2.0 to 3.0 mg/L at depths below 45 m.

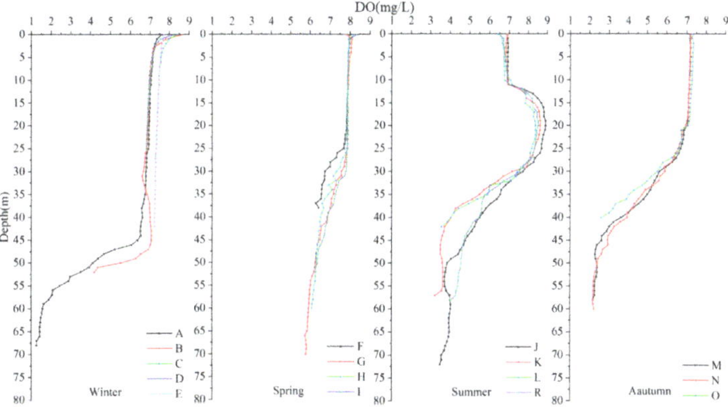

Figure 4. Seasonal variations of vertical dissolved oxygen (DO) section in Lake Lugu.

3.2.3. pH Value

The pH value of the Lake Lugu water body was generally slightly alkaline, and there was a clear stratification with obvious vertical and seasonal variations (Figure 5). Due to the high light intensity at the surface, many aquatic organisms thrived, photosynthesis was strong, and a large amount of CO_2 was consumed, resulting in a high surface water pH value. The photosynthesis of aquatic plants in deeper water was weak, and the decomposition of organic matter led to CO_2 and organic acids accumulation in the water. At the same time, due to the long retention time of the lower water and the slow molecular diffusion rate, the pH value decreased slowly. Thus, except in summer, the vertical change in pH value in Lake Lugu gradually decreased.

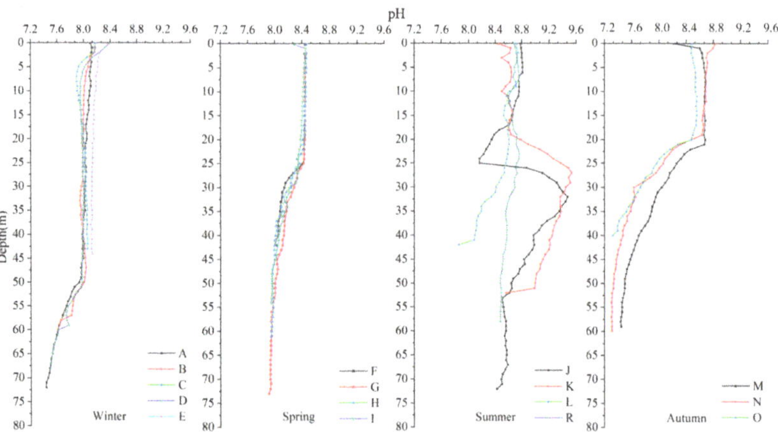

Figure 5. Seasonal variations of vertical pH section in Lake Lugu.

The vertical variation of pH value (Figure 5) shows that the pH value of the lake section is 7.6 to 8.4 in January due to the change of WT. pH value is very stable above 50 m water depth, varying around 8.2. Compared with January, the overall pH value of the water body appeared to increase in April, with a relatively high and evenly distributed pH value at 0~25 m. The pH value suddenly decreased in the 25 m water column. As the water depth increased, the pH was still 8.0 at the bottom of the water body. An increase in temperature and photosynthesis of aquatic plants in July led to the consumption of a large amount of CO_2 in the water body, which increased the pH value and alkalinity of the water body. At this time, due to various factors being most active, a diverse change in pH value was observed, especially at a water depth of 15 to 55 m, where the pH value could reach up to 9.6. Until October, the vertical distribution of pH values in the 0 to 20 m water body was relatively stable, ranged between 8.4 to 8.8, and it gradually decreased after the 20 m depth.

From a spatial point of view, the vertical pH change in the northern part of Lake Lugu in January was slightly more pronounced than in the southern part, but the difference was not significant. The vertical variation of pH in southern waters was stable. In April, pH variation was observed at the 25 m and 23 m in the northern (F and G) and southern (I) water area. In July, the pH of the north–south water area of Lake Lugu showed a distinct stratification in the vertical direction, and with the pH peak appearing in the northern area. In October, the pH values in the northern and southern water bodies were consistent and stratified but suddenly changed at the 19 m water layer. The pH value of surface waters in the northern area was slightly higher than in the south.

3.2.4. Chlorophyll-a (Chl-*a*)

Chl-*a* is an important indicator of phytoplankton biomass, and the content of Chl-*a* in waters reflects the number of algae in the water to some extent and is closely related to algae growth activity of algae, water transparency, nutrient salt concentration, and self-suspending characteristics [38]. Compared with EC, DO, and pH, the vertical stratification of Chl-*a* in Lake Lugu was not obvious, but the seasonal variation was significant. Seasonal analysis indicated that Chl-*a* content showed obvious seasonal variation. It peaked in summer and was lowest in winter. In January, the vertical distribution of Chl-*a* content was relatively uniform, and the concentration gradient at the 45 m water layer was greatest and remained almost unchanged in water layers below 55 m. Throughout April, the vertical change in Chl-*a* content was unclear and showed no obvious stratification. In July, minimum and maximum Chl-*a* content was recorded at surface water and 20 m water depth, respectively. In October, the average Chl-*a* content was higher than in January and April, and changed sharply in the 20 m water column. Overall, the Chl-*a* concentration gradient was larger in the thermocline, and comparison analysis showed that Chl-*a* fluctuated more vertically in October and the distribution was more evenly distributed in January and April.

According to the experimental results in January, April, and October 2015 (Figure 6), it could be seen that the Chl-*a* content differed to some extent in the spatial distribution, and there was a vertical change in Chl-*a* in the water layer. The change was more obvious in the northern part than the southern part of Lake Lugu, with the Chl-*a* content in the surface water being greater in the north than in the south.

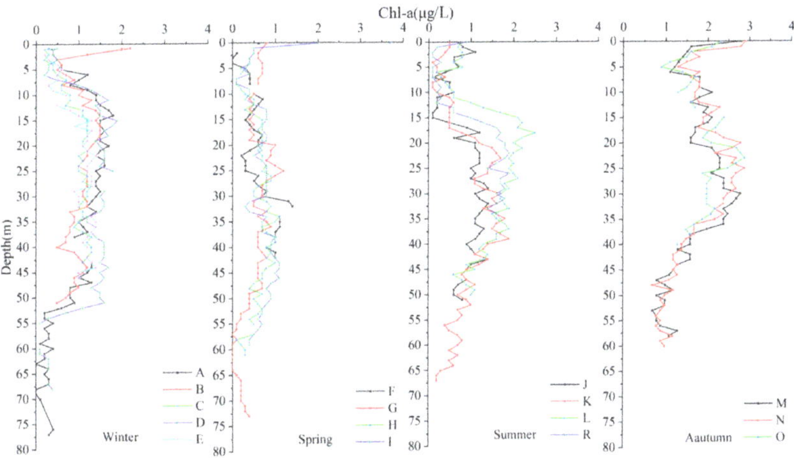

Figure 6. Seasonal variations in vertical chlorophyll- *a* (Chl-*a*) in Lake Lugu.

3.3. Relationship between Changes in Lake Lugu WT Stratification and Water Body Parameters

3.3.1. Water Mixing Type of Lake Lugu

The lake WT and temperature demonstrated obvious seasonal variation. Accordingly, the thermal stratification phenomenon of the lake water body also showed different seasonal patterns [38]. Therefore, most classifications of lakes are proposed based on thermal stratification and mixed types [39]. The vertical cross-section of WT in Lake Lugu revealed that the vertical distribution of WT in winter was close to the same temperature state, a positive temperature layer distribution in spring, summer, and autumn, and an obvious thermocline appeared in summer and autumn. Therefore, according to the lake classification scheme proposed by Lewis [40] based on revising previous work and the special geographical location and current situation of Lake Lugu, the lake-water mixing mode conformed to the characteristics of winter mixing, spring stratification formation, and stable stratification in summer and autumn, which indicate a warm single mixed lake.

3.3.2. Relationship between Seasonal Variation of Temperature Stratification and Spatial-Temporal Distribution of DO

The exponential method was used to determine the WT stratification type [41], if α = the total storage of volume flow divided by total storage capacity, when $\alpha < 10$, it represents a stable stratified type, and when $\alpha > 20$, it represents a completely mixed type. Lake Lugu has a low capacity of 1.953 billion m^3, and the annual volume of water entering the lake is 1.1 million m^3. Calculation results showed it had an $\alpha < 10$. It could be seen that the vertical WT distribution of the Lugu was stable, and the type of stratification matched the measurement results. Due to the thermal stratification of the water body, the exchange of material and energy between water layers was limited, causing an obvious response of water quality parameters. For deep-water lakes, a weaker hydrodynamics and longer retention time of water column refer to an uneven heat transfer from the water column, causing differences in the density of hot and cold-water columns, leading to differences in the physicochemical properties of different water layers and significant changes in the DO of the water column within the thermocline.

There are many factors that influence the vertical distribution of DO in a lake, such as depth, atmospheric temperature, basin shape, hydrothermal stratification, and biological effects. Lake Lugu is a deep-water arc, with its seasonal DO stratification greatly affecting WT stratification and vertical WT stratification greatly influencing the vertical DO distribution. Its upper water body was in direct contact with the atmosphere and was greatly disturbed by winds. Photosynthesis of aquatic plants in the temperature layer was strong, and the vertical mixing of DO was relatively uniform. However, the vertical mixing of DO was severely suppressed. Inside the isothermal bed, there was a very small number of aquatic plants, weak light, low rate of photosynthesis, and the decomposition of organic matter further increased the oxygen consumption of the water body. The presence of thermocline seriously blocked the exchange of substances and energy in the upper and lower bodies due to the isothermal water body. A long-term stagnation of water bodies at the mean temperature layer prevented the timely replenishment of DO in the upper layer of water bodies, causing the rate of oxygen consumption in the lower layer to exceed the rate of oxygen replenishment, resulting in the continuous decrease of DO content in the mean temperature layer of Lake Lugu and gradual development of anaerobic conditions.

The effect of temperature stratification of Lake Lugu on the vertical distribution of DO was comparable with that of Lake Wanfeng [42] and Korean Reservoir [43], and the seasonal stratification of WT resulted in seasonal stratification of the vertical distribution of DO. In winter, solar radiation was weak and the temperature was low, causing the lake water to release latent heat to the outside. The oxygen-rich water in the thermocline supplemented the consumption of DO in time, allowing DO to be evenly distributed in the vertical direction. The water level in the northern part of Lake Lugu was deeper, and the surface WT was higher. The thermocline was located at the 50 m water layer, where the temperature gradient changed greatly. At the same time, DO decreased sharply at 50 m, and the vertical change in WT in the south was uniform. The water layer below 55 m had a DO concentration <4.0 mg/L, indicating a state of hypoxia.

The thermal stratification initially formed in spring, and the existence of the thermocline effectively prevented the vertical convective mixing of water. The vertical distribution of DO in the thermocline was relatively uniform, and the DO in the thermocline gradually decreased and changed slowly. The depth of thermal rock formation in the northern water was lower than in the south and moved 5 m, shifting the vertical position of the DO in the northern water downwards compared with the southern water. The thermal stratification was stable in summer, DO vertical distribution in the temperature change layer was uniform, and the DO of the temperature change layer as a whole was lower than in April. The possible reason for such observations could be due to (1) decreased DO as temperature increased, (2) weakening of the hydrodynamic force of the temperature change layer, (3) day and night migration of aquatic organisms. In WT, DO changes were more complex and diverse. A peak was observed at the 20 m depth, and the DO decreased sharply above and below the 20 m depth. Factors causing extreme DO changes are very complex and can be broadly divided into physical factors, biological factors, and the combined effects of the two. In the case of Lake Lugu, physical factors were almost eliminated because of the physical influences and DO extremes due to WT or high-density water replenishment. The temperature difference between the Lake Lugu water body monitoring zones was small and mainly depended on precipitation. The position of the extreme value of DO cause by biological factors was consistent with the transparency of the water body [44]. Lake Lugu belongs to the Class I water quality with a transparency of 12 to 14 m. At the same time, Lake Lugu contained aquatic organisms that performed photosynthesis, because of its high altitude and the strong ultraviolet radiation in summer. To avoid the sun's ultraviolet radiation, the aquatic organisms mostly proliferated at about the 20 m water layer, which was in the thermocline. The presence of the thermocline severely inhibited the upward transport of O_2 produced by photosynthesis, resulting in low DO in the variable and mean temperature layers. In autumn, the DO decreased sharply at around 20 m depth. In the 0~20 m water layer, the aquatic organisms produced large amounts of DO by photosynthesis, and the strong turbulent mixing effects caused the DO to diffuse rapidly and made the vertical mixing more uniform in the 0~20 m water layer, leading to a very small DO concentration in the mean temperature layer and an anaerobic environment.

3.3.3. Influence of WT Stratification on the Temporal and Spatial Distribution of pH

The water pH was mainly controlled by the CO_2 content and HCO_3^- concentration, while the water CO_2 content was affected by WT, dissolved ions, microorganisms, and other factors [45]. Lake Lugu is a natural freshwater lake with a transparent water body. Therefore, a pH change in the water body was mainly related to the photosynthesis of aquatic organisms. The surface water body had sufficient light, a large plankton population, strong photosynthesis, and a large consumption of CO_2, which disrupted the equilibrium process of $CO_2-H_2CO_3-HCO_3^- -CO_3^{2-}$, causing the water to have an increased pH and weak alkalinity. Photosynthesis was very weak in the deep-water layer because light could not penetrate. In addition, mineralization and degradation of organic matter at the water–sediment interface generated a large amount of CO_2 and small molecular organic acids. Due to the cumulative effects of CO_2 and small molecular organic acids at layers closer to the bottom, their corresponding pH value was small. The existence of thermocline seriously blocked the exchange of materials and energy between the upper and lower water bodies. Comparatively, the lower water body had a long retention time and a slow molecular diffusion rate, resulting in a slower change in pH.

Similar to DO, the temporal and spatial pH distribution in Lake Lugu also demonstrated seasonal stratification characteristics. In winter, the water body released latent heat and was mixed vertically at the temperature change layer > 50 m, causing the vertical pH distribution to be uniform and more hydrodynamic conditions with deeper water depth. Thus, a weaker pH was observed in the northern water at a deeper water level, with a sudden change and considerable decrease observed at a water depth of 51 m. In spring, the pH value was slightly higher than in winter, which was related to the consumption of CO_2 in water due to the photosynthesis of aquatic organisms and the gradual decrease in the pH value of thermocline. Thermocline and weak stability in spring resulted in very weak pH change in the isothermal layer. In July, the pH value showed a low distribution trend at both ends, while it was high in the middle. Based on the vertical distribution characteristics of DO, it could be inferred that the pH value was higher at about 20 m. However, unlike this inference, the pH values of the northern and southern water bodies peaked at about 30 m and 24 m, respectively, and the pH value of the northern water body changed significantly, which may have been caused by the physical effects of warming and cooling of the water body, or the downward shift of the pH pole position due to gravity flow or density flow. In the north, the water level was deeper and the surface WT ranged between 20.5 °C to 21 °C, which was beneficial to the growth of aquatic organisms, leading to a greater amount of photosynthesis and CO_2 consumption, and a significant increase in pH. In October, the vertical distribution of pH in the 0 to 20 m water column was constant, and the higher pH was due to the cumulative effect of continuous CO_2 consumption by the photosynthesis of aquatic plants.

3.3.4. Effect of WT Stratification on the Spatiotemporal Distribution of EC

The EC of water was proportional to the concentration (or activity) of ions dissolved in the water, which mainly reflected the total amount of soluble ions in the water. The molecular formula of water (H_2O) shows that water is composed of electrically neutral molecules rather than ions, natural water is a good conductor of electricity, and it obeys Ohm's law. In lake water, a large number of substances dissolve, dissociate, and form electroactive ions, which can increase the conductivity of the water column, mainly controlled by soluble substances and temperature [46]. Figures 2 and 3 show that in the vertical direction, the seasonal stratification trend of EC and WT was consistent, suggesting that temperature was the main factor affecting EC. Based on its characteristics, the water quality of Lake Lugu is classified as Class I and has high transparency. The salinity of the lake water is determined by the total ion concentration in the water body. In the vertical section of the water body in lake Lugu, the salinity was almost constant (~0.10‰), ignoring the effects of ion concentration on conductivity and the relationship between EC and WT in each vertical section (Table 2).

Table 2. The relationship between electrical (EC) conductivity and water temperature (WT).

Sampling Point	Winter		Sampling Point	Spring	
	Equation	p		Equation	p
A	$C = -0.015T + 0.289$	-0.981 **	F	$C = 0.004T + 0.121$	0.999 **
B	$C = -0.004T + 0.184$	-0.464 **	G	$C = 0.004T + 0.122$	0.997 **
C	$C = -0.004T + 0.187$	-0.312 *	H	$C = 0.004T + 0.121$	0.996 **
D	$C = 0.004T + 0.103$	0.346 *	I	$C = 0.004T + 0.187$	0.999 **
E	$C = 0.004T + 0.096$	0.912 **			

Sampling Point	Summer		Sampling Point	Autumn	
	Equation	p		Equation	p
J	$C = 0.004T + 0.122$	0.997 **	M	$C = 0.003T + 0.143$	0.989 **
K	$C = 0.004T + 0.121$	0.999 **	N	$C = 0.003T + 0.144$	0.988 **
L	$C = 0.004T + 0.122$	0.999 **	O	$C = 0.003T + 0.146$	0.984 **
R	$C = 0.004T + 0.121$	0.999 **			

Notes: ** extremely significant correlation, $p < 0.01$; * Significant correlation, $p < 0.05$, C: EC (mS/cm), T: WT (°C).

Correlation analysis in Table 2 shows a significant and positive correlation between electrical EC and WT in winter (except in the water body at the northern part of Lake Lugu), spring, summer, and autumn, and that EC and temperature had a simple linear function. The thermal stratification of the water column was formed in April, and the stratification was stabilized in July. Due to the blocking effect of the thermocline, the EC fluctuated greatly. At the same time, when the solar radiation increased, the temperature increased, WT increased, and the molecular and ion movement rate accelerated, resulting in the exchange rate between the northern and southern part of Lake Lugu, causing the conductivity and the WT to be extremely similar, which could be summarized as the following function of conductivity and WT in April and July: $C = 0.004T + 0.12$. In October, a decrease in temperature, WT and difference in water density led to a downward shift of the thermocline position and a decrease in the slope between EC and WT, but no significant difference between the north and south water bodies. The functional relationship between EC and the WT on the vertical section of the water body was close to: $C = 0.003T + 0.14$. In winter, the vertical turbulence effect on the water body was enhanced and the electrical EC changed uniformly in the vertical direction. However, the EC of the water body in the northern part of Lake Lugu was inversely correlated with WT, showing a relationship: $C = -0.008T + 0.22$; while that in the south was positively correlated, showing a relationship: $C = 0.004T + 0.10$. In principle, EC was positively correlated with temperature and ion concentration, but this inverse correlation observed in the water of the northern part of Lake Lugu may have been due to an increase in the concentration of H^+ and soluble substances concentration in the northern water. The body temperature was lowered, the concentration of ions positively affected EC, and the effects of speed exceeded that of WT. The functional relationship between EC and WT in the vertical section of the Lake Lugu water body was concordant with the results of a previous study, $C\ (T) = aT + b$ [47].

The presence of thermocline effectively inhibited convection, turbulence, and molecular exchange in the upper and lower waters of the lake, seriously impeding the exchange of material and energy, thus affecting the vertical distribution of EC in the presence of thermal stratification. Therefore, thermal stratification in water also had a certain impact on the vertical distribution of EC. Taking the phenomenon of no stratification in winter as the reference to represent the relationship between the EC and temperature of the whole lake, it demonstrated a relatively high correlation between the EC–temperature, with some differences between the different layers (Table 3). The correlation analysis in Table 3 showed a significantly positive correlation between the EC and WT of the thermocline, epilimnion, and hypolimnion. The slope between EC–temperature increased gradually in the vertical direction of the monitored section, indicating that the influence of temperature on EC increased with increasing water depth.

Table 3. The functional relationship of epilimnion, thermocline, and hypolimnion between electrical conductivity (EC) and water temperature (WT).

Lake Stratification	Equation	p
Epilimnion	$C = 0.0065T + 0.079$	0.964 **
Thermocline	$C = 0.0082T + 0.063$	0.900 **
Hypolimnion	$C = 0.0334T - 0.176$	0.628 *

Notes: ** extremely significant correlation, $p < 0.01$; * Significant correlation, $p < 0.05$, C: EC (mS/cm), T: WT (°C).

3.3.5. Effect of WT Stratification on the Spatiotemporal Distribution of Chl-*a*

Chl-*a* is the pigment that makes plants green and is an important component of phytoplankton in water. Photosynthesis in water was mainly performed by phytoplankton [48]. Considering that phytoplankton contain Chl-*a*, it is commonly used to estimate the number of existing phytoplankton and photosynthesis and as a water quality monitoring indicator [49,50]. According to the China National Environmental Monitoring Center's evaluation method and technical classification regulations of lake (reservoir) eutrophication [51,52], the Chl-*a* of Lake Lugu was less than 3.09 μg/L in the four seasons of the year, suggesting that it is an oligotrophic lake. The water quality of Lake Lugu was classified as Class I in 2020, indicating good water quality. Further, its total nitrogen content based on the single evaluation index of lakes and reservoirs showed that Lake Lugu was classified as a Class I lake and had a nutritional status index of 13.2, further confirming it as an oligotrophic lake.

The Chl-*a* concentration in Lake Lugu varied significantly with the seasons, with a small growth peak in summer and autumn. The peak of Chl-*a* appeared at the surface water and after water stratification because WT and light conditions were suitable for algae growth. In the mean temperature layer, the Chl-*a* level was more uniform. In winter, Chl-*a* was higher when the surface WT of Lake Lugu dropped to its lowest level for the whole year, and during this period, Lake Lugu was controlled by rotating winds. The decrease in WT in the thermostat layer caused a thickening effect in the upper water body, increasing the instability of the water body in the thermostat layer. In addition, the continuous non-directional wind made the lower water body rich in nutrients and surpassed that of the upper water body. Overall, the vertical distribution of Chl-*a* in Lake Lugu fluctuated slightly. In summer, the southwest monsoon predominated. At this time, precipitation reached its highest levels of the year and the lowest Chl-*a* levels were recorded in the surface water. In summer, the temperature rose and due to strong solar ultraviolet radiation, algae were most active at a depth of 20 m, whereby Chl-*a* content peaked. On the whole, the variation of Chl-*a* content in the thermocline was more complex in summer, and the Chl-*a* content in the uniform temperature layer was more stable. In autumn, when the southwest monsoon retreated and algae growth was more prosperous, a higher Chl-*a* was recorded, causing a sudden jump at ~20 m. In the water layer below 40 m, Chl-*a* was evenly distributed vertically.

3.4. Limitations and Implications

First, this study analyzed the seasonal dynamics of WT and its vertical stratification structure based on monitoring data, and discussed the seasonal stratification characteristics of Lake Lugu water chemistry. However, long-term and high-frequency observations of the thermal stratification transition and its critical periods are lacking in terms of hydrodynamic profiles, nutrients, and phytoplankton. Second, due to the limited number of monitoring sites, the results may not be representative of the whole lake. With the improvement of monitoring systems and methods, this problem may be solved in the near future. Third, the impact of human activities on the water quality of Lake Lugu in the context of urbanization was not analyzed. In addition to climate change, human activities (population, GDP, impervious area, industrial structure, non-point pollution, etc.) also cause water quality changes in Lake Lugu. In the context of complex climate change and anthropogenic disturbances in the future, much work remains to be done to gain a more comprehensive and in-depth understanding of the thermal stratification characteristics of the water column in Lake Lugu and other similar lakes in the region and their ecological and environmental impacts (e.g., revealing their effects on changes in phytoplankton community structure and even the driving mechanisms of water blooms).

4. Conclusions

Lake Lugu is a typical warm single mixed lake. The stratification is characterized as mixed in winter and stratified in summer and autumn, and the lake is a single mixed lake. Lake Lugu thermal stratification controls the vertical distribution and variation of DO, pH, EC, resulting in the vertical stratification pattern of DO, pH, EC, and synergistic variation of thermocline. Meanwhile, WT stratification affects the spatial and temporal distribution of pH, EC, DO, and Chl-*a*. In winter, the water bodies are in the mixing period, and the water chemistry parameters EC, pH, and DO are evenly distributed vertically. In summer, the vertical stratification of EC, pH, and DO was more obvious, and the peak of pH and DO appear in the thermocline, and the trend was decreasing from the peak upward and downward.

The salinity of the Lake Lugu water column remains basically constant (about 0.10‰), and there is a simple linear function between EC and WT without considering the salinity effect, both in the vertical section and in the variable temperature layer, thermocline, and mean temperature layer. The results of correlation analysis showed that EC and WT were significantly and positively correlated in winter (except for Lake Lugu northern water body), spring, summer, and autumn, and the inverse correlation in Lake Lugu northern water body might be due to the increase of H^+ and soluble matter concentration in northern water body with the decrease of WT.

The spatial distribution pattern of Lake Lugu Chl-*a* content: south > north, which is due to the different effects of geographical location, lake current, and wind direction on Chl-*a* in the same lake at different depths. In addition to the above reasons, human activities are also one of the main reasons for the increase of Chl-*a* content in Lake Lugu, and the locations with the highest increase of Chl-*a* content are the most active tourist activities in Lake Lugu, which should be of great concern. With the seasonal formation and disappearance of the Lake Lugu thermocline, there will be an impact on the water quality of Lake Lugu. Therefore, to protect the ecosystem of Lake Lugu, water quality monitoring should be conducted in summer and autumn, and a rapid emergency mechanism should be developed in advance.

Author Contributions: H.Z. and P.H.: Conceptualization, Methodology, Software, Writing—Original draft preparation. F.C. and P.H.: Supervision, Writing—review & editing. H.Z.: Conceptualization, Supervision, Resources, Writing—Review & editing, Foundation's acquisition. L.D., X.W. and Y.Z.: Investigation, Data Curation. All authors have read and agreed to the published version of the manuscript.

Funding: This work was supported by the National Natural Science Foundation of China (Grant No. 41820104008) and Scientist workshop.

Data Availability Statement: The data that support the findings of this study are available from the corresponding author upon reasonable request.

Acknowledgments: We are grateful to the teachers and students who participated in the field collection of data.

Conflicts of Interest: The authors declare no conflict of interest.

References

1. Paerl, H.W.; Paul, V.J. Climate change: Links to global expansion of harmful cyanobacteria. *Water Res.* **2012**, *46*, 1349–1363. [CrossRef] [PubMed]
2. Bruesewitz, D.A.; Carey, C.C.; Richardson, D.C.; Weathers, K.C. Under-ice thermal stratification dynamics of a large, deep lake revealed by high-frequency data. *Limnol. Oceanogr.* **2015**, *60*, 347–359. [CrossRef]
3. Robertson, D.M.; Ragotzkie, R.A. Changes in the thermal structure of moderate to large sized lakes in response to changes in air temperature. *Aquat. Sci.* **1990**, *52*, 360–380. [CrossRef]
4. Zhang, Y.; Wu, Z.; Liu, M.; He, J.; Shi, K.; Zhou, Y.; Liu, X. Dissolved oxygen stratification and response to thermal structure and long-term climate change in a large and deep subtropical reservoir (Lake Qiandaohu, China). *Water Res.* **2015**, *75*, 249–258. [CrossRef]
5. Alexakis, D.; Kagalou, I.; Tsakiris, G. Assessment of pressures and impacts on surface water bodies of the Mediterranean. Case study: Pamvotis Lake, Greece. *Environ. Earth Sci.* **2013**, *70*, 687–698. [CrossRef]
6. Longyang, Q. Assessing the effects of climate change on water quality of plateau deep-water lake-A study case of Hongfeng Lake. *Sci. Total Environ.* **2019**, *647*, 1518–1530. [CrossRef]
7. Butcher, J.B.; Nover, D.; Johnson, T.E.; Clark, C.M. Sensitivity of lake thermal and mixing dynamics to climate change. *Clim. Chang.* **2015**, *129*, 295–305. [CrossRef]
8. Geller, W. The temperature stratification and related characteristics of Chilean lakes in midsummer. *Aquat. Sci.* **1992**, *54*, 37–57. [CrossRef]
9. Ryabov, A.B.; Rudolf, L.; Blasius, B. Vertical distribution and composition of phytoplankton under the influence of an upper mixed layer. *J. Theor. Biol.* **2010**, *263*, 120–133. [CrossRef]
10. Wang, S.; Qian, X.; Han, B.P.; Luo, L.C.; Hamilton, D.P. Effects of local climate and hydrological conditions on the thermal regime of a reservoir at Tropic of Cancer, in southern China. *Water Res.* **2012**, *46*, 2591–2604. [CrossRef]
11. Piccolroaz, S.; Toffolon, M.; Majone, B. The role of stratification on lakes' thermal response: The case of Lake S uperior. *Water Resour. Res.* **2015**, *51*, 7878–7894. [CrossRef]
12. Mullin, C.A.; Kirchhoff, C.J.; Wang, G.; Vlahos, P. Future projections of water temperature and thermal stratification in Connecticut reservoirs and possible implications for cyanobacteria. *Water Resour. Res.* **2020**, *56*, e2020WR027185. [CrossRef]
13. Kraemer, B.M.; Anneville, O.; Chandra, S.; Dix, M.; Kuusisto, E.; Livingstone, D.M.; McIntyre, P.B. Morphometry and average temperature affect lake stratification responses to climate change. *Geophys. Res. Lett.* **2015**, *42*, 4981–4988. [CrossRef]
14. Moser, K.A.; Baron, J.S.; Brahney, J.; Oleksy, I.A.; Saros, J.E.; Hundey, E.J.; Smol, J.P. Mountain lakes: Eyes on global environmental change. *Glob. Planet. Chang.* **2019**, *178*, 77–95. [CrossRef]
15. Adrian, R.; O'Reilly, C.M.; Zagarese, H.; Baines, S.B.; Hessen, D.O.; Keller, W.; Winder, M. Lakes as sentinels of climate change. *Limnol. Oceanogr.* **2009**, *54*, 2283–2297. [CrossRef] [PubMed]
16. Yang, X.; Lu, X. Drastic change in China's lakes and reservoirs over the past decades. *Sci. Rep.* **2014**, *4*, 6041. [CrossRef]
17. Ye, X.; Zhang, Q.; Liu, J.; Li, X.; Xu, C.Y. Distinguishing the relative impacts of climate change and human activities on variation of streamflow in the Poyang Lake catchment, China. *J. Hydrol.* **2013**, *494*, 83–95. [CrossRef]
18. Gleick, P.H. Methods for evaluating the regional hydrologic impacts of global climatic changes. *J. Hydrol.* **1986**, *88*, 97–116. [CrossRef]
19. Kalaimurugan, D.; Balamuralikrishnan, B.; Govindarajan, R.K.; Al-Dhabi, N.A.; Valan Arasu, M.; Vadivalagan, C.; Khanongnuch, C. Production and characterization of a novel biosurfactant molecule from Bacillus safensis YKS2 and assessment of its efficiencies in wastewater treatment by a directed metagenomic approach. *Sustainability* **2022**, *14*, 2142. [CrossRef]
20. Kamyab, H.; Yuzir, M.A.; Abdullah, N.; Quan, L.M.; Riyadi, F.A.; Marzouki, R. Recent Applications of the Electrocoagulation Process on Agro-Based Industrial Wastewater: A Review. *Sustainability* **2022**, *14*, 1985.

21. Leach, T.H.; Beisner, B.E.; Carey, C.C.; Pernica, P.; Rose, K.C.; Huot, Y.; Verburg, P. Patterns and drivers of deep chlorophyll maxima structure in 100 lakes: The relative importance of light and thermal stratification. *Limnol. Oceanogr.* **2012**, *63*, 628–646. [CrossRef]

22. Hou, P.; Chang, F.; Duan, L.; Zhang, Y.; Zhang, H. Seasonal Variation and Spatial Heterogeneity of Water Quality Parameters in Lake Chenghai in Southwestern China. *Water* **2022**, *14*, 1640. [CrossRef]

23. Liu, Y.; Wang, Y.; Sheng, H.; Dong, F.; Zou, R.; Zhao, L.; He, B. Quantitative evaluation of lake eutrophication responses under alternative water diversion scenarios: A water quality modeling based statistical analysis approach. *Sci. Total Environ.* **2014**, *468*, 219–227. [CrossRef]

24. Zohary, T.; Ostrovsky, I. Ecological impacts of excessive water level fluctuations in stratified freshwater lakes. *Inland Waters* **2011**, *1*, 47–59. [CrossRef]

25. Noori, R.; Ansari, E.; Bhattarai, R.; Tang, Q.; Aradpour, S.; Maghrebi, M.; Haghighi, A.T.; Bengtsson, L.; Kløve, B. Complex dynamics of water quality mixing in a warm mono-mictic reservoir. *Sci. Total Environ.* **2021**, *777*, 146097. [CrossRef]

26. Zhao, X.; Wang, Q.; Liu, Z. Seasonal variations of thermal stratification in Lake Lugu. *Highlights Sci. Online* **2014**, *7*, 2441–2448.

27. Wang, Q.; Yang, X.; Anderson, N.J.; Ji, J. Diatom seasonality and sedimentation in a subtropical alpine lake (Lugu Hu, Yunnan-Sichuan, Southwest China). *Arct. Antarct. Alp. Res.* **2015**, *47*, 461–472. [CrossRef]

28. Chen, D. A Preliminary Numerical Simulation of the Thermodynamic Conditions of Lugu Lake in Recent Years. Ph.D. Thesis, Jinan University, Guangzhou, China, 2015.

29. Wu, G.; Zhang, Q.; Zheng, X.; Mu, L.; Dai, L. Water quality of Lugu Lake: Changes, causes and measurements. *Int. J. Sustain. Dev. World Ecol.* **2008**, *15*, 10–17. [CrossRef]

30. Sheng, E.; Yu, K.; Xu, H.; Lan, J.; Liu, B.; Che, S. Late holocene Indian summer monsoon precipitation history at Lake Lugu, northwestern Yunnan Province, southwestern China. *Palaeogeogr. Palaeoclimatol. Palaeoecol.* **2015**, *438*, 24–33. [CrossRef]

31. Yang, K.; Yu, Z.; Luo, Y.; Zhou, X.; Shang, C. Spatial-temporal variation of lake surface water temperature and its driving factors in Yunnan-Guizhou Plateau. *Water Resour. Res.* **2019**, *55*, 4688–4703. [CrossRef]

32. Wang, Q.; Yang, X.; Anderson, N.J.; Zhang, E.; Li, Y. Diatom response to climate forcing of a deep, alpine lake (Lugu Hu, Yunnan, SW China) during the Last Glacial Maximum and its implications for understanding regional monsoon variability. *Quat. Sci. Rev.* **2014**, *86*, 1–12. [CrossRef]

33. State Environmental Protection Administration (SEPA). GB3838-2002. In *Environmental Quality Standard for Surface Water*; Standards Press: Beijing, China, 2002; pp. 1–8. (In Chinese)

34. Podobnik, B.; Stanley, H.E. Detrended cross-correlation analysis: A new method for analyzing two non-stationary time series. *Phys. Rev. Lett.* **2008**, *100*, 084102. [CrossRef] [PubMed]

35. Austin, J.; Colman, S. A Century of Temperature Variability in Lake Superior. *Limnol. Oceanogr.* **2008**, *53*, 2724–2730. [CrossRef]

36. Dong, C.Y.; Yu, Z.M.; Wu, Z.X.; Wu, C.J. Study on seasonal characteristics of thermal stratification in lacustrine zone of Lake Qiandao. *Huan Jing Ke Xue = Huanjing Kexue* **2013**, *34*, 2574–2581. [PubMed]

37. Kraus, E.B.; Turner, J.S. A one-dimensional model of the seasonal thermocline II. The general theory and its consequences. *Tellus* **1967**, *19*, 98–106. [CrossRef]

38. Woolway, R.I.; Meinson, P.; Nõges, P.; Jones, I.D.; Laas, A. Atmospheric stilling leads to prolonged thermal stratification in a large shallow polymictic lake. *Clim. Chang.* **2017**, *141*, 759–773. [CrossRef]

39. Kirillin, G.; Shatwell, T. Generalized scaling of seasonal thermal stratification in lakes. *Earth-Sci. Rev.* **2016**, *161*, 179–190. [CrossRef]

40. Lewis, W.M., Jr. A revised classification of lakes based on mixing. *Can. J. Fish. Aquat. Sci.* **1983**, *40*, 1779–1787. [CrossRef]

41. Magee, M.R.; Wu, C.H. Response of water temperatures and stratification to changing climate in three lakes with different morphometry. *Hydrol. Earth Syst. Sci.* **2017**, *21*, 6253–6274. [CrossRef]

42. Wen, X.; Zhang, H.; Chang, F.; Li, H.; Duan, L.; Wu, H.; Ouyang, C. Seasonal stratification characteristics of vertical profiles of water body in Lake Lugu. *Adv. Earth Sci.* **2016**, *31*, 858–869.

43. Lee, Y.G.; Kang, J.H.; Ki, S.J.; Cha, S.M.; Cho, K.H.; Lee, Y.S.; Kim, J.H. Factors dominating stratification cycle and seasonal water quality variation in a Korean estuarine reservoir. *J. Environ. Monit.* **2010**, *12*, 1072–1081. [CrossRef] [PubMed]

44. Read, J.S.; Hamilton, D.P.; Jones, I.D.; Muraoka, K.; Winslow, L.A.; Kroiss, R.; Gaiser, E. Derivation of lake mixing and stratification indices from high-resolution lake buoy data. *Environ. Model. Softw.* **2011**, *26*, 1325–1336. [CrossRef]

45. Berge, J.A.; Bjerkeng, B.; Pettersen, O.; Schaanning, M.T.; Øxnevad, S. Effects of increased sea water concentrations of CO_2 on growth of the bivalve *Mytilus edulis* L. *Chemosphere* **2006**, *62*, 681–687. [CrossRef] [PubMed]

46. Vega, M.; Pardo, R.; Barrado, E.; Debán, L. Assessment of seasonal and polluting effects on the quality of river water by exploratory data analysis. *Water Res.* **1998**, *32*, 3581–3592. [CrossRef]

47. Boehrer, B.; Schultze, M. Stratification of lakes. *Rev. Geophys.* **2008**, *46*, 7. [CrossRef]

48. Søndergaard, M.; Larsen, S.E.; Jørgensen, T.B.; Jeppesen, E. Using chlorophyll a and cyanobacteria in the ecological classifi-cation of lakes. *Ecol. Indic.* **2011**, *11*, 1403–1412. [CrossRef]

49. Mignot, A.; Claustre, H.; Uitz, J.; Poteau, A.; D'Ortenzio, F.; Xing, X. Understanding the seasonal dynamics of phytoplank-ton biomass and the deep chlorophyll maximum in oligotrophic environments: A Bio-Argo float investigation. *Glob. Bio-Geochem. Cycles* **2014**, *28*, 856–876. [CrossRef]

50. Hou, P.; Luo, Y.; Yang, K.; Shang, C.; Zhou, X. Changing characteristics of chlorophyll a in the context of internal and external factors: A case study of Dianchi lake in China. *Sustainability* **2019**, *11*, 7242. [CrossRef]

51. Dodds, W.K.; Jones, J.R.; Welch, E.B. Suggested classification of stream trophic state: Distributions of temperate stream types by chlorophyll, total nitrogen, and phosphorus. *Water Res.* **1998**, *32*, 1455–1462. [CrossRef]
52. Paerl, H.W.; Xu, H.; McCarthy, M.J.; Zhu, G.; Qin, B.; Li, Y.; Gardner, W.S. Controlling harmful cyanobacterial blooms in a hyper-eutrophic lake (Lake Taihu, China): The need for a dual nutrient (N & P) management strategy. *Water Res.* **2011**, *45*, 1973–1983.

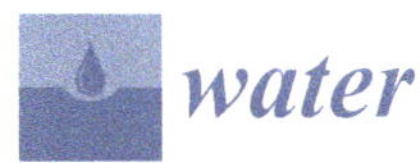

Article

Seasonal Variation and Spatial Heterogeneity of Water Quality Parameters in Lake Chenghai in Southwestern China

Pengfei Hou, Fengqin Chang *, Lizeng Duan, Yang Zhang and Hucai Zhang *

Institute for Ecological Research and Pollution Control of Plateau Lakes, School of Ecology and Environmental Science, Yunnan University, Kunming 650500, China; houyu8806@163.com (P.H.); duanlizeng2019@ynu.edu.cn (L.D.); zhangyang9106@ynu.edu.cn (Y.Z.)
* Correspondence: changfq@ynu.edu.cn (F.C.); zhanghc@ynu.edu.cn (H.Z.)

Abstract: Seasonal dynamics and the vertical stratification of multiple parameters, including water temperature (WT), dissolved oxygen (DO), pH, and chlorophyll-*a* (Chl-*a*), were analyzed in Lake Chenghai, Northern Yunnan, based on monitoring data collected in 2015 (October), 2016 (March, May, July), 2017 (March, June, October), 2018 (August), and 2020 (June, November). The results indicate that the lake water was well mixed in winter and spring when the water quality was stable. However, when WT becomes stratified in summer and autumn, the Chl-*a* content and pH value changed substantially, along with the vertical movement of the thermocline. With rising temperature, the position of the stratified DO layer became higher than the thermocline, leading to a thickening of the water body with a low DO content. This process induced the release of nutrients from lake sediments and promoted eutrophication and cyanobacteria bloom. The thermal stratification structure had some influence on changes in DO, pH, and Chl-*a*, resulting in the obvious stratification of DO and pH. In summer, with an increase in temperature, thermal stratification was significant. DO and pH achieved peak values in the thermocline, and exhibited a decreasing trend from this peak, both upward and downward. The thermocline was anoxic and the pH value was low. Although Chl-*a* maintained a low level below the thermocline and was not high, there was a sudden increase in the surface layer, which should be urgently monitored to prevent large-scale algae reproduction and even local outbreaks in Lake Chenghai. Moreover, Lake Chenghai is deeper in the north and shallower in the south: this fact, together with the stronger wind–wave disturbance in the south, results in surface WT in the south being lower than that in the north year-round. This situation results in a gradual diminution of aquatic plants from north to south. Water quality in the lake's southern extent is better than that in the north, exhibiting obvious spatial heterogeneity. It is recommended that lake water quality monitoring should be strengthened to more fully understand lake water quality and take steps to prevent further deterioration.

Keywords: Lake Chenghai; Yunnan; water quality parameters; seasonality; spatial heterogeneity

Citation: Hou, P.; Chang, F.; Duan, L.; Zhang, Y.; Zhang, H. Seasonal Variation and Spatial Heterogeneity of Water Quality Parameters in Lake Chenghai in Southwestern China. *Water* **2022**, *14*, 1640. https://doi.org/10.3390/w14101640

Academic Editor: Roohollah Noori

Received: 14 March 2022
Accepted: 27 April 2022
Published: 20 May 2022

Publisher's Note: MDPI stays neutral with regard to jurisdictional claims in published maps and institutional affiliations.

1. Introduction

Eutrophication has been a major problem affecting water resources and the environment around the world, which has profound effects on water quality and safety, as well as aquatic ecosystem health [1]. In the context of global change, the study of the mechanism of the effect of increasing temperature and eutrophication on the thermal stratification of lakes and reservoirs, as well as its ecological and environmental effects, has become one of the most prevalent issues of current international research [2,3]. Lake thermal stratification and thermal cycling are important factors governing various physicochemical processes (such as dissolved oxygen (DO) distribution, nutrient exchange, microbial activity, nutrient release from the bottom sediment, etc.) and dynamical phenomena (such as upper and lower water mixing and convection in lakes), which are important indicators affecting the biological production and ecosystem evolution of lakes [4,5]. For deep-water lakes (including reservoirs), with extreme and persistent temperature differences, the vertical distribution and variation patterns of water temperature determine the vertical stratification and mixing exchange of chemical factors as well as biological factors (phytoplankton, animals, etc.), which in turn profoundly affect the lake ecosystem [6,7]. Therefore, a deep understanding of the significance of deep-water lake hydrochemical parameters requires an in-depth study of the seasonal thermal stratification of lakes [8]. Meanwhile, the vertical variation of lake thermal stratification affects the vertical distribution of water chemistry parameters, such as lake water temperature (WT), DO, chlorophyll-*a* (Chl-*a*), pH, and electrical conductivity (EC), and this seasonal stratification of water chemistry parameters caused by water temperature changes is a typical feature of deep-water lakes. Therefore, an in-depth study of the thermodynamic stratification of lakes can help to improve the understanding of physical, chemical, and bioecological processes in lakes, and thus help to improve the health of lake ecosystems.

Over the last few decades, the hydrological conditions of many lakes have changed to the extent that the lakes have fundamentally altered in appearance [9] or in their stratification patterns. However, the effects of climate change on stratification phenology remain largely unexplored on a global scale [10]. Some scholars have conducted in-depth systematic studies of the seasonal stratification and hydrochemical characteristics of many natural deep-water and shallow-water lakes and large artificial reservoirs, such as Lakes Taihu, Tianmu, Fuxian, Wanfeng, Qiandao, Lugu, and Hongfeng [11,12]. Studies have shown that WT, DO, Chl-*a*, pH, EC, the cell densities of cyanobacteria, turbidity, and other parameters are prone to vertical seasonal stratification in summer, especially in deep-water lakes and reservoirs [13,14]; the vertical variation of lake thermal stratification affects the top-to-bottom distribution of DO, pH, Chl-*a*, and other hydrochemical parameters, and the seasonal stratification of hydrochemical parameters caused by variation in WT is a feature typical of deep-water lakes [15–17]. Stratification occurs quickly, causing hypoxia in the uniform temperature layer of water bodies, and algae grows and reproduces rapidly in the temperature change layer, disrupting balanced aquatic ecosystems and worsening the water quality. However, studies of the seasonal thermal stratification and the hydrochemical parameters of plateau deep-water lakes are few. Therefore, understanding changes in hydrochemical parameters, such as lake temperature, Chl-*a* and DO concentration, and pH value, is not only of practical significance for lake eutrophication prevention and water quality protection, but also has very important implications for local and even global climate change research.

Lake Chenghai is an important freshwater resource, supporting the productivity and lives of people around the lake. It is rich in animal and plant resources, and more importantly, Lake Chenghai is one of only three natural growth areas of *Spirulina* cyanobacteria in the world and is famous for producing high-quality *S. (Arthrospira) platensis* [18,19]. However, with the increasing human populations around the lake and the gradual development of industry and agriculture, the water quality of Lake Chenghai has deteriorated [20]. Especially since the beginning of the artificial cultivation of *Spirulina*, the eutrophication of the water body has become increasingly serious, and the perennial outbreak of algal blooms has a significant impact on the ecology and environment of the lake [21]. Most of the previous studies analyzed the eutrophication of Lake Chenghai from the perspective of aquatic plants and zooplankton, and the analytical studies of water quality parameters were only based on short-term and small-scale monitoring data, while long-term sentinel monitoring data of water quality parameters were rarely published or reviewed [22,23]. As an important deep-water lake in the southwest monsoon region, we still lack a detailed understanding of Lake Chenghai and lack detailed studies on the seasonal stratification of its temperature and vertical variation of water chemistry parameters. Faced with such a situation, we monitored Lake Chenghai from October 2015 in order to reveal the seasonal stratification characteristics and patterns of Lake Chenghai's WT, and to further explore the environmental effects brought about by the seasonal vertical stratification of the lake's

WT, providing a scientific basis for the conservation of high-quality freshwater lakes in the plateau.

2. Data and Methods

2.1. Background of Lake Chenghai

Lake Chenghai is located on the contiguous Qinghai–Tibet and Yunnan–Guizhou Plateaus at $100°33'$ to $100°45'$ E, $26°25'$ to $26°40'$ N, and in Yongsheng County, Yunnan Province. Lake Chenghai is located in the Chenghai Fracture Zone, and is a tectonically fractured lake with a small catchment area and no distant river input or outflow [24]. The results of water chemistry analysis of the lake water show that the total ions of the lake water reach 933.2 mg/L [25], and that the lake water is yellowish green; is weakly alkaline; has high hardness; is rich in Ca^{2+}, CO_3^{2-}, and HCO_3^- plasma; and can easily form autogenous carbonate mineral precipitation [26]. The soil around the lake is dominated by red loam and red-brown loam, and the basin has a wide variety of bedrock, including basalt, sandstone, dolomitic limestone, and shale [27,28]. The lake lies at an elevation of 1502 m a.s.l., its watershed area is 318.3 km^2, and the lake area itself is 75.97 km^2. Oval Lake Chenghai is oriented roughly north–south with a long axis measuring 19.15 km, a maximum east–west width of 5.21 km, and a maximum water depth of 35.87 m, with an average of 24.98 m. Lake Chenghai is one of the typical plateau deep-water lakes and has a water capacity of 1.98 million m^3. The lake was formed in the middle Quaternary period (ca. 1.2 Ma) in a fault graben formed by the Himalayan orogeny [29]. The drainage area has a subtropical climate, with an average temperature of 18.7 °C and without frost throughout the year. Lake Chenghai is an inland closed plateau deep-water lake, where surface evaporation is approximately three times greater than watershed precipitation [30]. Lake Chenghai used to be an outflow lake; the water flows 30 km southward into the Jinsha River through the Cheng River, but it is a closed lake at present and mainly recharged by groundwater and precipitation. Due to its location in the dry and hot valley of the Jinsha River, the evaporation is approximately three times greater than the precipitation in the basin, resulting in a continuous decline in the water level.

Lake Chenghai is the fourth largest among nine plateau lakes in Yunnan, one of the only alkaline lakes in the world, and one of only three lakes in which *Spirulina* can grow naturally [31,32]. The water level of Lake Chenghai has dropped by 3.97 m in the last decade due to high evaporation in the basin and agricultural water use surrounding the lake. Human activities have long aimed to obtain economic benefits and have often overlooked the fragility of lacustrine ecosystems. In recent years, due to the continuous demographic and economic development of the Lake Chenghai watershed, rapid population growth, and water pollution, eutrophication has become increasingly serious. In addition to the "point source", i.e., pollution from *Spirulina* farming wastewater, the most important source of contamination is domestic sewage from villages, agricultural non-point sources, and soil erosion [33]. At present, the environmental health of Lake Chenghai is rated as Class IV, its water quality is mesotrophic, and it faces the threat of eutrophication.

2.2. Sampling

Controlled by the tectonic background, the water depth of Lake Chenghai is characterized by a north–south oriented feature, and is deeper on the western side than the eastern side. Therefore, in this study, four sampling points were set up along the deep-water axis of the lake from north to south, marked as A ($100°39'07''$ E~$26°35'26''$ N), B ($100°38'41''$ E~$26°33'23''$ N), C ($100°38'37''$ E~$26°31'33''$ N), and D ($100°38'45''$ E~$26°29'30''$ N) (Figure 1). On-site monitoring was carried out in 2015 (October), 2016 (March, May, July), 2017 (March, June, October), 2018 (August), and 2020 (June, November). Sampling locations were established by means of a satellite-based global positioning system (GPS) and water quality parameters (including WT, DO, and Chl-*a* concentration; pH value; and phycocyanin concentration; among others) were measured with a Xylem Analytics YSI-6600 multi-parameter sonde. One vertical line was established at each site to monitor water quality at different depths. Data were first collected between 0.1 and 1 m below the water surface and the last data were monitored 0.5 m above the lake bottom, with additional data collected at one-meter intervals. To ensure accuracy, each depth was measured six times.

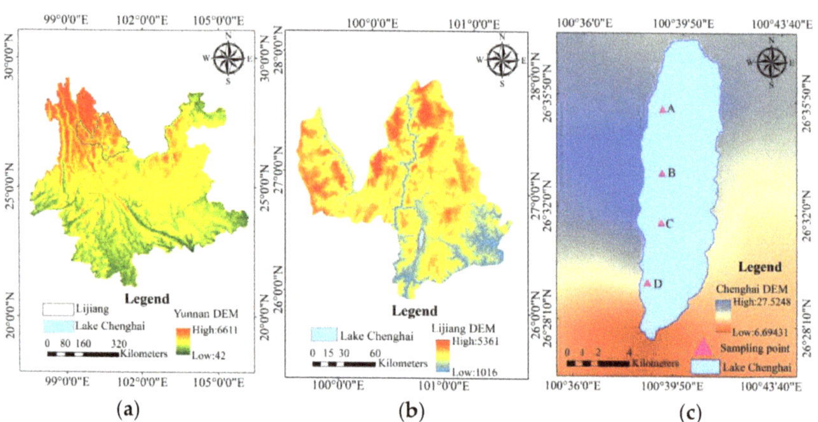

Figure 1. Distribution of sampling locations in Lake Chenghai: (**a**) Yunnan Province; (**b**) Lijiang City; (**c**) Sampling point of Lake Chenghai.

2.3. Analysizing Methods

2.3.1. Comprehensive Nutrition State Index

According to lake and reservoir eutrophication evaluation methods and grading technical regulations, total phosphorus (TP), permanganate index (COD_{Mn}), DO, Chl-*a*, and transparency are the main indicators of nutrient levels in water [34]. To systematically understand water quality and the eutrophication level of Lake Chenghai, the comprehensive nutrition state index was used to evaluate lake eutrophication levels under various scenarios. The comprehensive nutrition state index (*TLI*) is given as:

$$TLI\left(\sum\right) = \sum_{j=1}^{m} w_j \cdot TLI(j) \tag{1}$$

where *TLI* ($\sum$) is the integrated trophic level index, *TLI*(*j*) is the trophic level index of *j*, and W_j is the correlative weighted score for the trophic level index of *j*.

$$W_j = \frac{r_{ij}^2}{\sum_{j=1}^{m} r_{ij}^2} \tag{2}$$

where W_j is the correlative weighted score for the trophic level index of *j* and r_{ij} is a relative coefficient.

In Chinese lakes and reservoirs, the correlation coefficient for Chl-*a* to other parameters is presented in Table 1 [35]:

Table 1. Correlation coefficient for Chl-*a* to other parameters in Chinese lakes and reservoirs.

Parameter	Chl-*a*	TP	TN	SD	COD_{Mn}
r_{ij}	1	0.84	0.82	−0.83	0.83
r_{ij}^2	1	0.7056	0.6724	0.6889	0.6889

The computational formula for each eutrophication index is:

$$TLI(Chla) = 10(2.5 + 1.086 \ln Chla) \tag{3}$$

$$TLI(TP) = 10(9.435 + 1.624 \ln TP) \tag{4}$$

$$TLI(TN) = 10(5.435 + 1.694 \ln TN) \tag{5}$$

$$TLI(CODMn) = 10(0.109 + 2.66 \ln CODMn) \tag{6}$$

$$TLI(SD) = 10(5.118 - 1.94 \ln SD) \tag{7}$$

A series of 0~100 continuous values were adopted for grading the eutrophication level: trophic level index $TLI (\sum) < 30$, oligotrophic; $30 \leq TLI (\sum) \leq 50$, mesotrophic; $TLI (\sum) > 50$, eutrophic; $50 < TLI (\sum) \leq 60$, weak eutrophic; $60 < TLI (\sum) \leq 70$, middle eutrophic; and $TLI (\sum) > 70$ hypo-eutrophic.

2.3.2. Lake Quality Level

River water quality classification was based on national quality standards (GB 3838-2002) [36]. According to the environmental functions and protection objectives of surface waters, it is divided into five categories on a functional level (Table 2).

Table 2. Water function and standard classification.

Water Quality Classification	Scope of Application
Class I	Mainly applicable to source waters and national nature reserves.
Class II	Mainly applicable to centralized drinking water, surface water sources, first-class protected areas, etc.
Class III	Mainly applicable to secondary protection zones, fisheries, and swimming areas of centralized drinking water surface water sources.
Class IV	Mainly applicable to general industrial water use areas and recreational areas where the human body is not in direct contact with water.
Class V	Mainly applicable to agricultural water use areas and general landscape requirements.

2.3.3. Correlation Analysis Method

Pearson's correlation coefficient is a metric used to describe relationships among variables. This method uses the covariance matrix of data to evaluate the strength of the relationship between two vectors. Normally, the Pearson's correlation coefficient between two variables, β_i and β_j, can be calculated as shown in Equation, where $cov(\beta_i, \beta_j)$ is the covariance, $var(\beta_i)$ is the variance of β_i, and $var(\beta_j)$ is the variance of β_j [37].

$$R(\beta_i, \beta_j) = \frac{cov(\beta_i, \beta_j)}{\sqrt{var(\beta_i) \times var(\beta_j)}} \tag{8}$$

3. Results and Discussion

3.1. Vertical Stratification and Seasonal Temperature Fluctuations

In large deep-water lakes, WT changes vertically due to inconsistent warming and cooling of the upper and lower water bodies. The vertical distribution of WT in March, May, July, and October at each sampling locus in Lake Chenghai clearly shows that seasonal variation of WT in the lake is excellent, and the vertical temperature gradient of the water body is obvious in May, July, and October. The depth of the thermocline gradually changes over time (Figure 2). With the increase in solar radiation and air temperature, the surface water body rapidly warmed up, the temperature difference with the deep-water body gradually increased, and the water body appeared as a stratification phenomenon. According to the standard definition, a thermocline is a water layer whose temperature gradient is more than 0.2 °C/m [38]. The air temperature of Lake Lugu reaches the highest in May, which makes the external heat continuously transfer downward from the surface layer, the maximum temperature difference decreases, the depth of the temperature leap layer moves up, and the temperature stratification phenomenon is obvious. The WT of Lake Chenghai reaches its highest in summer and autumn, which leads to the continuous transfer of external heat from the surface layer to the deeper layer, a decrease in the maximum temperature difference, and an increase in the depth of the thermocline, showing a clear temperature stratification. Changes in temperature always lead to the thermal stratification of lakes, during which period the WT decreases slowly in the epilimnion and hypolimnion and sharply in the thermocline with increasing depth [39,40].

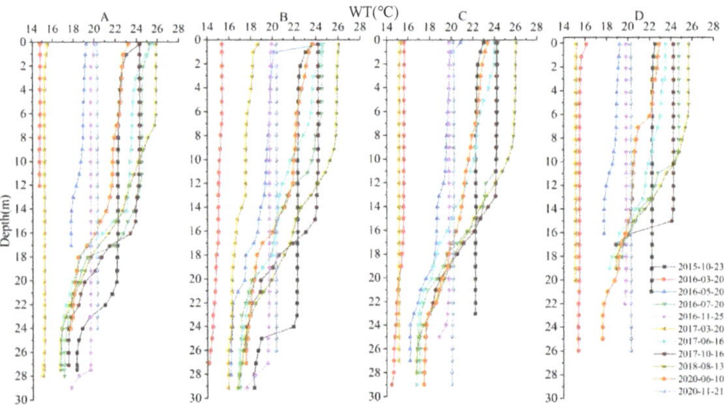

Figure 2. Vertical profile of water temperature (WT) in Lake Chenghai, Yunnan.

The overall temperature of the lake in October 2015 was relatively high, with the temperature in the south higher than that in the north. Vertically, there was an obvious thermocline of 20 to 26 m depth at Locus A, with a concomitant temperature increase of around 4 °C. At Locus B, there was a clear thermocline 23 to 27 m in depth and a temperature rise again of 4 °C. There was no obvious thermocline at Locus C and D, where the water is shallow. The WT in March 2016, the lowest monitored, was higher in the north than in the south, which may be related to the larger thermocline in the deeper lake area. There was no apparent stratification at the four sampling loci from north to south, and the temperature only slowly fell with increasing depth. In May 2016, the lake WT was evenly distributed horizontally, and the WT in the north and south was relatively consistent. Vertically, temperature increases were recorded at all four monitoring points. In the north, in the deep-water area, a 3 °C temperature jump appeared at Locus A between 21 and 24 m. In the deep-water area, a 2.4 °C temperature increase appeared at Locus B between 16 and 24 m in depth. The thickness of this layer reached 8 m. In the south, a layer of increased temperature measuring 2.4 °C appeared at Locus C at 17 to 24 m depth. At Locus D, between 8 and 14 m depth, a modest 1.2 °C temperature increase was recorded. The average temperature of the lake water in July 2016 was the highest at all four monitoring locations, especially in the upper water layer, with a maximum value of 25.5 °C. Horizontally, surface WT in the north was higher than that in the south, while vertically all four loci showed obvious temperature increases. Between 13 and 22 m depth, a temperature increase layer of 6.8 °C formed at Locus A, while below 22 m, the temperature decreased slowly. A thermocline appeared at 7 to 23 m depth at Locus B, reflected by a temperature increase of 7.1 °C. Below 23 m, the temperature decreased slowly. At Locus C, the temperature barely changed in the 0 to 11 m depth range, and small temperature changes were recorded between 11 and 15 m, with a clear thermocline apparent at 15 to 23 m. At 5.5 °C, the temperature below 23 m decreased slowly with increasing depth, and the temperature at Locus D hardly changed in the top 10 m of the water column. Below 10 m to the bottom of the lake (ca. 18 m), the temperature dropped sharply.

The WT decreased between 0 and 5 m by 0.4 °C at Locus A in March 2017. The trend of change was smoother between 5 and 20 m, with WT remaining at around 15.3 °C, and an intensified decreasing trend from 25 to 30 m. In June 2020, Locus A exhibited a temperature increase between 0 and 15 m, with a temperature increase of 2.1 °C. Below 15 m, the temperature dropped sharply by 3.4 °C. At Locus B, the WT increased with depth as the temperature decreased by 0.89 °C between 0 and 5 m. From 5 to 10 m, the change trend was smooth, and from 10 to 30 m, the temperature decreased by 1.69 °C. In June, Locus A yielded a WT of 25.59 °C, the highest in 2017. From 0 to 15 m, WT remained stable, near 24 °C, but WT decreased sharply to 16.8 °C with increasing depth between 15 and 30 m, a decline of 8.79 °C. In October 2017, Locus B exhibited a thermocline from 7 to 23 m depth, reflecting a temperature increase of 5.96 °C, while WT decreased slowly below 23 m. The October temperature trend at Locus A, as with June, showed a significant thermocline between 14 and 25 m depth with a temperature jump of 6.4 °C, while the WT between 0 and 30 m dropped slowly by 6.8 °C. In 2020, the temperature at Locus A tended to be stable from 0 to 7 m in June, and decreased by 4.06 °C below 10 m. The WT at Locus B was nearly constant from 0 to 11 m depth, and there was a significant thermocline indicated by an increased temperature of 6.4 °C between 15 and 27 m.

From the change trend of Lake Chenghai's WT (Figure 2), it can be seen that the seasonal stratification feature of Lake Chenghai's WT is obvious. The stratification of water bodies is characterized by mixing in winter and stratification in summer and autumn, and the lake belongs to a single mixed type of lake. With the rapid decline in temperature in winter, the temperature of the upper water layer decreases, the upper and lower water layers are mixed, and the stratification phenomenon disappears; the temperature difference between surface water and bottom water is small, forming a more uniform water temperature. Lake Chenghai enters the mixing period in winter, and the water temperature gradient changes minimally, especially in the range of water depth below 10 m. In Lake Chenghai, in the summer, after the temperature rose rapidly with the enhancement of the surface layer temperature, it began to gradually increase; the surface layer and the bottom of the water temperature gap began to increase, the lower layer of water temperature change was minimal, and the water body gradually demonstrated a water temperature stratification phenomenon. After October, the surface layer temperature difference gradually became smaller, and the variable temperature layer became thicker, which is due to the fall as the temperature decreases; the reservoir surface temperature gradually decreased, the higher temperature of the bottom layer of water received the cooling of the upper layer of water, and the temperature stratification gradually weakened.

3.2. Seasonal Variation Characteristics of the Water Quality Profile

3.2.1. Dissolved Oxygen (DO)

DO is a proxy of primary productivity and hydrodynamic conditions in lacustrine environments, and it is inversely proportional to salinity and temperature but positively proportional to the intensity of lake wave action; additionally, the level of DO is related to the photosynthesis of algae [41–43]. Seasonal variations in DO in Lake Chenghai are obvious, exhibiting distinct characteristics in different months (Figure 3). Our monitoring data show that the DO content of Lake Chenghai was not high in October 2015. The DO content at Locus A, the deep area in the north, increased slightly and then decreased at a depth of 0 to 7 m, with a maximum of 9.6 mg/L at a depth of 1 m. The DO content decreased slowly from 7 to 16 m, maintained a value of 5.3 to 4.6 mg/L, and then decreased rapidly from 16 to 21 m and finally maintained a concentration of approximately 0.3 mg/L. At Locus B, DO decreased between 0 and 7 m, and then remained stable at 5 mg/L to a depth of 22 m. Between 22 and 25 m, the DO content dropped sharply. The DO content at Locus C and D in the southern lake area decreased slowly with increased depth, but no sharp changes were observed. Horizontally, the DO content in the north was much higher than the south. The average DO content at 0 to 12 m depth was 6.5 mg/L at Locus A, 6.1 mg/L at Locus B, 5.7 mg/L at Locus C, and 4.5 mg/L at Locus D, showing a general decreasing trend from north to south.

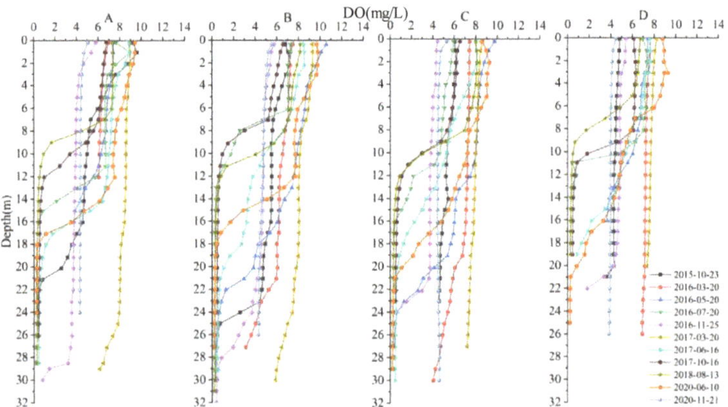

Figure 3. Vertical profile of dissolved oxygen (DO) in Lake Chenghai, Yunnan.

The amount of DO in Lake Chenghai in March 2016 was relatively high, with the amount at Locus A decreasing slowly as the depth increased. A maximum of 7.58 mg/L was recorded at the surface, and a minimum value of 6.92 mg/L at the bottom. At Locus B, DO also decreased slowly from 0 to 19 m, and rapidly below 19 m in depth. The maximum DO content was recorded on the surface (7.43 mg/L), and the minimum at the bottom (4.23 mg/L). The DO content between 0 and 20 m at Locus C decreased slowly and then increased sharply. A maximum value was recorded at the surface (7.47 mg/L) and a minimum value at the bottom (3.12 mg/L). Due to the shallow depth at Locus D, the amount of DO slowly decreased with increased depth. The maximum value (6.96 mg/L) occurred in the surface layer, and the minimum value (6.07 mg/L) was recorded in the lower layer. Horizontally, the DO content in the north was much higher than in the south. The average DO content at a depth of 0 to 12 m was 7.3 mg/L at Locus A, 7.3 mg/L at Locus B, 6.9 mg/L at Locus C, and 6.3 mg/L at Locus D.

The highest DO content in Lake Chenghai was observed in May 2016, and the DO content of the upper water column was higher than that of the lower water column at Locus A. The DO content first increased and then decreased from 0 to 6 m in water depth, reaching a maximum of 9.71 mg/L at a depth of 2 m. The DO content decreased slowly from 6 to 19 m, and then decreased sharply at 19 m to 0.3 mg/L. The DO content at Locus B decreased rapidly between 0 and 4 m, and maintained a stable state from 7 to 13 m. Below 13 m, the DO content decreased continuously, forming a stable hypoxic environment below 22 m. At Locus C, the DO content decreased gradually with increasing water depth from 0 to 11 m, and increased sharply between 11 and 14 m. The DO content remained stable from 14 to 19 m, and decreased sharply below 19 m, creating a stable hypoxic environment below 27 m. The amount of DO between 0 and 10 m depth at Locus D decreased steadily as the water depth increased, and the amount of DO at depths greater than 10 m decreased sharply. The DO content in the north was much higher than that in the south. The average DO content between 0 and 12 m was 8.8 mg/L at Locus A, 8.7 mg/L at Locus B, 8.4 mg/L at Locus C, and 6.9 mg/L at Locus D. The DO content of Lake Chenghai in July 2016 was low, but higher between 0 and 3 m at Locus A and almost unchanged between 3 and 11 m water depth. The DO content decreased sharply from 11 to 16 m and remained below 0.3 mg/L at a depth of 16 m. The DO content in the depth range of 0 to 6 m at Locus B was basically maintained above 7 mg/L, and the DO content in the depth range of 6 to 11 m decreased sharply and stabilized below 0.3 mg/L below 11 m. The DO content in the depth range of 0 to 11 m at Locus C decreased slowly with increasing water depth and stabilized below 0.3 mg/L below 17 m. The DO content in the depth range of 0 to 9 m at Locus D in the southern shallow lake area was basically stable at approximately 7 mg/L, and then decreased sharply from 9 to 12 m. However, the DO content below 12 m slowly decreased as the water depth increased. The DO content was relatively high at Locus A and D and comparatively low at Locus B and C in the horizontal direction. In winter, the DO and WT in Chenghai water were characterized by uniform mixing at the same time. The reason for this change can be inferred from the fact that, in winter, the DO in Chenghai water is reduced due to the vertical exchange of water bodies, the consumption of DO in water by reducing substances, and the upwelling of anoxic water bodies in the lower layer, which consumes more oxygen than reoxygenation, causing the DO to decrease and reach anoxia in severe cases. In summer and autumn, the DO concentration decreases with the increase in water depth and decreases sharply in the depth range of the thermocline; below the thermocline, the DO is basically constant.

Generally, it can be seen that in the vertical direction, the variation trend of DO is obvious (Figure 3), and the lake shows a generally decreasing trend with the increase in water depth. This might be attributed to the photosynthesis of planktonic algae. When photosynthesizing in the surface layer, O_2 is released, which increases the oxygen content in the water body; when respiration occurs in the bottom layer, O_2 is consumed in the water body, which results in decreased oxygen content in the water body. The DO in the rainy season is always higher than that in the dry season, which might be due to the increase in algae activity when the WT increases. On the other hand, with the increase in precipitation in the rainy season, the enhanced hydrodynamics and lake wave action cause the DO content in the surface water column to increase rapidly, and at the same time, along with the water level rising, the respiration of bottom plants increases, and so does the oxygen consumption.

3.2.2. pH Value

The pH of water bodies is one of the most important indicators of chemical and biological changes in lake systems and is a major influence on inland water bodies and associated ecosystems [44]. Our monitoring data indicate that the lake experiences obvious seasonal changes (Figure 4). There is a general lacustrine phenomenon in which, due to the strong photosynthesis of algae in surface water and the high consumption of CO_2, the pH values are high. On the other hand, the photosynthesis of bottom-dwelling algae is weak. High algal die-off causes CO_2 to accumulate in large quantities. In addition, the decomposition of organic matter produces aids, which reduce the pH values [45,46].

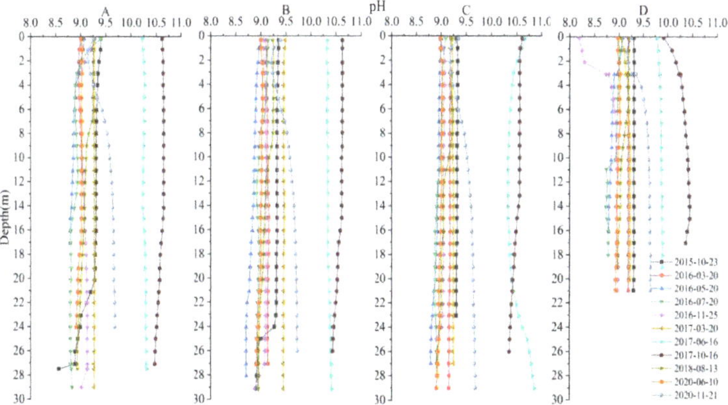

Figure 4. Vertical profile of pH values in Lake Chenghai, Yunnan.

In October 2015, the pH of Lake Chenghai was relatively high, with significant vertical reductions near the thermocline at Locus A and B in the northern deep-water region, and a jump gradient of approximately 0.4. Locus C and D, located in the southern lake with shallow depth, exhibited small vertical changes, and the pH values of the samples taken in the south were higher those that in the north. The pH value of lake water in March was relatively stable and there was almost no detectable change vertically, and the pH values in the northern lake were higher than those in the southern lake. The pH value of the lake water dropped significantly near the thermocline at Locus A and B in May 2016, with a jump gradient of approximately 0.4 and a steady drop at Locus C. The pH decreased sharply at Locus D between 0 and 4 m water depth, which is the shallowest location in the south. The vertical pH values at Locus A dropped sharply from the surface to a depth of 3 m in July 2016, with a jump amplitude of 0.6. The pH value first increased slightly and then decreased slowly between 3 and 17 m water depth, finally becoming stable. The pH values at Locus B decreased slowly with increasing depth between 0 and 14 m, and then stabilized. At Locus C, the pH values decreased slowly with increasing depth from 0 to 17 m, and then increased slowly with growing water depth. The pH values at Locus D decreased slowly between 0 and 11 m water depth. Horizontally, the pH values in the northern lake area were higher than those in the south. In general, the pH change in Chenghai is not a significant change. The changes in pH in the Lake Chenghai water column are mainly related to the photosynthesis of the algae growth process; the surface water algae consume CO_2 in water through photosynthesis. The accumulation of CO_2 under the thermocline in the stratified water column slightly reduces the pH of the water column, because the light cannot penetrate the thermocline and the CO_2 in the lower water column is not consumed by photosynthesis.

From the pH variation curves in Figure 4, it can be seen that the pH varies very little in the vertical direction in Lake Chenghai, and all data measured at the four sites are between 8.18 and 10.65. The overall decreasing trend with increasing water depth may be related to the respiration of planktonic algae. The surface planktonic algae photosynthesize and absorb CO_2 from the water, thus increasing the pH value; the bottom algae mainly respire and exhale CO_2, thus decreasing the pH value due to the decrease in sunlight transmission.

3.2.3. Chlorophyll-*a* (Chl-*a*)

The Ch-*a* content is an important proxy of phytoplankton biomass in lakes [47]. Our study results show that there are 175 species of algae in Lake Chenghai, and mainly diatoms, green algae, and cyanobacteria dominated quantitatively, with cyanobacteria showing typical eutrophic cyanobacterial characteristics. The spatio-temporal content of Chl-*a* in Lake Chenghai varied greatly (Figure 5).

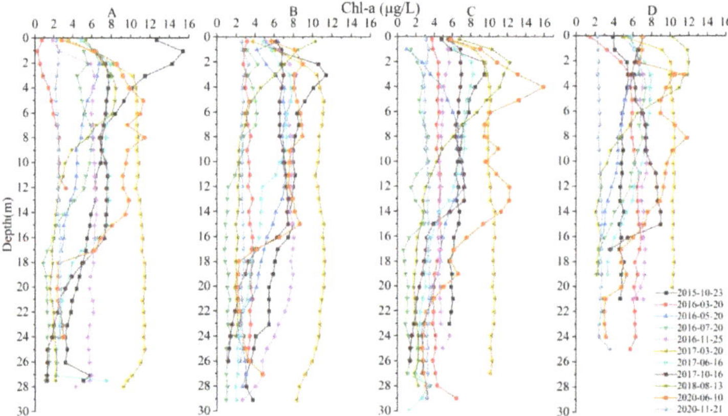

Figure 5. Vertical profile of chlorophyll-*a* (Chl-*a*) in Lake Chenghai, Yunnan.

Our monitoring results show that the Chl-*a* content in Lake Chenghai was relatively high in October 2015, exhibiting a top-down increase and then a sharp decrease between 0 and 7 m at Locus A in the northern deep waters of the lake, with a maximum value (15.4 µg/L) observed at 1 m depth. The Chl-*a* content decreased slowly with depth between 7 and 26 m and showed a slight increase below 26 m, mainly due to algal death and sedimentation. The Chl-*a* content between 7 and 26 m decreased slowly with depth, and showed a slight increase below 26 m. This may be due to the death and sedimentation of some algae, and a maximum value of 11.4 µg/L was recorded at Locus B between 0 and 6 m. The Chl-*a* content stabilized at 8 µg/L between 6 and 16 m, decreased again between 16 and 19 m, and stabilized from 19 to 24 m water depth. The Chl-*a* content at a depth of 29.5 m changed three times at Locus B, indicating that the algal species in this position were relatively abundant and the population size was relatively large. At Locus C, Chl-*a* first increased at a depth of 0 to 6 m and then decreased sharply, reaching a maximum value of 9.5 µg/L at a depth of 2 to 3 m, and then decreased slowly below 6 m. The Chl-*a* content gradually increased at Locus D between 0 and 3 m and then slowly decreased below 3 m. The maximum recorded Chl-*a* value was 5.6 µg/L at 3 m. Horizontally, the Chl-*a* content gradually decreased from north to south, and the average Chl-*a* content in the 0 to 12 m depth range was 9.4 µg/L at Locus A, 8.4 µg/L at Locus B, and 7.3 µg/L at Locus C. The average Chl-*a* content at the 0 to 12 m water depth was 9.4 µg/L at Locus A, 8.4 µg/L at Locus B, 7.3 µg/L at Locus C, and 4.9 µg/L at Locus D.

The Chl-*a* content in Lake Chenghai was relatively low in March 2016; the surface content of Locus A was low and increased rapidly between 0 and 3 m water depth, after which changes were small. The Chl-*a* content increased slowly between 3 and 16 m water depth, and decreased from 16 to 26 m, with a maximum value (7 µg/L) recorded at 16 m. The Chl-*a* content between 0 and 12 m decreased and then increased again with depth, with the lowest value (0.3 µg/L) occurring at a depth of 1 m. The Chl-*a* content gradually decreased from north to south, with the average Chl-*a* content between 0 and 12 m at Locus A being 0.3 µg/L. The average content of Chl-*a* was 5.4 µg/L at Locus A, 4.4 µg/L at Locus B, 3.2 µg/L at Locus C, and 1.8 µg/L at Locus D. The vertical content of Chl-*a* varied greatly in May 2016, with Chl-*a* content between 0 and 2 m rising rapidly from 3.1 µg/L to 8.3 µg/L at Locus A, and then remaining constant around 8 µg/L from 2 to 20 m water depth. Below 20 m, the Chl-*a* content decreased gradually, finally reaching a minimum of 3 µg/L. The Chl-*a* content from 0 to 4 m at Locus B first decreased and then increased rapidly, finally stabilizing at around 7 µg/L between 4 and 13 m water depth. The Chl-*a* content from 13 to 24 m water depth continued to decrease, finally achieving a minimum value of 3 µg/L between 24 and 28 m. The Chl-*a* content between 0 and 4 m decreased and then increased rapidly, finally maintaining a value of approximately 7 µg/L at the depth ranging from 4 to 13 m. The Chl-*a* content between 13 and 24 m decreased continuously with increasing depth, and then increased slowly from 24 to 28 m, with a minimum value of 2.2 µg/L recorded. At Locus C, the Chl-*a* content between 0 and 12 m increased continuously downward, with a larger increase between 0 and 4 m. At Locus C, the Chl-*a* content from 0 to 22 m increased continuously with water depth. From 12 to 22 m, the Chl-*a* content continued to decrease with water depth, and below 22 m, the Chl-*a* content was essentially stable at 2.8 µg/L. In the southern part of the lake at Locus D, the Chl-*a* content increased rapidly with water depth between 0 and 2 m, and then decreased rapidly between 2 and 6 m, and continued to decrease steadily below 6 m. The average content of Chl-*a* between 0 and 12 m decreased with water depth. The average content of Chl-*a* at this depth was 7.9 µg/L at Locus A, 6.6 µg/L at Locus B, 5.4 µg/L at Locus C, and 5.3 µg/L at Locus D. The average Chl-*a* content in the water depth range of 0 to 12 m was 7.9 µg/L at Locus B, 5.4 µg/L at Locus C, and 5.3 µg/L at Locus D.

The largest changes in the Chl-*a* content were observed in the vertical water column in July 2016. For example, the Chl-*a* content increased sharply from 9.9 µg/L to 17.4 µg/L from 0 to 2 m water depth at Locus A, and then decreased to 13.7 µg/L between 2 and 4 m water depth. From 4 to 12 m water depth, the Chl-*a* content decreased slowly. The Chl-*a* content decreased from 10.2 µg/L to 2.2 µg/L at 12 to 17 m water depth, and remained stable at 1.7 µg/L below 17 m depth. The Chl-*a* content was the maximum measured at 19.9 µg/L between 0 and 4 m water depth at Locus B, and remained stable at 16 µg/L from 4 to 6 m, while it decreased sharply to 1.9 µg/L between 6 and 15 m. The Chl-*a* content increased and then decreased between 0 and 5 m at Locus C. From 1 to 3 m water depth, the Chl-*a* content was maintained at a high level of 17 µg/L. The Chl-*a* content decreased with depth from 5 to 10 m, and also from 10 to 17 m. However, below 17 m water depth, the Chl-*a* content decreased sharply from 12.7 µg/L to 2.2 µg/L with increasing depth, and remained at a low level of around 2 µg/L. The Chl-*a* content of the surface layer at Locus D was 4.1 µg/L, which jumped to 12.5 µg/L at 1 m water depth, and remained at approximately 12 µg/L between 1 and 10 m water depth. However, the Chl-*a* content increased to 15 µg/L at 15 m depth and decreased sharply from 15 µg/L to 2 µg/L between 10 and 15 m. Horizontally, the Chl-*a* content was the highest at Locus C in the south, followed by Locus A and B in the north, and the lowest appeared at Locus D, again in the south, exhibiting strong spatial heterogeneity.

From the Chl-*a* curves in Figure 5, it can be seen that the trend of Lake Chenghai's Chl-*a* content is obvious and might be influenced by the wave and WT. The Chl-*a* content showed a decreasing trend with increasing water depth. The results of correlation analysis revealed that the Chl-*a* content had a significant positive correlation with WT. Meanwhile, the differences in the Chl-*a* content in Lake Chenghai water bodies were larger in the dry season and smaller in the rainy season, while the Chl-*a* content in the rainy season was lower than that in the dry season. This may be attributed to the fact that the concentration of toxins produced by the massive outbreak of Microcystis aeruginosa reached a certain level during the rainy season, which had a lysis effect on other algae. On the other hand, it may be related to the pH changes, as the increased water volume of Lake Chenghai in the rainy season had a dilution effect, which reduced the Chl-*a* content, and the enhanced hydrodynamics of the lake made the water body more homogeneous.

3.3. Correlation Analysis of Water Quality in Lake Chenghai

From the above discussion, it can be concluded that the spatial and seasonal changes in the water quality are significant in Lake Chenghai. Based upon the correlation analysis of the water quality parameters, including WT, pH, Chl-*a*, and DO (Table 3), there is a significant positive correlation between various parameters. The correlation coefficient between pH and DO is 0.55, the correlation coefficient between WT and Chl-*a* is 0.39, and the correlation coefficient between pH and Chl-*a* is 0.39. This verifies that WT affects the Chl-*a* content, while WT and Chl-*a* jointly influence both pH and DO.

Table 3. Vertical profile of chlorophyll-*a* (Chl-*a*) in Lake Chenghai, Yunnan.

	pH	WT	Chl-*a*
DO	0.55 **	0.24 **	0.28 **
pH		0.24 **	0.39 **
WT			0.39 **

** Significant correlation at 0.01 level (bilateral).

Based on the analysis of monitoring data from Lake Chenghai and meteorological data from Yongsheng County, it can be concluded that the feedback sensitivity of the lake water to the temperature is low, the temperature change in the water body always lags behind the air temperature change, and the precipitation has a strong influence on WT. The average monthly air temperature was 2.3 °C lower in October 2015 than one year later due to the larger number of rainy days in October 2016. Throughout the year, the average temperature of the southern lake area was higher than that of the northern lake area in both summer and autumn; however, during winter and spring, the average WT of the northern lake area was higher than that of the southern. The surface water temperature of the northern lake area was higher than that of the southern area. This is mainly due to the perennial prevalence of a southerly wind in Lake Chenghai. The mixing of shallow lake water in the southern area was uniform and heat exchange was frequent, while the northern deep-water lake area was stable. At Lake Chenghai, during summer and autumn, due to the obvious stratification in the deep-water area in the north, convection exchange between the upper and lower water bodies is hindered, and the distribution of nutrients such as N and P and light in the water body is affected, thus impacting the DO concentration and the vertical distribution of aquatic organisms, which then affects the pH, resulting in changes in water quality. In 2020, the water quality of Lake Chenghai was listed as Class IV, meaning slightly polluted, and this does not meet the requirements of the water environment function [48]. The lake's nutritional status index was 45.9, and the nutritional status was medium (Figure 6).

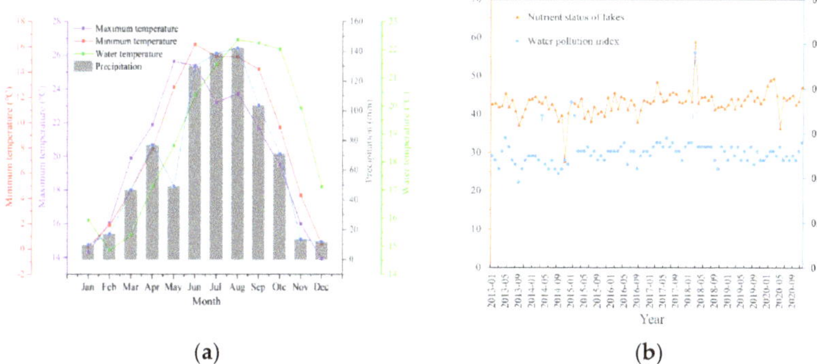

(a) (b)

Figure 6. (a) Changes in monthly average temperature and precipitation in Yongsheng County, Yunnan from 2015–2020; (b) Lake Chenghai water pollution index and lake nutrition state index.

The Chl-*a* content in clear water correlates positively with WT. As WT rises, the amount of phytoplankton increases and multiple peaks occur in different depth ranges, indicating the diversity and complexity of aquatic plant species [49,50]. Differences in pH occur principally because of changes in CO_3^{2-} and HCO_3^{-}; thus, when the temperature rises in the spring, the pH values increase with the growth of phytoplankton. However, in Lake Chenghai, a different situation occurred. In May and July, the WT was significantly higher than in March, but the pH value was lower than in March, which may be due to the end of multiple algal blooms occurring during this period and the decomposition of algae, such as cyanobacteria of the genus *Anabaena*.

3.4. Limitations and Implications

Human activities are also one of the main reasons for the large changes in the water quality in Lake Chenghai, which should be given special attention in the future. First, in this study, the seasonal dynamics of WT and its vertical stratification structure were analyzed based on monitoring data, and the seasonal stratification characteristics of the water chemical properties in Lake Chenghai were discussed. However, there was a lack of long-term and high-frequency observations on the thermal stratification transformation and its critical period regarding the hydrodynamic profile, nutrients, and phytoplankton. Second, due to the limited number of monitoring sites, the results possibly cannot represent the situation in the whole lake. With improvements in the monitoring system and methods, this problem may be solved in the near future. Third, the impact of human activities on the water quality of Lake Chenghai in the context of urbanization was not analyzed. In addition to climate change, human activities (population, GDP, impervious area, industrial structure, non-point pollution, etc.) also cause water quality changes in Lake Chenghai. In the context of complex climate change and anthropogenic disturbances in the future, there is still much work to be carried out in order to gain a more comprehensive and in-depth understanding of the thermal stratification characteristics of the water column in Lake

Chenghai and other similar lakes in the region, and their ecological and environmental effects (e.g., to reveal their effects on changes in the phytoplankton community structure and even the driving mechanisms of water blooms).

4. Conclusions

The stratification of Lake Chenghai seawater is characterized by mixing in winter and stratification in summer and autumn, with the lake being a single mixed lake. In winter, the lake water of Lake Chenghai is in a state of mixing, and complete water convection exchange causes the bottom water to maintain a high level of DO. Thermal stratification in summer and autumn directly hinders the exchange of substances between the temperature change layer and the stagnant water layer, resulting in the formation of stratified DO levels. The higher the lacustrine WT, the closer stratification of DO is to the water surface. In other words, with a thicker and more anoxic water body, coupled with the long-term effects of reducing substances and microorganisms, such as sulfides, nitrites, and ferrous ions in the sediment, an anaerobic zone is formed near the sediment, leading to the accelerated release of nitrogen and phosphorus from the sediments, deteriorating the bottom water quality. With the seasonal formation and disappearance of the thermal stratification of Lake Chenghai, there will be an impact on the water quality of the lake. Therefore, to protect the ecosystem of Lake Chenghai, water quality monitoring should be conducted in summer and autumn, and a rapid emergency response mechanism should be developed in advance.

In terms of Lake Chenghai, though the lake remains at a mesotrophic level, it will still face the risk of accelerated eutrophication and algal bloom outbreaks. Especially under the influence of complex climate change and human activities in the future, a great deal of work is still needed, and it would be worthwhile to understand the thermal stratification characteristics of Lake Chenghai's water column and its ecological effects (e.g., to reveal its driving mechanism on the structural changes of the phytoplankton community and even the occurrence of water blooms) in a more comprehensive way. Therefore, in a subsequent study, the monitoring of water quality parameters in Lake Chenghai needs to be continued and supplemented with observations from other seasons and months in order to fully reflect the thermal stratification in the lake in different seasons and months, as well as the spatial differences and potential influencing factors of the stratification parameters. Meanwhile, the seasonal variations in lake thermal stratification (depth, intensity, and thickness), nutrient salinity, sediment phytoplankton biomass, and their influencing factors will be explored.

Author Contributions: H.Z. and P.H.: conceptualization, methodology, software and writing—original draft preparation. F.C.: supervision, writing—review and editing. H.Z.: conceptualization, supervision, resources, writing—review and editing, and foundation's acquisition. L.D. and Y.Z.: investigation and data curation. All authors have read and agreed to the published version of the manuscript.

Funding: This work was supported by the Special Project for Social Development of Yunnan Province (202103AC100001) and the Yunnan Provincial Government Leading Scientist Program (No. 2015HA024).

Data Availability Statement: The data that support the findings of this study are available from the corresponding author upon reasonable request.

Acknowledgments: We are grateful to the teachers and students who participated in the field collection of data.

Conflicts of Interest: The authors declare no conflict of interest.

References

1. Lu, Y.; Wang, R.; Zhang, Y.; Su, H.; Wang, P.; Jenkins, A.; Ferrier, R.C.; Bailey, M.; Squire, G. Ecosystem health towards sustainability. *Ecosyst. Health Sustain.* **2015**, *1*, 1–15. [CrossRef]
2. Paerl, H.W.; Paul, V.J. Climate change: Links to global expansion of harmful cyanobacteria. *Water Res.* **2012**, *46*, 1349–1363. [CrossRef] [PubMed]

3. Bruesewitz, D.A.; Carey, C.C.; Richardson, D.C.; Weathers, K.C. Under-ice thermal stratification dynamics of a large, deep lake revealed by high-frequency data. *Limnol. Oceanogr.* **2015**, *60*, 347–359. [CrossRef]
4. Zhang, Y.; Wu, Z.; Liu, M.; He, J.; Shi, K.; Zhou, Y.; Wang, M.; Liu, X. Dissolved oxygen stratification and response to thermal structure and long-term climate change in a large and deep subtropical reservoir (Lake Qiandaohu, China). *Water Res.* **2015**, *75*, 249–258. [CrossRef] [PubMed]
5. Alexakis, D.; Kagalou, I.; Tsakiris, G. Assessment of pressures and impacts on surface water bodies of the Mediterranean. Case study: Pamvotis Lake, Greece. *Environ. Earth Sci.* **2013**, *70*, 687–698. [CrossRef]
6. Kraemer, B.M.; Anneville, O.; Chandra, S.; Dix, M.; Kuusisto, E.; Livingstone, D.M.; Rimmer, A.; Schladow, S.G.; Silow, E.; Sitoki, L.M.; et al. Morphometry and average temperature affect lake stratification responses to climate change. *Geophys. Res. Lett.* **2015**, *42*, 4981–4988. [CrossRef]
7. Alexakis, D.; Tsihrintzis, V.A.; Tsakiris, G.; Gikas, G.D. Suitability of Water Quality Indices for Application in Lakes in the Mediterranean. *Water Resour. Manag.* **2016**, *30*, 1621–1633. [CrossRef]
8. Butcher, J.B.; Nover, D.; Johnson, T.E.; Clark, C.M. Sensitivity of lake thermal and mixing dynamics to climate change. *Clim. Chang.* **2015**, *129*, 295–305. [CrossRef]
9. Foley, B.; Jones, I.D.; Maberly, S.C.; Rippey, B. Long-term changes in oxygen depletion in a small temperate lake: Effects of climate change and eutrophication. *Freshw. Biol.* **2012**, *57*, 278–289. [CrossRef]
10. Magee, M.R.; Wu, C.H. Response of water temperatures and stratification to changing climate in three lakes with different morphometry. *Hydrol. Earth Syst. Sci.* **2017**, *21*, 6253–6274. [CrossRef]
11. Zou, R.; Zhang, X.; Liu, Y.; Chen, X.; Zhao, L.; Zhu, X.; He, B.; Guo, H. Uncertainty-based analysis on water quality response to water diversions for Lake Chenghai: A multiple-pattern inverse modeling approach. *J. Hydrol.* **2014**, *514*, 1–14. [CrossRef]
12. Liu, X.; Lu, X.; Chen, Y. The effects of temperature and nutrient ratios on Microcystis blooms in Lake Taihu, China: An 11-year investigation. *Harmful Algae* **2011**, *10*, 337–343. [CrossRef]
13. Leach, T.H.; Beisner, B.E.; Carey, C.C.; Pernica, P.; Rose, K.C.; Huot, Y.; Brentrup, J.A.; Domaizon, I.; Grossart, H.-P.; Ibelings, B.W.; et al. Patterns and drivers of deep chlorophyll maxima structure in 100 lakes: The relative importance of light and thermal stratification. *Limnol. Oceanogr.* **2012**, *63*, 628–646. [CrossRef]
14. Zhang, Y.; Wu, Z.; Liu, M.; He, J.; Shi, K.; Wang, M.; Yu, Z. Thermal structure and response to long-term climatic changes in Lake Qiandaohu, a deep subtropical reservoir in China. *Limnol. Oceanogr.* **2014**, *59*, 1193–1202. [CrossRef]
15. Liu, Y.; Wang, Y.; Sheng, H.; Dong, F.; Zou, R.; Zhao, L.; Guo, H.; Zhu, X.; He, B. Quantitative evaluation of lake eutrophication responses under alternative water diversion scenarios: A water quality modeling based statistical analysis approach. *Sci. Total Environ.* **2014**, *468-469*, 219–227. [CrossRef]
16. Zohary, T.; Ostrovsky, I. Ecological impacts of excessive water level fluctuations in stratified freshwater lakes. *Inland Waters* **2011**, *1*, 47–59. [CrossRef]
17. Kirillin, G.; Shatwell, T. Generalized scaling of seasonal thermal stratification in lakes. *Earth-Sci. Rev.* **2016**, *161*, 179–190. [CrossRef]
18. Zhang, L.; Xu, K.; Wang, S.; Wang, S.; Li, Y.; Li, Q.; Meng, Z. Characteristics of dissolved organic nitrogen in overlying water of typical lakes of Yunnan Plateau, China. *Ecol. Indic.* **2018**, *84*, 727–737. [CrossRef]
19. Yang, K.; Yu, Z.; Luo, Y. Analysis on driving factors of lake surface water temperature for major lakes in Yunnan-Guizhou Plateau. *Water Res.* **2020**, *184*, 116018. [CrossRef]
20. Yan, D.; Xu, H.; Yang, M.; Lan, J.; Hou, W.; Wang, F.; Zhang, J.; Zhou, K.; An, Z.; Goldsmith, Y. Responses of cyanobacteria to climate and human activities at Lake Chenghai over the past 100 years. *Ecol. Indic.* **2019**, *104*, 755–763. [CrossRef]
21. Song, Y.; Qi, J.; Deng, L.; Bai, Y.; Liu, H.; Qu, J. Selection of water source for water transfer based on algal growth potential to prevent algal blooms. *J. Environ. Sci.* **2021**, *103*, 246–254. [CrossRef] [PubMed]
22. Yang, K.; Yu, Z.; Luo, Y.; Zhou, X.; Shang, C. Spatial-Temporal Variation of Lake Surface Water Temperature and Its Driving Factors in Yunnan-Guizhou Plateau. *Water Resour. Res.* **2019**, *55*, 4688–4703. [CrossRef]
23. Yu, Y.; Zhang, M.; Qian, S.; Li, D.; Kong, F. Current status and development of water quality of lakes in Yunnan-Guizhou Plateau. *J. Lake Sci.* **2010**, *22*, 820–828.
24. Wan, G.J.; Chen, J.A.; Wu, F.C.; Xu, S.Q.; Bai, Z.G.; Wan, E.Y.; Wang, C.; Huang, R.; Yeager, K.; Santschi, P. Coupling between 210Pbex and organic matter in sediments of a nutrient-enriched lake: An example from Lake Chenghai, China. *Chem. Geol.* **2005**, *224*, 223–236. [CrossRef]
25. Du, L.N.; Jiang, Y.E.; Chen, X.Y.; Yang, J.X.; Aldridge, D. A family-level macroinvertebrate biotic index for ecological assessment of lakes in Yunnan, China. *Water Resour.* **2017**, *44*, 864–874. [CrossRef]
26. Sun, W.; Zhang, E.; Liu, E.; Ji, M.; Chen, R.; Zhao, C.; Shen, J.; Li, Y. Oscillations in the Indian summer monsoon during the Holocene inferred from a stable isotope record from pyrogenic carbon from Lake Chenghai, southwest China. *J. Southeast Asian Earth Sci.* **2017**, *134*, 29–36. [CrossRef]
27. Wu, F.; Xu, L.; Liao, H.; Guo, F.; Zhao, X.; Giesy, J.P. Relationship between mercury and organic carbon in sediment cores from Lakes Qinghai and Chenghai, China. *J. Soils Sediments* **2013**, *13*, 1084–1092. [CrossRef]
28. Zhang, H.; Huo, S.; Yeager, K.M.; Xi, B.; Zhang, J.; He, Z.; Ma, C.; Wu, F. Accumulation of arsenic, mercury and heavy metals in lacustrine sediment in relation to eutrophication: Impacts of sources and climate change. *Ecol. Indic.* **2018**, *93*, 771–780. [CrossRef]

29. Wang, Y.; Peng, J.; Cao, X.; Xu, Y.; Yu, H.; Duan, G.; Qu, J. Isotopic and chemical evidence for nitrate sources and transformation processes in a plateau lake basin in Southwest China. *Sci. Total Environ.* **2020**, *711*, 134856. [CrossRef]
30. Wang, Y.; Tang, Y.; Xu, Y.; Yu, H.; Cao, X.; Duan, G.; Bi, L.; Peng, J. Isotopic dynamics of precipitation and its regional and local drivers in a plateau inland lake basin, Southwest China. *Sci. Total Environ.* **2021**, *763*, 143043. [CrossRef]
31. Yu, H.; Qi, W.; Liu, C.; Yang, L.; Wang, L.; Lv, T.; Peng, J. Different stages of aquatic vegetation succession driven by environmental disturbance in the last 38 years. *Water* **2019**, *11*, 1412. [CrossRef]
32. Liu, W.; Li, S.; Bu, H.; Zhang, Q.; Liu, G. Eutrophication in the Yunnan Plateau lakes: The influence of lake morphology, watershed land use, and socioeconomic factors. *Environ. Sci. Pollut. Res.* **2012**, *19*, 858–870. [CrossRef] [PubMed]
33. Zan, F.; Huo, S.; Xi, B.; Zhang, J.; Liao, H.; Wang, Y.; Yeager, K.M. A 60-year sedimentary record of natural and anthropogenic impacts on Lake Chenghai, China. *J. Environ. Sci.* **2012**, *24*, 602–609. [CrossRef]
34. China Environmental Status Bulletin. 2002. Available online: http://www.mee.gov.cn/ (accessed on 3 May 2003).
35. China National Environmental Monitoring Centre. *Lake (Reservoir) Eutrophication Evaluation Method and Classification Technical Provisions, 2001*; China National Environmental Monitoring Centre: Beijing, China, 2002.
36. State Environmental Protection Administration (SEPA). GB3838-2002. *Environmental Quality Standard for Surface Water*; Standards Press: Beijing, China, 2002; pp. 1–8. (In Chinese)
37. Podobnik, B.; Stanley, H.E. Detrended cross-correlation analysis: A new method for analyzing two non-stationary time series. *Phys. Rev. Lett.* **2008**, *100*, 084102. [CrossRef] [PubMed]
38. Lei, H.; Junbo, W.; Liping, Z.; Jianting, J.; Yong, W.; Qingfeng, M. Water temperature and characteristics of thermal stratification in Nam Co, Tibet. *J. Lake Sci.* **2015**, *27*, 711–718. [CrossRef]
39. Noori, R.; Ansari, E.; Bhattarai, R.; Tang, Q.; Aradpour, S.; Maghrebi, M.; Haghighi, A.T.; Bengtsson, L.; Kløve, B. Complex dynamics of water quality mixing in a warm mono-mictic reservoir. *Sci. Total Environ.* **2021**, *777*, 146097. [CrossRef]
40. Li, Y.; Tian, L.; Bowen, G.J.; Wu, Q.; Luo, W.; Chen, Y.; Wang, D.; Shao, L.; Cai, Z.; Tao, J. Deep lake water balance by dual water isotopes in Yungui Plateau, southwest China. *J. Hydrol.* **2021**, *593*, 125886. [CrossRef]
41. Yu, Z.; Yang, K.; Luo, Y.; Shang, C.; Zhu, Y. Lake surface water temperature prediction and changing characteristics analysis-A case study of 11 natural lakes in Yunnan-Guizhou Plateau. *J. Clean. Prod.* **2020**, *276*, 122689. [CrossRef]
42. Schwefel, R.; Gaudard, A.; Wüest, A.; Bouffard, D. Effects of climate change on deepwater oxygen and winter mixing in a deep lake (Lake Geneva): Comparing observational findings and modeling. *Water Resour. Res.* **2016**, *52*, 8811–8826. [CrossRef]
43. Missaghi, S.; Hondzo, M.; Herb, W. Prediction of lake water temperature, dissolved oxygen, and fish habitat under changing climate. *Clim. Chang.* **2017**, *141*, 747–757. [CrossRef]
44. Chen, X.; Liu, X.; Peng, W.; Dong, F.; Chen, Q.; Sun, Y.; Wang, R. Hydroclimatic influence on the salinity and water volume of a plateau lake in southwest China. *Sci. Total Environ.* **2019**, *659*, 746–755. [CrossRef] [PubMed]
45. Cao, X.; Wang, Y.; He, J.; Luo, X.; Zheng, Z. Phosphorus mobility among sediments, water and cyanobacteria enhanced by cyanobacteria blooms in eutrophic Lake Dianchi. *Environ. Pollut.* **2016**, *219*, 580–587. [CrossRef] [PubMed]
46. Ni, L.; Li, D.; Su, L.; Xu, J.; Li, S.; Ye, X.; Geng, H.; Wang, P.; Li, Y.; Li, Y.; et al. Effects of algae growth on cadmium remobilization and ecological risk in sediments of Taihu Lake. *Chemosphere* **2016**, *151*, 37–44. [CrossRef] [PubMed]
47. Filazzola, A.; Mahdiyan, O.; Shuvo, A.; Ewins, C.; Moslenko, L.; Sadid, T.; Blagrave, K.; Imrit, M.A.; Gray, D.K.; Quinlan, R.; et al. A database of chlorophyll and water chemistry in freshwater lakes. *Sci. Data* **2020**, *7*, 310. [CrossRef]
48. Pu, J.; Song, K.; Lv, Y.; Liu, G.; Fang, C.; Hou, J.; Wen, Z. Distinguishing Algal Blooms from Aquatic Vegetation in Chinese Lakes Using Sentinel 2 Image. *Remote Sens.* **2022**, *14*, 1988. [CrossRef]
49. Zhang, M.; Yu, Y.; Yang, Z.; Kong, F. Deterministic diversity changes in freshwater phytoplankton in the Yunnan–Guizhou Plateau lakes in China. *Ecol. Indic.* **2016**, *63*, 273–281. [CrossRef]
50. Zhou, Q.; Wang, W.; Huang, L.; Zhang, Y.; Qin, J.; Li, K.; Chen, L. Spatial and temporal variability in water transparency in Yunnan Plateau lakes, China. *Aquat. Sci.* **2019**, *81*, 36. [CrossRef]

Article

Spatiotemporal Changes in Water Quality Parameters and the Eutrophication in Lake Erhai of Southwest China

Kun Chen [1], Lizeng Duan [2],*, Qi Liu [2], Yang Zhang [2], Xiaonan Zhang [2], Fengwen Liu [2] and Hucai Zhang [2],*

[1] Institute for International Rivers and Eco-Security, Yunnan University, Kunming 650500, China
[2] Institute for Ecological Research and Pollution Control of Plateau Lakes,
School of Ecology and Environmental Science, Yunnan University, Kunming 650500, China
* Correspondence: duanlizeng2019@ynu.edu.cn (L.D.); zhanghc@ynu.edu.cn (H.Z.)

Abstract: To understand the lake status and reasons of eutrophication at Lake Erhai in recent years, water quality, including water temperature (T), pH, dissolved oxygen (DO), total nitrogen (TN), total phosphorus (TP) and chlorophyll-a (Chl-a) from 2016 to 2020 was monitored and analyzed. The results showed no obvious thermocline in the vertical direction at Lake Erhai, while Chl-a demonstrated obvious spatiotemporal distribution characteristics in Lake Erhai. Chl-a concentrations increased to a maximum in summer in August with the low TN:TP value, leading to algal blooms, most notably in the southern lakes. Low pH and DO appeared due to the thermocline of Erhai Lake (August 2016). A large area of algae distribution due to the increase of total phosphorus appeared in the northern lake area of Lake Erhai in December 2016, with a tendency of mesotrophic to light eutrophic in summer by the nutritional evaluation of Lake Erhai, especially in the central lake area and the northern lake area. Pearson's correlation coefficient and principal component analysis showed a significant positive correlation between Chl-a and T (r = 0.34, $p \leq 0.01$) and TP (r = 0.31 $p \leq 0.01$) in the mesotrophic Lake Erhai, indicating that TP content was one of the triggering factors for the algal blooming. Based on the spatiotemporal changes in water quality parameters and their relationship with eutrophication, scientific agencies should implement management strategies to protect Lake Erhai, supplemental to the costly engineering measurements.

Keywords: Lake Erhai; temperature; Chl-a; seasonal changes; eutrophication

Citation: Chen, K.; Duan, L.; Liu, Q.; Zhang, Y.; Zhang, X.; Liu, F.; Zhang, H. Spatiotemporal Changes in Water Quality Parameters and the Eutrophication in Lake Erhai of Southwest China. *Water* **2022**, *14*, 3398. https://doi.org/10.3390/w14213398

Academic Editor: Roohollah Noori

Received: 8 September 2022
Accepted: 24 October 2022
Published: 26 October 2022

Publisher's Note: MDPI stays neutral with regard to jurisdictional claims in published maps and institutional affiliations.

1. Introduction

The study of lakes eutrophication is important to understand the changes in their ecological environment under the influence of natural and human activities [1]. Research on the seasonal vertical distribution characteristics of physical and chemical parameters of plateau lakes helps identify corresponding environmental indicators and assess changes in the ecosystems of plateau lakes, indicating the importance of long-term and continuous monitoring and research [2,3]. The heat distribution of the lake water affects its stratification and mixing, resulting in changes in its physical and chemical parameters [4,5]. T measurement shows that the temperature of lakes has a strong seasonal variation, while stratification shows that they have a seasonal pattern, suggesting that lakes can be classified based on T [6]. The research results indicated that typical eutrophic non-aquaculture water had mean concentrations of Chl-a of higher than 10 µg/L, and significant positive correlations were found between pH, DO and Chl-a. When the mean concentration of Chl-a was less than 10 µg/L, no correlation was found between DO and Chl-a for waters with a high exchange rate or heavily organically polluted natural waters [7].

External nutrients inputs above a critical level may drive shallow lakes to shift from clear water to a turbid state, enhanced eutrophication process, and the biomass of submerged macrophyte and dominance community changed by nutrient input [8,9]. The algal growth was often within a certain range in lakes caused by nutrient input. The study of

Hou et al. showed that the lake water in Chenghai appeared temperature stratification in summer. Induces nutrient release from lake sediments and promotes eutrophication resulting from cyanobacterial reproduction in the upper thermocline [10]. The eutrophication of lakes is usually affected by total nitrogen or total phosphorus. When different types of exogenous polluted water flows into the lake, the N/P ratio will change, resulting in changes in the nitrogen and phosphorus limitation of the lake. It has been reported that the nutrient status of Jianhu has gradually shifted from N-limited (2000–2010) to P-limited in recent years (2010–2018) [11]. There have a study assessed the actual water situation in the estuarine area of Lake Taihu, China, based on eutrophication levels and status of water quality using the trophic level index (TLI) and water quality index (WQI) methods. In the wet (August 2017) and dry (March 2018) seasons, the average TLI and WQI values in the wet season were worse than that in the dry season (TLI: 57.40, WQI: 65.74), and P may be the main factor in the dry season [12].

Previous studies on Lake Erhai mostly focused on the analysis of imported phosphorus species in Lake Erhai [13], the role of sediment bioavailability phosphorus in algal growth [14], historical changes in water level and responses to human activities and climate change [15]. Comparatively, this study aimed to analyze the monitoring data of Lake Erhai from 2016 to 2020, investigate the vertical and horizontal distribution characteristics of the physical and chemical parameters in recent years, as well as the seasonal change processes, and analyze their relationship with eutrophication to provide a scientific basis for the environmental restoration and eutrophication mitigation of Lake Erhai.

2. Materials and Methods

2.1. Overview of the Study Area

Lake Erhai is located in the central part of the Dali Bai Autonomous Prefecture in Yunnan (100°05′ E 100°17′ E, 25°36′ N 25°58′ N) and is a shallow plateau lake formed by tectonic and rifting movement (Figure 1). It belongs to the Lancang River Basin, originating from Cibi Lake in Eryuan County. Its water source is from Bagu Mountain via the Mizhi River in the north and Xier River in the west and relies on surface runoff and lake precipitation. The annual water volume of the lake is 13.78×10^8 m^3, with an area of 256.5 km^2, water storage capacity of 270×10^8 m^3, average water depth of about 10.5 m, deepest depth reaching up to 20.9 m, and a water retention time of about 2.75 years [16]. Lake Erhai has a subtropical plateau monsoon climate, with mild four seasons, low average temperature, large daily range, long daylight hours, and clear dry and rainy seasons.

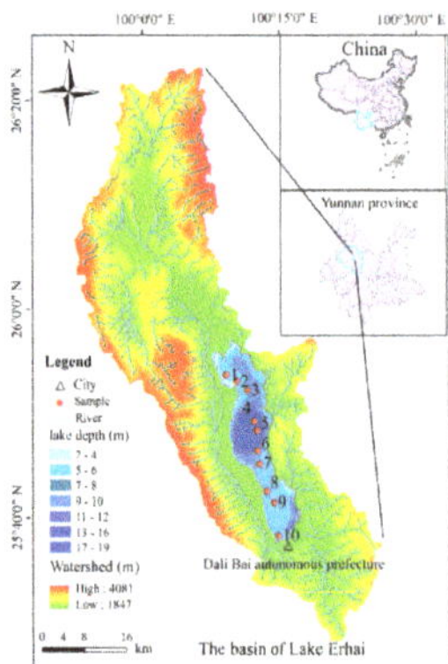

Figure 1. Map showing the location of Lake Erhai.

Erhai Basin has a total population of 883,900, and the urban population is mainly concentrated in the western and southern regions. The western basin of Lake Erhai is the main agricultural area and the high-value areas of pollution discharge in the basin are concentrated in the northern area of Lake Erhai. From 2007 to 2017, the non-point source pollution discharge in Erhai Basin accounted for 94.10% of the total discharge in the basin [17].

2.2. Data Collection and Research Methods

To comprehensively and systematically investigate Lake Erhai's water quality status and seasonal changes, water samples were collected from January 2016 to June 2020. A total of 10 sites were set up from north to south for fixed-point monitoring, of which sites 1–3 were located in the northern lake area of Lake Erhai, sites 4–7 in the middle lake area and sites 8–10 in the southern lake area. The water quality parameters, including temperature, pH, DO and Chl-a, were measured simultaneously, using YSI multiparameter sonde (product model: EXO2, United States, sampling frequency: 0.2 s) at the sampling site at 1 m interval.

Because water depth in most of the Lake Erhai is less than 5 m, we selected the measured date at 1 m and 5 m depth to discuss and the data at 1 m and 5 m from January, April, August and October 2017 were used to draw a heat map for seasonal comparison (representing winter, spring, summer, and autumn). The depth data at 1 m, 5 m, 10 m and 15 m from January 2016 to June 2020 (averaged data of each month) were used to analyze the change in trends and investigate long-term series change. Sites 2, 7 and 10 (representing the northern, central and southern parts of the lake) were used to assess TN and TP, and analyze the changing trends of different depths and long-term series in the three lake areas.

The water quality evaluation of surface layer mainly used the lake eutrophication level scoring standard of Aizaki Morihiro to score the three sampling site 2, 7 and 10 (Table 1), and used the Carson index method to classify the monitoring site [18].

Table 1. Grading and classification standard.

Score	Eutrophication Level
0~30.0	Oligotropher
30.1~50.0	Mesotropher
50.1~60.0	Light eutropher
60.1~70.0	Middle eutropher
70.1~100.0	Hyper eutropher

From the six water quality parameters, we selected all corresponding data (at the same depth and time: total 287 groups) at sites 2, 7, and 10 for Pearson correlation analysis to determine the causes of the higher Chl-a content in the water environment of Lake Erhai. At the same time, the 1 m, 5 m and 10 m depth data (at the same depth and time: total 76 groups) at sites 2, 7 and 10 were selected for principal component analysis to investigate the correlation between temperature, water quality parameters and TP and Chl-a. The above analysis was conducted by origin (version: 2021b).

3. Results

3.1. Spatiotemporal Changes in Water Quality Parameters

3.1.1. Spatiotemporal Changes in T

According to the meteorological data from 1981 to 2010 (data source: www.nmc.cn (accessed on 1 May 2022), the annual average precipitation at Dali Bai autonomous prefecture is about 1078.9 mm, with monthly average precipitation in the rainy season from May to October greater than 76 mm, accounting for 85–96% of the annual precipitation (Figure 2). Years of meteorological data as the basis for the division of dry and rainy seasons. Precipitation and temperature has obvious distribution characteristics in dry and rainy seasons, of which May-October is the rainy season, with an average temperature >15 °C. The dry season is from November to April and has an average temperature <15 °C. Thus, the dry season is from January to April and the rainy season is from May to October.

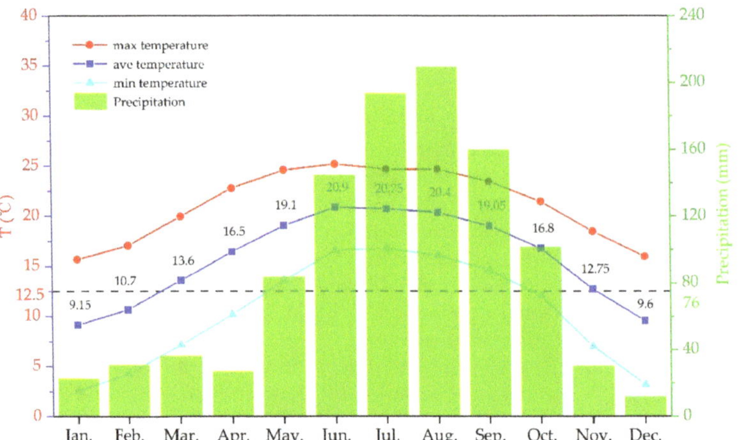

Figure 2. Monthly average temperature and precipitation from 1981 to 2010 (Dali Bai autonomous prefecture).

Based on the monthly changes in the T, our monitoring data from 2016 to 2020 (Figure 3) showed that the T of Lake Erhai was higher than the average temperature of Dali Bai autonomous prefecture during the same period but lower than the highest temperature in that same period. Since the monitoring time was mainly concentrated during the daytime when the temperature was higher than the average temperature.

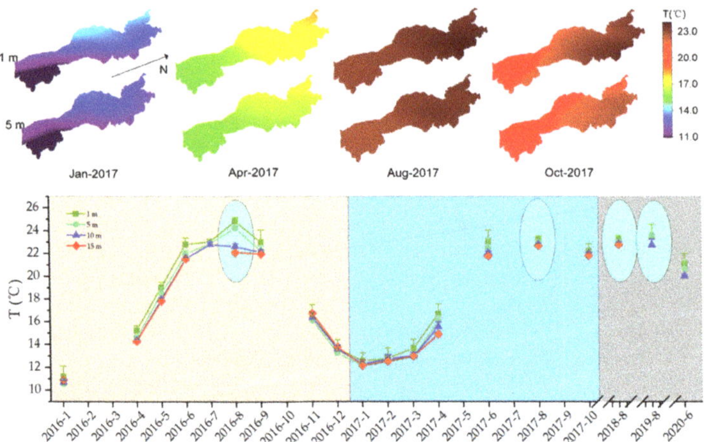

Figure 3. Spatiotemporal changes in T.

According to the research of Taihu Lake, the key factors affecting the growth of algae are lake T, light energy [19]. From the perspective of different lake areas, the T in the northwestern lake area was slightly higher than in its southeast area, especially in January and April 2017. In different lake layers, we observed that the overall T decreased as the water depths increased. The T and local temperature had the same trend in different seasons.

From the temporal change of Lake Erhai's temperature, the T in the rainy season was found to be significantly higher than in the dry season, with the highest temperature recorded in summer and starting to drop in autumn. The lake T did not change much in the four seasons from 2016 to 2017. From 2017 to 2020, several high-temperature rainy months were observed. A large temperature difference was observed between the upper and lower layers of the lake (5–10 m) during the rainy August seasons in 2016. Studies have shown that the process of thermal stratification will lead to oxygen deficiency ($O_2 < 1$ mg/L), and the oxygen consumption rate of lakes indicates the more eutrophic nature of water bodies [20].

3.1.2. Spatiotemporal Changes in pH

Algae growth and bacteria decomposing organic matter affect pH. The study found that high correlation values were obtained for pH and algae cellular growth [21]. From the temporal change of Lake Erhai's pH, the pH of the entire lake was low alkaline, higher than 7.5 per all monitoring months throughout the year (Figure 4).

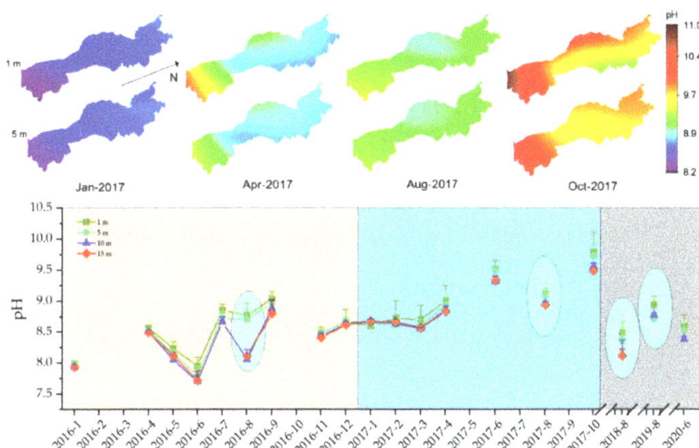

Figure 4. Spatiotemporal changes in pH.

In regard to seasons varies of pH, it was lowest in winter and started to rise until it was maximum in autumn. The data at different lake areas showed that the pH in the southwestern lake area was slightly higher than in the northeast area. In different lake layers, the pH decreased with increased water depths. The lake's pH has increased from 2016 to 2017. From the summer of 2017 to 2020, the pH of Lake Erhai drops in summer. Depth change in pH reached a maximum in summer, with low pH appeared in August 2016 between 5–10 m. The low pH value may be caused by the decomposition of organic matter and algae by microorganisms after the thermocline appears.

3.1.3. Spatiotemporal Changes of DO

The DO in lakes was important to maintain the dynamic balance of the ecological environment of a water body for the survival of aquatic organisms. DO also participates in the transformation of some substances [22]. Relevant studies have shown that the activities of plankton directly or indirectly determine the scope and degree of the hypoxic (less than 4 mg/L) less zone in the Qiandao Lake area. At the same time, the low oxygen area will affect the growth of aquatic organisms [23].

From the seasonal changes of DO in Lake Erhai in 2017, the DO increased slightly from winter to spring (from 7 mg/L to about 8 mg/L), and from spring to summer, the DO in water dropped sharply (from 8 mg/L to 8 mg/L). Dropped to about 4 mg/L and the DO at a depth of 15 m was as low as 2 mg/L, forming a hypoxic area. From summer to autumn, the DO content in water increases (from 4 mg/L to about 10 mg/L). This may be related to the decrease in algae density in the water and the fact that photosynthesis produces more oxygen than respiration consumes, and photosynthesis is stronger on the lake surface, especially in the northeastern lake area (Figure 5).

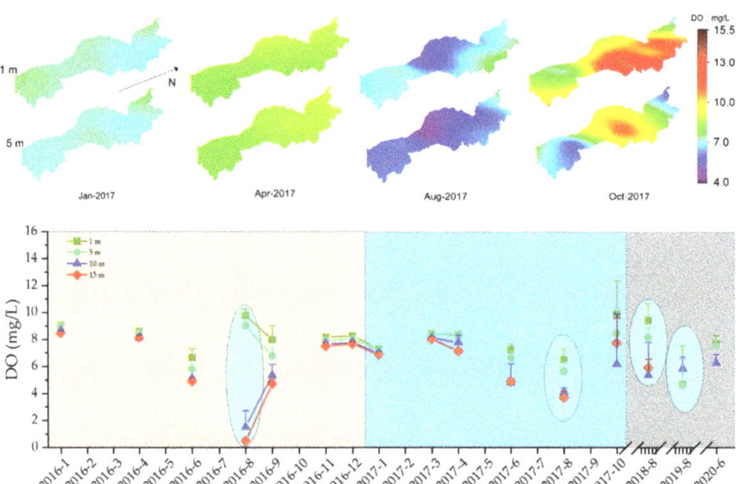

Figure 5. Spatiotemporal changes in DO.

From the long-term sequence of 2017–2020, the phenomenon appeared in August in summer when the DO in the lake water at a depth of 1–5 m was significantly greater than that in a depth of 10–15 m, of which 2016 was the most obvious, followed by 2018. This phenomenon may be related to the fact that a large number of algae in the upper layer of photosynthesis is stronger than respiration to increase the oxygen concentration, while the lower layer of algae is decomposed by bacteria due to the shading of the upper layer of algae and lack of light. The bacterial decomposition process consumes oxygen and produces organic acid species, resulting in a drop in DO and pH.

3.1.4. Spatiotemporal Changes in Chl-a

Chl-a is the main parameter characterizing the existing amount of phytoplankton and is one of the important indicators of lake water quality [24]. A change in Chl-a concentration can reflect the nutritional status of the water body. Previous research results showed significant seasonal changes in Chl-a in Lake Erhai, with the overall content beginning to rise gradually in April, increasing sharply in July, reaching a peak in October, and dropping sharply in January the following year [25].

From the temporal change of Lake Erhai Chl-a, the months with higher Chl-a content appeared from June to September. The distribution characteristics of Chl-a concentration at a depth of 1–5 m in different lake areas in four seasons in 2017 were southern area > northern area > central area (Figure 6). Combined with the long-term series, Chl-a appeared higher in four periods, namely July-September 2016, December 2016, June-August 2017, and August 2018. The content of Chl-a in the summer and winter of 2016 showed that the content in the lower layer was higher than that in the upper layer. In August 2017 and 2018, the Chl-a content of the upper layer was higher than that of the lower layer. This reflects the increasing tendency of algal growth space to the surface layer of 1–5 m in summer.

3.1.5. Temporal Changes in TN and TP

Assessment of changes in TN and TP concentrations with depth at indicated sites of Lake Erhai showed no obvious change trend in TN content with depth, while the TP content slightly decreased with increased depth (Figure 7). In terms of overall season, the TN content was lowest in August 2017, while that of TP was highest. The characteristics of TP concentration in different seasons were as follows: summer > autumn > spring > winter.

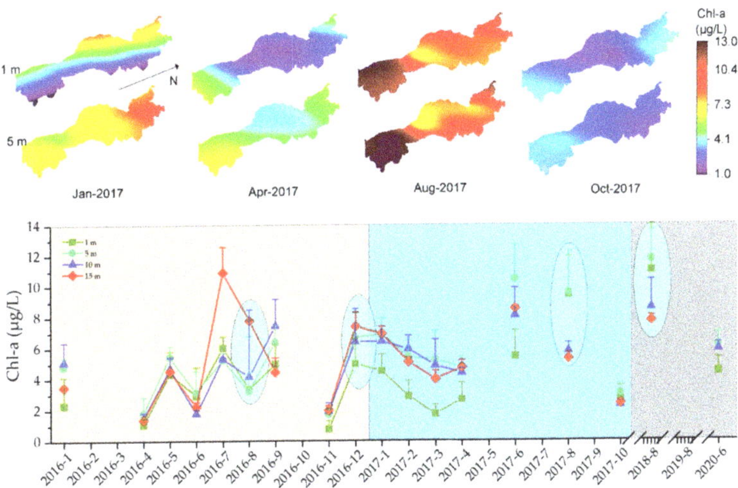

Figure 6. Spatiotemporal changes in Chl-a.

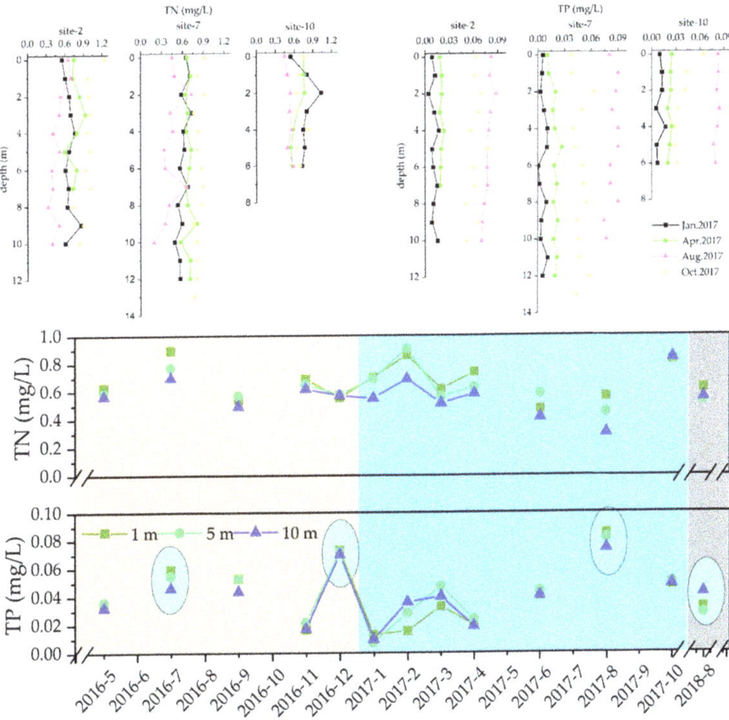

Figure 7. Spatiotemporal changes in TN and TP.

From the long-term observation of TN, we noticed a decreasing trend in TN in Lake Erhai, with its value fluctuating between 0.3–0.9 mg/L. A higher value was mainly recorded in February and August. Its content at the surface layer was generally higher than that of the lower layer. From the long-term observation of TP, we noticed significant variation in TP content in Lake Erhai during different months. Higher values were mainly recorded in

July and December 2016, August 2017 and August 2018, and TP and Chl-a have the same seasonal distribution trend.

3.2. Trophic Status Evaluation

A study used the comprehensive trophic state index to evaluate Dianchi Lake, Lake Erhai, and Fuxian Lake in 2017. The results showed that the medians of the comprehensive trophic state index were 63.21, 31.69, and 18.55, respectively. Lake Erhai was evaluated as mesotrophic [26].

According to the evaluation of eutrophic level of surface water at three points in Lake Erhai from 2016 to 2018 (Table 2). The evaluation results showed that Lake Erhai is mainly in a mesotrophic state, while only the northern and central lake regions in 2018 were in a light eutrophic state. The north lake and central lakes showed a fast transition to light eutrophic, while the southern lakes showed a slower trend. Moreover, the TP score increased while the TN score decreased, which means that the value of N: P was decreasing, indicating that the contribution of TP to eutrophication was increasing.

Table 2. Eutrophication level evaluation in Lake Erhai.

Sample	Year-Month	Chl-a Score1	TN Score2	TP Score3	Average Score	Eutrophication Level
Site 2	2016-7	17.5	50.9	43.5	37.3	Mesotropher
	2016-12	45.7	41.6	43.1	43.5	Mesotropher
	2017-8	47.9	48.5	42.7	46.3	Mesotropher
	2018-8	47.3	44.9	59.8	50.7	Light eutropher
Site 7	2016-7	27.0	53.9	47.7	42.9	Mesotropher
	2016-12	39.5	45.1	44.2	42.9	Mesotropher
	2017-8	42.3	42.3	41.2	41.9	Mesotropher
	2018-8	50.0	48.0	53.3	50.4	Light eutropher
Site 10	2016-7	33.6	49.3	46.3	43.1	Mesotropher
	2016-12	41.1	46.6	44.5	44.1	Mesotropher
	2017-8	52.6	42.6	42.2	45.8	Mesotropher
	2018-8	47.2	46.2	51.1	48.1	Mesotropher

3.3. Correlation and Principal Component Analysis

The correlation analysis of Chl-a at different depths showed (Figure 8a): (1) a positive correlation with temperature (r = 0.36, $p \leq 0.01$) and TP (r = 0.31, $p \leq 0.01$), (2) a negative correlation with TN (r = -0.31, $p \leq 0.01$).

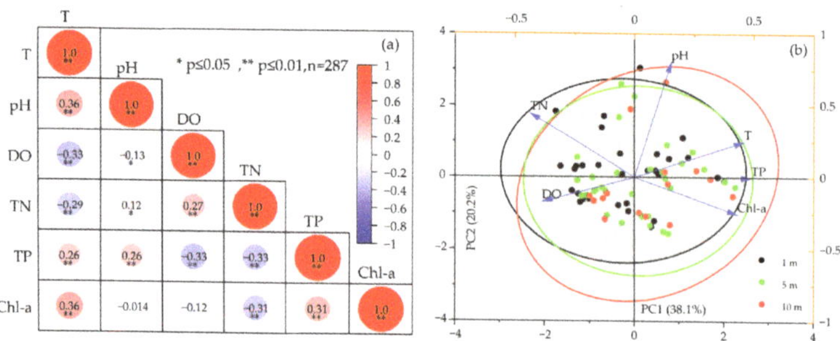

Figure 8. Correlation (**a**) and principal component (**b**) analysis of six water quality parameters.

Chl-a reflects the density of photosynthetic aquatic plants in lake water, and the vertical distribution of aquatic plants in lakes changes regional DO and pH. This explains the decomposition of large amounts of algae at a depth of 10–15 m in August 2016, leading to the formation of anoxic and acidic environments in this area.

In our study, temperature and TP in Lake Erhai were the main promoting factors for algal growth in Lake Erhai. TN had a downward trend in summer, and was negatively correlated with Chl-a, which was not the limiting factor for algal growth in Lake Erhai in summer. According to the trend of evaluation scores of TN and TP, TP has become one of the main limiting nutrients for aquatic photosynthetic plants in Lake Erhai in summer. This is consistent with Li Donglin's research results in Qilu Lake [27]. From the principal component analysis (Figure 8b), it can be seen that the water quality parameters's differences of Lake Erhai at depths of 1, 5 and 10 m are relatively low, indicating that the spatial mixing of the lake water is relatively high. The principal component PC1 (38.1%) may represent a nutrient factor that promotes water eutrophication. It can be seen that when it is reduced to only three depths, it also has a relatively consistent correlation with Figure 8a.

Based on the above analysis results, the temperature and TP content of Lake Erhai demonstrated a significant and positive correlation with Chl-a content. Using 95% regression linear fitting (Figure 9a,b), we found that the linear regression coefficients for temperature and TP were (R1 = 0.121, R2 = 0.099; $p > 0.05$). From the scatter plot, when the temperature was lower in winter, the recorded Chl-a value was higher, which may be related to the sudden increase of TP in December 2016. The increase of TP significantly reduced the distribution of Chl-a in the low value area. TP was the main limiting nutrient for elevated Chl-a concentration. This result is consistent with that of Chl-a in many studies [28–31].

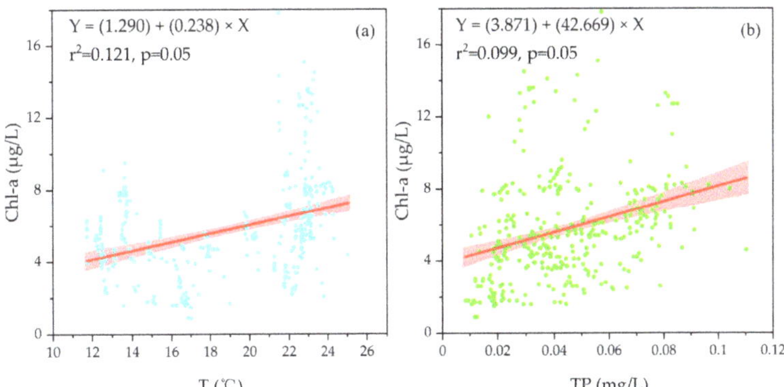

Figure 9. Linear regression relationship of T to Chl-a. (**a**) and TP to Chl-a. (**b**).

4. Discussion

4.1. Water Quality Parameters

T is an important factor affecting Chl-a concentration, and it is a key factor for the growth of phytoplanktons [32,33]. Similar to the local monthly average temperature at Lake Erhai, obvious seasonal changes were observed in the lake T. The monthly average air temperature directly affected the T of the lake. However, as a large shallow lake, Lake Erhai has a large specific heat capacity. Since sampling was performed during the day, the recorded surface T of the lake was slightly higher than the local monthly average air temperature. The spatial distribution of Lake Erhai's temperature was mainly affected by wind, the southwest wind was the main wind direction, The wind speed of Lake Erhai was significantly different at different locations of the lake, with the southern lake region > central lake region > northern lake region [34]. The southern lake area of Lake Erhai was the shallowest with strong wind disturbance. The southwest wind blew the surface water of the southern lake area to the northern lake area, so the lower water level at the northern lake area flowed to the southern lake area as a supplement. At the same time, the strong wind in the southern part of Lake Erhai cooled the shallow lake faster. Therefore, the T in the northern lake area was observed to be higher than in the central and southern lake areas. The temperature of Lake Erhai demonstrated little vertical variation. The lake water was completely mixed under the influence of wind, and there was no obvious thermocline in the T in most months. In the summer of August 2016, thermocline was observed at a depth of 5–10 m in Lake Erhai. At this time, the wind makes the lake water less mixed. the algae die and settle down and are decomposed by bacteria, resulting in the release of a large amount of Chl-a, and the DO and pH in the water body decrease, which is the most likely condition for eutrophication in this month. Studies have shown that the location of the thermocline and the euphotic depth can create a functional niche for diazotrophic cyanobacteria, resulting the upward transport of nitrate into the euphotic zone is reduced by a subjacent thermocline changed of nitrogen and phosphorus limiting factors on algal blooms [35].

From the perspective of the spatiotemporal changes in the lake's pH value, the photosynthesis of aquatic plants and algae consuming a large amount of CO_2 is the main reason for the increase in pH value. In terms of seasons, in summer and autumn, the total photosynthetic rate of aquatic organisms was greater than the respiration rate, which reduced the dissolved CO_2 content and increased the pH value of the lake water. On the contrary, the light intensity was weakest in winter, and the photosynthesis intensity of aquatic plants and algae was lowest. The photosynthesis rate was lower than the respiration rate, making the dissolved CO_2 content higher and the lake water pH lower. The pH value was also related to the lake's depth, which was the main reason for the spatial differentiation in pH values. In summer and autumn, the light and intensity of aquatic plants in the shallow lake area of southern Lake Erhai were greater than the respiration intensity, which was the reason for the higher pH value of the southern shallow lake area. In winter, a lower pH value was recorded in the shallow lake area of southern Lake Erhai because the light and intensity of aquatic plants were less than the respiration intensity. In the summer of August 2016, the temperature was higher and the wind was lower, resulting in a decrease in pH value in the area below 5 m. Studies have investigated the relationships between early summer partial pressures of CO_2 and dissolved organic carbon concentration in the surface waters of 27 northern Wisconsin lakes. CO_2 had a strong positive relationship with dissolved organic carbon concentration [36]. Other studies show that bacteria decomposed the organic matter on the bottom of the lake, oxygen was consumed, and a large amount of acidic substances were produced [37].

DO in Lake Erhai showed significant seasonal variation. It was highest in winter and lowest in summer. This could have been related to the low temperature in winter due to strong wind and weak solar radiation, which increased the contact of the water body from the surface to the bottom with the atmosphere, increasing the frequency of oxygen exchange. In summer, the temperature was high and precipitation was heavy. The death and decomposition of algae and aquatic plants consume a lot of oxygen, and the effect of microorganisms in the water below the mixed layer to decompose organic matter in the water is strengthened. As a result, the DO content of the lower water body decreases. We found that the content of Chl-a is affected by lake depth, wind and DO, and studies have shown that when DO increases, aquatic plants increase, resulting in an increase in Chl-a content [38]. Warmer summer temperatures result in high algal biomass. The photosynthesis of the algae on the surface is greater than the respiration to produce oxygen. The algae in the lower layer are decomposed to produce Chl-a, so that the oxygen concentration and Chl-a content have opposite trends in the vertical direction. Combining the above two situations, the DO and Chl-a in Lake Erhai were negatively correlated ($r = -0.12$). It was reported that DO stratification remarkably influence N species and transformation pathways in different water columns by high frequency sampling during summers in Longjing Lake, China. Results showed that Oxycline (4–11 m) was the major place for N transformations [39]. It was explained that the change of DO in Erhai Lake caused nitrogen to be removed and phosphorus to become a limiting nutrient factor.

Previous studies have investigated the Chl-a concentration in Lake Erhai from 2009 to 2013, and found that the annual average Chl-a concentration in Lake Erhai was 5–10.0 μg/L. Compared with this study, the content of Chl-a in Lake Erhai from 2016 to 2018 showed an increasing trend [40]. In addition, analysis of monitoring data showed that the summer temperature and TP in 2016–2018 also led to an upward trend in Chl-a. Due to a large number of algal blooms in summer, the temperature of Lake Erhai began to drop in autumn, and the DO content began to rise. After a large number of phytoplankton were decomposed by bacteria the Chl-a concentration also began to decline. This observation is consistent with the use of models to simulate seasonal changes in water quality parameters in Lake Erhai [41]. In December 2016, the lake area in the northwest of Lake Erhai became higher in TP content, even with fast wind speed and high DO content in December. However, the T in the northwestern lake area is higher, and a large amount of phosphorus is conducive to the growth of aquatic plants and algae, resulting in higher Chl-a content in the middle and lower layers (5–15 m) of the lake area in the northwestern Lake Erhai. The phenomenon that Chl-a in the middle and lower layers is higher than that in the surface layer may be related to the fact that the surface temperature is lower than that in the middle and lower layers in December.

4.2. Trophic Status and Changes

From the eutrophication score, it can be seen that the nutrient score changes: northern lake area > central lake area > southern lake area, which may be related to the fact that the outflow river Xi'er is in the south and the main wind direction is southwesterly. The gradual increase of TP drives the increase of Chl-a content, which leads to the change of Lake Erhai from light eutrophic state in summer. The sudden increase in TP in December 2016 also increased the score. The TN score is not high in summer, and phosphorus gradually becomes the main nutritional factor of algae blooming. The buffer capacity and hydrodynamic conditions of Lake Erhai should be improved and the input of exogenous phosphorus in the northwestern lake area should be controlled to prevent the massive growth of algae [42].

4.3. Trophic Causes and Effects

The changes in water quality in Lake Erhai are complex, and the trend of increasing nutrient levels is due to various factors. Although Lake Erhai is still in the mesotrophic, principal component analysis and correlation analysis show that temperature and phosphorus elements promote the growth of algae in the lake, and the excessive growth of algae changes DO and pH. This is consistent with the nutrition level analysis of the Sabalan Dam Reservoir in northwest Iran that implied TN:TP value was lower in summer than in winter, and therefore phosphorus became the main limiting factor of eutrophication [43]. Lake Erhai shows similar nutrient characteristics in summer, in addition that the range and trend of Erhai algae controlled by wind and lake currents.

5. Conclusions

Water quality parameters of Lake Erhai measured from 2016 to 2020 provided a deepening understanding of the lake's eutrophication features and their changes. The monitoring data analysis shows that there is no obvious thermocline occured in the lake, indicating that Lake Erhai belongs to a shallow lake with a high degree of water mixed state.

The Chl-a in Lake Erhai showed obvious spatiotemporal distribution characteristics. It is higher in the southern lake area in August and higher in the northern lake area in December. A high Chl-a values appeared around August in summer and December in 2016. T (r = 0.36) and TP (r = 0.31) are the main promoting factors for the increase of Chl-a content in Lake Erhai. The phenomenon of low T, DO, pH, and high Chl-a appeared in the middle and lower water depths (<5 m) of the central Lake Erhai in August 2016, which may be caused by the decomposition of a large number of algae by microorganisms. The thermocline and low TN:TP values may cause the risk of water quality deterioration in summer. The abnormal high Chl-a content appeared in Lake Erhai in winter in December 2016 is directly related to the particularly high TP content this month. The nutritional evaluation of the main months with high Chl-a content in Lake Erhai showed that has a trend of mesotrophic to light eutrophic.

Author Contributions: Methodology, Software, Writing—Original draft preparation, K.C.; Conceptualization, Supervision, Resources, Writing—review & editing, Foundations acquisition, H.Z.; Investigation, Data Curation, L.D., Q.L., Y.Z., X.Z. and F.L. All authors have read and agreed to the published version of the manuscript.

Funding: This research was funded by the Yunnan Provincial Government Scientist workshop and the Special Project for Social Development of Yunnan Province (Grant No. 202103AC100001).

Data Availability Statement: The data that support the findings of this study are available from the corresponding author upon reasonable request.

Acknowledgments: We thank all the graduate students whom participated the field works and laboratory analysis.

Conflicts of Interest: The authors declare no conflict of interest.

References

1. Li, Y.P.; Tang, C.Y.; Yu, Z.B.; Acharya, K. Correlations between algae and water quality: Factors driving eutrophication in Lake Taihu, China. *Int. J. Environ. Sci. Technol.* **2014**, *11*, 169–182. [CrossRef]
2. Xu, W.; Duan, L.; Wen, X.; Li, H.; Li, D.; Zhang, Y.; Zhang, H. Effects of Seasonal Variation on Water Quality Parameters and Eutrophication in Lake Yangzong. *Water* **2022**, *14*, 2732. [CrossRef]
3. Chang, F.Q.; Hou, P.F.; Wen, X.Y.; Duan, L.Z.; Zhang, Y.; Zhang, H.C. Seasonal Stratification Characteristics of Vertical Pro-files and Water Quality of Lake Lugu in Southwest China. *Water* **2022**, *14*, 2554. [CrossRef]
4. Liu, M.; Zhang, Y.L.; Shi, K.; Zhu, G.W.; Wu, Z.X.; Liu, M.L.; Zhang, Y.B. Thermal stratification dynamics in a large and deep subtropical reservoir revealed by high-frequency buoy data. *Sci. Total Environ.* **2019**, *651*, 614–624. [CrossRef]
5. Chen, J.; Lyu, Y.; Zhao, Z.; Liu, H.; Zhao, H.; Li, Z. Using the multidimensional synthesis methods with non-parameter test, multiple time scales analysis to assess water quality trend and its characteristics over the past 25 years in the Fuxian Lake, China. *Sci. Total Environ.* **2018**, *655*, 242–254. [CrossRef] [PubMed]
6. Yang, K.; Yu, Z.; Luo, Y.; Zhou, X.; Shang, C. Spatial-Temporal Variation of Lake Surface Water Temperature and Its Driving Factors in Yunnan-Guizhou Plateau. *Water Resour. Res.* **2019**, *55*, 4688–4703. [CrossRef]
7. Zang, C.; Huang, S.; Wu, M.; Du, S.; Scholz, M.; Gao, F.; Lin, C.; Guo, Y.; Dong, Y. Comparison of Relationships Between pH, Dissolved Oxygen and Chlorophyll a for Aquaculture and Non-aquaculture Waters. *Water Air Soil Pollut.* **2011**, *219*, 157–174. [CrossRef]
8. He, L.; Zhu, T.; Cao, T.; Li, W.; Zhang, M.; Zhang, X.; Ni, L.; Xie, P. Characteristics of early eutrophication encoded in submerged vegetation beyond water quality: A case study in Lake Erhai, China. *Environ. Earth Sci.* **2015**, *74*, 3701–3708. [CrossRef]
9. Qi, J.; Deng, L.; Song, Y.J.; Qi, W.X.; Hu, C.Z. Nutrient Thresholds Required to Control Eutrophication: Does It Work for Natural Alkaline Lakes? *Water* **2022**, *14*, 2674. [CrossRef]
10. Hou, P.; Chang, F.; Duan, L.; Zhang, Y.; Zhang, H. Seasonal Variation and Spatial Heterogeneity of Water Quality Parameters in Lake Chenghai in Southwestern China. *Water* **2022**, *14*, 1640. [CrossRef]
11. Zhang, Y.; Chang, F.; Zhang, X.; Li, D.; Liu, Q.; Liu, F.; Zhang, H. Release of Endogenous Nutrients Drives the Transformation of Nitrogen and Phosphorous in the Shallow Plateau of Lake Jian in Southwestern China. *Water* **2022**, *14*, 2624. [CrossRef]
12. Wang, J.; Fu, Z.; Qiao, H.; Liu, F. Assessment of eutrophication and water quality in the estuarine area of Lake Wuli, Lake Taihu, China. *Sci. Total Environ.* **2018**, *650*, 1392–1402. [CrossRef] [PubMed]
13. Ji, N.; Liu, Y.; Wang, S.; Wu, Z.; Li, H. Buffering effect of suspended particulate matter on phosphorus cycling during transport from rivers to lakes. *Water Res.* **2022**, *216*, 118350. [CrossRef]
14. Ni, Z.K.; Wang, S.R.; Cai, J.J.; Li, H.; Jenkins, A.; Maberly, S.C.; May, L. The potential role of sediment organic phosphorus in algal growth in a low nutrient lake. *Environ. Pollut.* **2019**, *255*, 113235. [CrossRef] [PubMed]
15. Wen, Z.; Ma, Y.; Wang, H.; Cao, Y.; Yuan, C.; Ren, W.; Ni, L.; Cai, Q.; Xiao, W.; Fu, H.; et al. Water Level Regulation for Eco-social Services Under Climate Change in Erhai Lake Over the Past 68 years in China. *Front. Environ. Sci.* **2021**, *9*, 196. [CrossRef]
16. Wang, S.M.; Dou, H.S. *Chronicles of Chinese Lakes*; Science Press: Beijing, China, 1998. (In Chinese)
17. Wang, M.J.; Yu, B.; Zhuo, R.R.; Wang, Y. Correlation analysis of water environmental pollution and water environmental change in Erhai Basin. *J. Huazhong Norm. Univ.* **2020**, *54*, 700–710. (In Chinese)
18. Zou, J.Y.; Yan, C.Q.; Wang, Y.W.; Liu, H.; Gu, F.; Hao, Y.Y.; Niu, W.B.; Zhang, Y.; Yu, S. Analysis of water quality and eu-trophication safety of Luoma Lake. *Eng. J. Heilongjiang Univ.* **2015**, *6*, 52–55. (In Chinese)
19. Zhao, Q.H.; Sun, G.D.; Wang, J.J.; Yu, Z.G.; Jiang, B. Coupling effects of T and light energy on algal growth in Taihu Lake in spring. *Lake Sci.* **2018**, *30*, 385–393. (In Chinese)
20. Yaseen, T.; Bhat, S.U. Assessing the Nutrient Dynamics in a Himalayan Warm Monomictic Lake. *Water Air Soil Pollut.* **2021**, *232*, 1–21. [CrossRef]
21. Acuña-Alonso, C.; Lorenzo, O.; Álvarez, X.; Cancela, A.; Valero, E.; Sánchez, A. Influence of Microcystis sp. and freshwater algae on pH: Changes in their growth associated with sediment. *Environ. Pollut.* **2020**, *263*, 114435. [CrossRef]

22. Crossman, J.; Futter, M.; Elliott, J.; Whitehead, P.; Jin, L.; Dillon, P. Optimizing land management strategies for maximum improvements in lake dissolved oxygen concentrations. *Sci. Total Environ.* **2018**, *652*, 382–397. [CrossRef] [PubMed]
23. Yu, Y.; Liu, D.F.; Yang, Z.J.; Zhang, J.L.; Xu, Y.Q.; Liu, J.G.; Yan, G.H. The vertical stratification characteristics of DO and phytoplankton in Qiandao Lake and their influencing factors. *Environ. Sci.* **2017**, *38*, 1393–1402. (In Chinese)
24. Li, D.; Wu, N.; Tang, S.; Su, G.; Li, X.; Zhang, Y.; Wang, G.; Zhang, J.; Liu, H.; Hecker, M.; et al. Factors associated with blooms of cyanobacteria in a large shallow lake, China. *Environ. Sci. Eur.* **2018**, *30*, 27. [CrossRef] [PubMed]
25. Yang, W.; Deng, D.G.; Zhang, S.; Xie, P.; Yu, L.G.; Wang, S.R. Seasonal dynamics and spatial distribution of Chl-a concentra-tion. *Lake Sci.* **2012**, *24*, 858–864. (In Chinese)
26. Zhou, Q.; Wang, W.; Huang, L.; Zhang, Y.; Qin, J.; Li, K.; Chen, L. Spatial and temporal variability in water transparency in Yunnan Plateau lakes, China. *Aquat. Sci.* **2019**, *81*, 36. [CrossRef]
27. Li, D.L.; Chang, F.Q.; Wen, X.Y.; Duan, L.Z.; Zhang, H.C. Seasonal Variations in Water Quality and Algal Blooming in Hy-pereutrophic Lake Qilu of Southwestern China. *Water* **2022**, *14*, 2611. [CrossRef]
28. Lin, S.-S.; Shen, S.-L.; Zhou, A.; Lyu, H.-M. Assessment and management of lake eutrophication: A case study in Lake Erhai, China. *Sci. Total Environ.* **2021**, *751*, 141618. [CrossRef] [PubMed]
29. An, G.Y.; Guo, Z.C.; Ye, P. Climatic changes and impacts on water quality of Lake Erhai in Dali area, Yunnan province over the period from 1989 to 2019. *Geoscience* **2022**, *36*, 406.
30. Yu, G.; Jiang, Y.; Song, G.; Tan, W.; Zhu, M.; Li, R. Variation of Microcystis and microcystins coupling nitrogen and phosphorus nutrients in Lake Erhai, a drinking-water source in Southwest Plateau, China. *Environ. Sci. Pollut. Res.* **2014**, *21*, 9887–9898. [CrossRef]
31. Chen, X.K.; Liu, X.B.; Li, B.G.; Peng, W.Q.; Dong, F.; Hang, A.P.; Wang, W.J.; Cao, F. Water quality assessment and spatial–temporal variation analysis in Erhai Lake, southwest China. *Open Geosci.* **2021**, *13*, 1643–1655. [CrossRef]
32. Adama, H.; Ye, J.; Persaud, B.; Slowinski, S.; Pour, H.K.; Cappellen, P.V. Chlorophyll-a growth rates and related environmental variables in global temperate and cold-temperate lakes. *Earth Syst. Sci. Data Discuss.* **2021**, *329*, 1–30.
33. Woolway, R.I.; Kraemer, B.M.; Zscheischler, J.; Albergel, C. Compound hot temperature and high chlorophyll extreme events in global lakes. *Environ. Res. Lett.* **2021**, *16*, 124066. [CrossRef]
34. Wang, X.; Deng, Y.; Tuo, Y.; Cao, R.; Zhou, Z.; Xiao, Y. Study on the temporal and spatial distribution of chlorophyll a in Erhai Lake based on multispectral data from environmental satellites. *Ecol. Inform.* **2020**, *61*, 101201. [CrossRef]
35. Ehrenfels, B.; Bartosiewicz, M.; Mbonde, A.S.; Baumann, K.B.L.; Dinkel, C.; Junker, J.; Kamulali, T.; Kimirei, I.A.; Odermatt, D.; Pomati, F.; et al. Thermocline depth and euphotic zone thickness regulate the abundance of diazotrophic cyanobacteria in Lake Tanganyika. *Biogeosci. Discuss.* **2020**, *214*, 1–21.
36. Diane, H.; Timothy, K.K.; Joan, L.R. Relationship between P CO 2 and Dissolved Organic Carbon in Northern Wisconsin Lakes. *J. Environ. Qual.* **1996**, *25*, 1442–1445.
37. Zhang, L.; Zhang, S.; Lv, X.; Qiu, Z.; Zhang, Z.; Yan, L. Dissolved organic matter release in overlying water and bacterial community shifts in biofilm during the decomposition of Myriophyllum verticillatum. *Sci. Total Environ.* **2018**, *633*, 929–937. [CrossRef] [PubMed]
38. Islam, S.; Ali, Y.; Kabir, H.; Zubaer, R.; Meghla, N.T.; Rehnuma, M.; Hoque, M.M.M. Assessment of Temporal Variation of Water Quality Parameters and the Trophic State Index in a Subtropical Water Reservoir of Bangladesh. *Grassroots J. Nat. Resour.* **2021**, *4*, 164–184. [CrossRef]
39. Su, X.; He, Q.; Mao, Y.; Chen, Y.; Hu, Z. Dissolved oxygen stratification changes nitrogen speciation and transformation in a stratified lake. *Environ. Sci. Pollut. Res.* **2019**, *26*, 2898–2907. [CrossRef] [PubMed]
40. Chen, X.H.; Qian, X.Y.; Li, X.P.; Wei, Z.H.; Hu, S.Q. Temporal change characteristics of eutrophication in Lake Erhai (1988–2013) and analysis of social and economic drivers. *Lake Sci.* **2018**, *30*, 70–78. (In Chinese)
41. Wu, Y.; Zhang, J.P.; Hou, Z.Y.; Tian, Z.B.; Chu, Z.S.; Wang, S.R. Seasonal Dynamics of Algal Net Primary Production in Re-sponse to Phosphorus Input in a Mesotrophic Subtropical Plateau Lake, Southwestern China. *Water* **2022**, *14*, 835. [CrossRef]
42. Guo, Y.; Dong, Y.; Chen, Q.; Wang, S.; Ni, Z.; Liu, X. Water inflow and endogenous factors drove the changes in the buffering capacity of biogenic elements in Erhai Lake, China. *Sci. Total Environ.* **2021**, *806*, 150343. [CrossRef]
43. Noori, R.; Ansari, E.; Jeong, Y.-W.; Aradpour, S.; Maghrebi, M.; Hosseinzadeh, M.; Bateni, S.M. Hyper-Nutrient Enrichment Status in the Sabalan Lake, Iran. *Water* **2021**, *13*, 2874. [CrossRef]

Article

Effects of Seasonal Variation on Water Quality Parameters and Eutrophication in Lake Yangzong

Weidong Xu [1], Lizeng Duan [2], Xinyu Wen [3], Huayong Li [4], Donglin Li [1], Yang Zhang [2] and Hucai Zhang [2,*]

[1] Institute for International Rivers and Eco-Security, Yunnan University, Kunming 650504, China
[2] Institute for Ecological Research and Pollution Control of Plateau Lakes, School of Ecology and Environmental Science, Yunnan University, Kunming 650504, China
[3] School of Geography and Engineering of Land Resources, Yuxi Normal University, Yuxi 653100, China
[4] School of Resources Environment and Tourism, Anyang Normal University, Anyang 455000, China
* Correspondence: zhanghc@ynu.edu.cn

Abstract: Understanding the seasonal variation characteristics and trends in water quality is one of the most important aspects for protecting and conserving lakes. Lake Yangzong water quality parameters and nutrients, including water temperature, dissolved oxygen (DO), pH, conductivity, Chlorophyll-*a*, phycocyanin, total nitrogen (TN) and total phosphorus (TP), were monitored in different seasons from 2015 to 2021. Based on the monitoring data, the temporal and spatial variations of various parameters were analyzed. The results showed that Lake Yangzong is a warm monomictic lake. The Pearson correlation coefficient and correlation analysis showed water quality parameters were significantly correlated and probably affected by temperature. Cyanobacteria were at risk of blooming in spring and autumn. The contents of TN and TP in winter were significantly higher than in summer, especially TN, with both reaching a peak at the epilimnion and hypolimnion in December 2020 (TN = 1.3 mg/L, TP = 0.06 mg/L). We also observed a dual risk of endogenous release and exogenous input. Therefore, strengthening the supervision for controlling eutrophication caused by human activities and endogenous release is urgently needed.

Keywords: Lake Yangzong; water quality parameters; temporal and spatial variations; cyanophyte relative quantity index; nutrient reduction

Citation: Xu, W.; Duan, L.; Wen, X.; Li, H.; Li, D.; Zhang, Y.; Zhang, H. Effects of Seasonal Variation on Water Quality Parameters and Eutrophication in Lake Yangzong. *Water* **2022**, *14*, 2732. https://doi.org/10.3390/w14172732

Academic Editor: Anas Ghadouani

Received: 18 July 2022
Accepted: 29 August 2022
Published: 1 September 2022

Publisher's Note: MDPI stays neutral with regard to jurisdictional claims in published maps and institutional affiliations.

1. Introduction

Lakes have a variety of functions, including water supply, flood prevention, aquaculture, transport and tourism [1]. As the water exchange of lake water is long, it has weak self-purification abilities [2]. Lakes are easily contaminated because they receive water from surrounding areas. The spatial and temporal changes in the water quality parameters can directly reflect the environmental water conditions of a lake, as variations in water quality parameters can lead to changes in lake trophic status [3]. Therefore, it is of practical significance to strengthen the monitoring and analysis of water quality parameters, especially for preventing and controlling eutrophication and protecting water quality and safety.

Mazhar et al. studied the changes of water bodies in different regions under different conditions. The results show that the river water profile has different laws in different climatic regions, which has an impact on DO dynamics [4]. The Ara Waterway, located in the wet regions, has a higher water quality variation in seasonal scale than that of the Yamuna Waterway, which is in the dry region [5]. In the southern estuarine water ecosystem of the Boseong County in Korea, there is a high Carlson Trophic State Index in the cropping area and land settlements in summer and autumn [6]. Through the groundwater monitoring data in Pakistan, it can be found that excessive groundwater abstraction has caused adverse impacts on groundwater quality [7].

Thermal stratification is the result of geographic location and summer–winter climatological differences and the depth of the lake. The seasonal vertical distribution of nutrients accompanies the thermal structure dynamics [8,9]. The thermal stratification of lakes may lead to a lack of oxygen in the hypolimnion and a boom in algae growth in the epilimnion in summer. This would destroy the balance of aquatic ecosystems and damage water quality.

Algal growth rate and other physiological characterization of living organisms are responding temperature and not vice versa. Chlorophyll-*a* is one of the important parameters of the ecological water system. Its content reflects phytoplankton biomass in water and is the basis of determining eutrophication. Phycocyanin is a kind of phycobiliprotein usually found in the body of cyanophyte, red alga, cryptophyta, and dinoflagellate. In these kinds of algae, red algae are usually found in the matrix of an oligotrophic stream. Cryptomonas mainly live in water with poor or moderate levels of nutrition. It is a common alga in tropical and polar waters, contributing less to the abundance of phytoplankton species. The amount of dinoflagellate is also lower in fresh water. However, cyanobacteria have remarkable heterogeneous structures and functions and a relatively broad ecological niche, with an absorption peak near a wavelength of 620 nm [10]. Cyanobacteria include many species, which are often the dominant species in lakes. The algal toxins contained in it will have a serious impact on organisms [11]. According to a study by Xie et al. [12], there are 13 genera (33.3%) of cyanophytes, 4 genera (10.3%) of dinoflagellate and 1 genus of cryptophyta in the water body of Lake Yangzong in 2013. Based on the above, we can preliminarily determine that cyanobacteria are the main algae in Lake Yangzong. Chlorophyll-*a* is usually used as an indicator to evaluate the trophic status of the water in most previous studies, but the use of phycocyanin is rare. Normally, eutrophication is often accompanied by the outbreak of cyanobacteria bloom; therefore, strengthening the estimation of cyanobacteria quantity is of great significance for preventing eutrophication.

Since cyanobacteria contain both Chlorophyll-*a* and phycocyanin, the ratio of the two pigments in lakes may indicate the relative amount of cyanobacteria in the lake water. The previous Cyanophyte Relative Quantity Index (CRQI) estimation mainly relied on the data from remote sensing rather than the water quality data measured in situ. According to the calculation formula established by Zhang et al. [13], Chlorophyll-*a* and phycocyanin were measured in situ to estimate the relative quantity index of cyanobacteria in this study. This method could avoid the estimation error caused by the separate use of Chlorophyll-*a* as an indicator. Similarly, TN and TP contents are also important conventional indicators for measuring water quality as they are associated with lake water eutrophication which can lead to cyanobacteria bloom and other consequences. The distinctness in release intensity of N and P could modify the N/P limitation in the lake, which affects algae growth and nutrient control [14], such as normalized cyanobacteria outbreaks formed in lakes such as Lake Dian (Dianchi) [15] and Lake Jian. Therefore, the accurate determination and analysis of TN and TP in lakes are of great significance for the management of lakes.

Lake Yangzong is the third deepest lake in the Yunnan province of southwestern China. It has a variety of social and economic functions, and its water quality is of high concern. Previous studies on Lake Yangzong mainly focused on the analysis of phytoplankton biomass and their population structure [12,16,17], the dynamic characteristics and inflow capacity of TN and TP [18,19], the source of arsenic pollution [20,21], water quality change process [22] and environmental risks caused by heavy metal and arsenic pollution [23]. However, studies on water quality parameters were limited to monitoring data for a short time and on a small scale [24]. Therefore, this study combined multi-season, high-frequency, large-scale and continuous water quality monitoring data to provide a scientific basis for the evaluation of the nutritional status of Lake Yangzong and water quality management.

2. Material and Methods

2.1. Physical Geographical Background of Lake Yangzong

Lake Yangzong is a freshwater lake located in Yiliang county, Yunnan Province (24°51′–24°58′ N, 102°55′–103°02′ E). Its average water depth is 20 m with a maximum depth of 31 m. The lake has an average altitude of about 1800 m above sea level, with a lake area of 31 km^2 and a water volume of 6.04×10^8 m^3. The catchment of Lake Yangzong belongs to the northern subtropical monsoon climate zone, covering an area of 192 km^2. The annual average temperature of Lake Yangzong is 15.2 °C, the average maximum temperature is 21.5 °C, and the average minimum temperature is 12.4 °C [25], with an obvious difference in wet and dry seasons and strong evaporation in dry seasons [26]. The main rivers flowing into the lake include the Yangzong River, Qixing River and Luxichong River in Chengjiang County, Baiyi River, and Tangchi River in Yiliang county [27]. The annual average surface rainfall of Lake Yangzong is 824.7 mm. The rainy season is from May to October, during which precipitation accounts for 85% of the total annual rainfall, forming a typical low-latitude regional climate. The main water source of the lake is atmospheric precipitation. Rainfall is sufficient in summer and scarce in winter, forming two transition periods in spring and autumn [22].

The basin economy is dominated by industry and supplemented by tourism. For a long time, the vulnerability of the lake's ecosystem has been neglected, despite the effects of human activities. Chemical fertilizers have induced non-point pollution and long-distance water diversion into the lake, and combined with aquaculture, rural domestic sewage, solid wastes and soil and water loss caused by vegetation damage, they are the main sources of pollution of the lake [27]. Lake Yangzong is the water supply for industry, agriculture and tourism.

In recent years, the water supply for real estate development and land replacement around the lake has depended mainly on water extraction. However, because of the catastrophic artificial arsenic pollution in 2008 and subsequent chemical treatments, the water quality of Lake Yangzong was seriously affected. Although the water quality has improved after treatment, the arsenic content was remained at a moderate level (in 2010, it fluctuated around 0.05 mg/L > 0.01 mg/L), and the nutrition level gradually rose [16,18].

2.2. Sampling

Based on the shape of the lake, three monitoring sites were set up in the southern (S1), middle (S2) and northern (S3) parts of Lake Yangzong (Figure 1). In situ monitoring was performed in April and November of 2015, January, June and November of 2016, April, June, July and September of 2017, December 2020 and August 2021. The whole lake monitoring was carried out in May 2018 and 2019, December 2020 and August 2021 (in total, 17 points were sampled and measured). The sampling sites were marked and located by a GPS satellite navigator, and the water quality parameters, including water temperature (WT), dissolved oxygen (DO) concentration, Chlorophyll-*a* (Chl-*a*) concentration, pH value and conductivity, were measured with a multi-parameter water quality monitoring instrument (YSI6600V2). A vertical line was set to monitor the water quality (including WT, DO, Chl-a, pH, conductivity and phycocyanin) at different depths at each site. The first data were measured about 0.5 m below the water surface, the last data were monitored 0.5 m above the bottom of the lake, and other data were collected at one-meter intervals. To ensure the accuracy of the data, each depth was measured six times.

Similarly, TN and TP samples from the vertical section of the lake water were collected in 1 L brown polyethylene bottles in September, October, November and December 2016, June, July, September and October 2017, March, May, June, August and October 2018, May 2019, December 2020 and August 2021. The samples were sent to a laboratory and placed in a refrigerator at 4 °C for chemical analysis within 2 h after sample collection.

2.3. Analysis Methods

The TP and TN levels were measured using an ultraviolet spectrophotometer (UV-2600) to determine the absorbance at the wavelengths 200 nm, 275 nm (TN) and 700 nm (TP). Then, the absorbance of the standard sample was corrected to obtain the specific TN and TP contents.

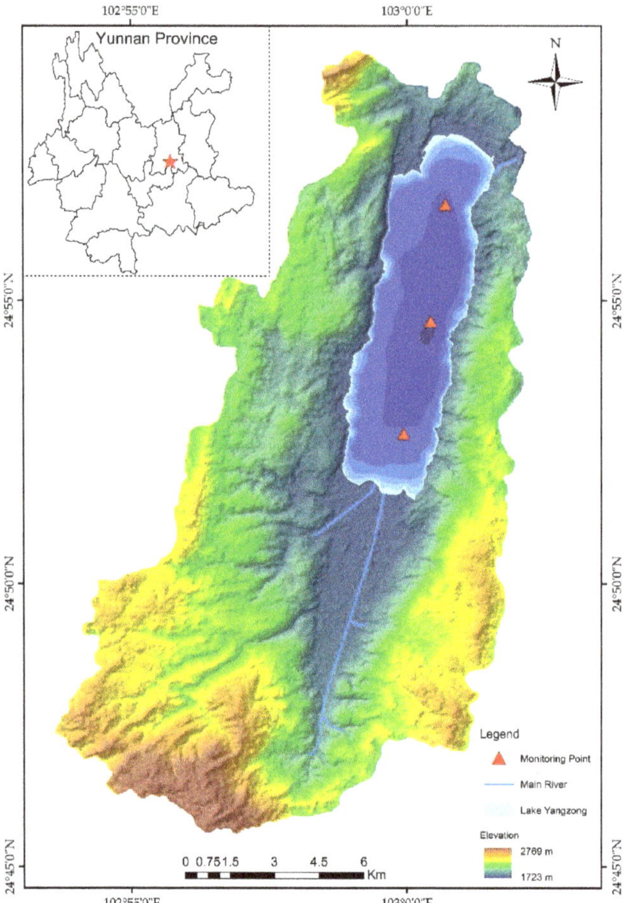

Figure 1. Study area and sampling sites of Lake Yangzong.

2.4. Data Processing

Microsoft Excel 2016 was used to assess the recorded data. The vertical profile diagram, column chart and correlation analysis diagram of each parameter were drawn using Origin 2020b (Origin Lab, Ltd., Northampton, MA, USA). In the column chart, the ratio of the total concentration of Chl-*a* to phycocyanin at the vertical section of each monitoring site was calculated using Microsoft Excel 2016.

3. Results

3.1. Seasonal Variation of Water Temperature

The temperature of deep water plateau lakes is affected by changes in air temperature [8,28], with changes in temperature leading to the thermal stratification of lakes. Increasing depths of water lakes cause a slow decrease in water temperature at the epilimnion and hypolimnion, leading to a sharp decrease in the thermocline. Similar to other deep-water plateau lakes (high mountains), the water temperature of Lake Yangzong presents a stratification and mixing phenomenon in the vertical profile. The lake is layered in spring, summer and autumn, and mixed in winter (Figure 2).

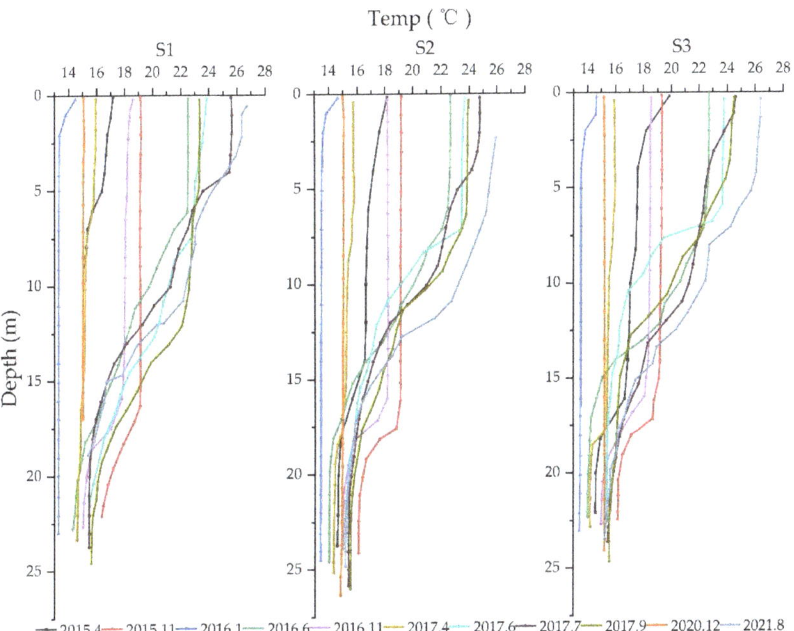

Figure 2. Vertical profile of the water temperature in Lake Yangzong.

Our findings showed that the temperature change pattern could be divided into four seasons, which included April 2015 and 2017 in spring (between 14.2 °C and 19.8 °C); June 2016, June, July and September 2017 and August 2021 in summer (between 14.1 °C and 26.6 °C); November 2015 and 2016 in autumn (between 15 °C and 19 °C); and January 2016 and December 2020 in winter (between 13.3 °C and 15.1 °C). These grouping modes represented the division of the four seasons. In winter, the lake belonged to the mixed period. Except for slight changes in the epilimnion from 0 m to 2 m (January), the other water depths were evenly mixed. In spring, epilimnion, thermocline and hypolimnion appeared but were not stable and significant. In summer, the water temperature increased dramatically in the epilimnion (water depth: 0 m to 6 m), while the variation in hypolimnion (water depth: ~18 m) was not obvious. There was a significant temperature gradient and a wide range in the thermocline (water depth: 6 m to 18 m). The hierarchical structure was stable during this time. The water temperature dropped obviously from surface to bottom in autumn. Temperature changes dropped into the deep waters. Moreover, a relatively higher temperature formed in the hypolimnion before the end of summer, which lasted until autumn. The thermocline (water depth: 16 m to 20 m) was in a lower depth and had a smaller range, indicating the vanishing stage of thermal stratification. As soon as the temperature of epilimnion dropped to the temperature of hypolimnion, the lake became mixed. Among the four seasons, the surface temperature range was 14.49 °C to 25.58 °C, the maximum temperature difference was 11.09 °C, the bottom temperature range was 13.31 °C to 16.38 °C and the maximum temperature difference was 3.07 °C. The whole year's maximum temperature and maximum temperature difference appeared in summer.

The difference among the three monitoring sites was not significant horizontally. In April 2015, the water temperature in the northern part of the lake was higher than the others. In July 2017, the water temperature in the southern part of the lake was higher than the others. In November 2015, according to the strict definition of thermocline [10], the thermocline in the south part of the lake disappeared, and the mixing period started.

3.2. Seasonal Variation Characteristics of Water Quality Profile

3.2.1. Dissolved Oxygen (DO)

There were obvious seasonal stratification and mixing phenomena in the variation of the DO concentration in Lake Yangzong (Figure 3).

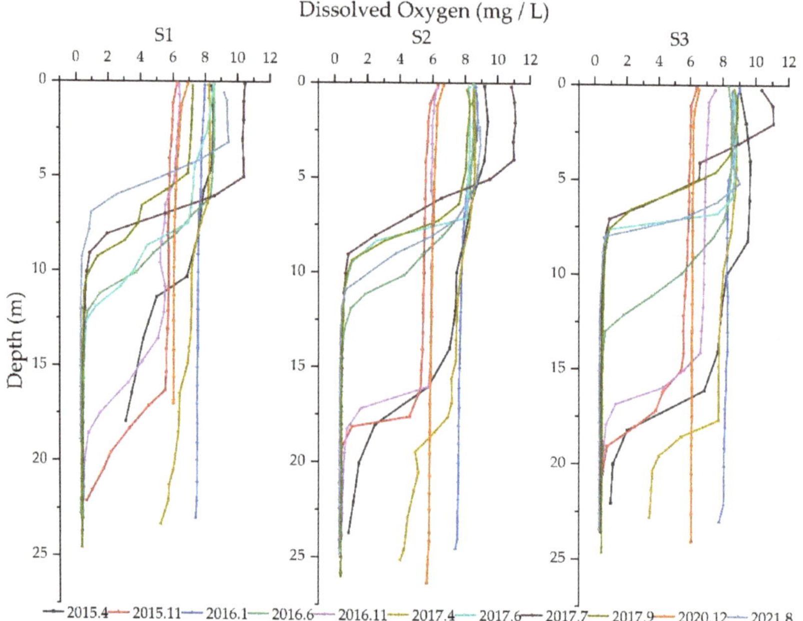

Figure 3. Vertical profile of dissolved oxygen (DO) in Lake Yangzong.

Similar to the variation trend of water temperature, the higher the temperature differences are, the stronger the stratification stability and DO distribution are. The curve of dissolved oxygen concentration could also be divided into four seasons. The highest value of surface DO was recorded (about 10.69 mg/L) in July 2017. The DO in the vertical direction was similar in winter. The DO stratification phenomenon began in spring, and the concentration of DO decreased with the water depth. Even during this time, the stratification was not stable. The stratification phenomenon was obvious in the middle and north parts of the lake. In summer, an obvious gradient of DO formed within the range of 4 m to 12 m below the water surface, where the content of DO decreased sharply. In this period, the stable stratification of DO was formed, leading to a lack of DO in deeper waters. In autumn, the DO stratification phenomenon began to vanish, and the water above 15 m was evenly mixed. In the horizontal direction, the seasonal changes in the central and northern regions were more obvious. The change trends of DO in all groups were consistent with the trend of temperature change.

3.2.2. pH Values and Their Variations

The water in Lake Yangzong was alkaline, and the change in pH is obvious (Figure 4).

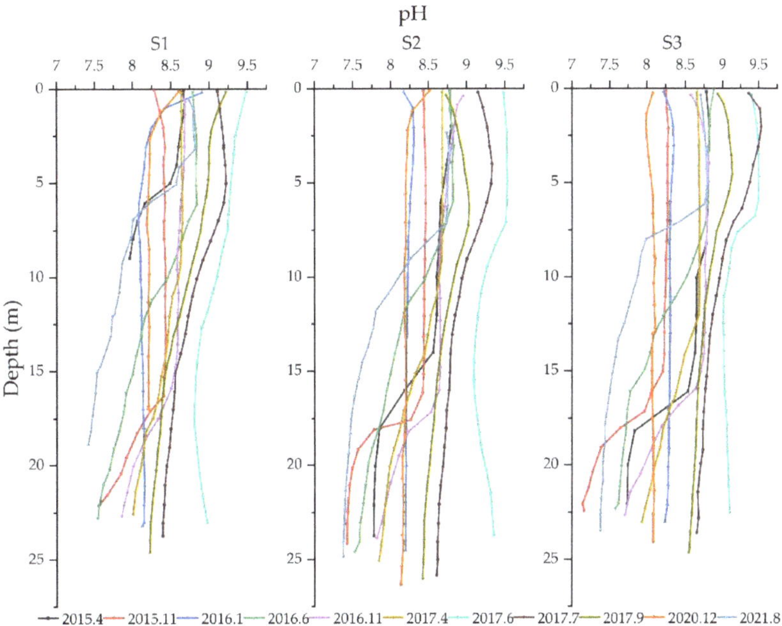

Figure 4. Vertical pH profile in Lake Yangzong.

In spring, summer and autumn, with the stratification of temperature and DO, the pH was also stratified. The same four seasons were analyzed according to the grouping mode of temperature and DO. In winter, the variation range of pH was small, with some differences found at the surface. In January 2016, in the range of 0 m to 2 m, the pH decreased sharply in the southern part of the lake (from 8.91 to 8.07). Conversely, in the northern and central parts of the lake, the pH increased within the top 2 m. In spring, an increase in variation range was observed, with the difference between surface and bottom indicating the stratification of pH. In April 2015, there was a dramatically changing layer in the vertical profile, but in April 2017, the pH gradient at each depth was similar. Since the depths differed in the three monitoring sites, sharp changes in the layers were located at different depths. In summer, except in June 2017, the pH decreased with depth and formed a stable pH stratification. During this period, the changes mainly appeared in the southern part of the lake. In autumn, a variation trend in pH identical to the trend of temperature and DO was observed. In the central part of the lake, the pH value at the epilimnion was relatively low.

3.2.3. Conductivity

Conductivity refers to the ability to transmit electricity. It is mainly affected by salinity, dissolved solids, temperature and water supply. In Lake Yangzong, seasonal variation in conductivity was obvious, i.e., the vertical change in conductivity in spring and winter was not significant, while it was clearly stratified in summer and autumn (Figure 5).

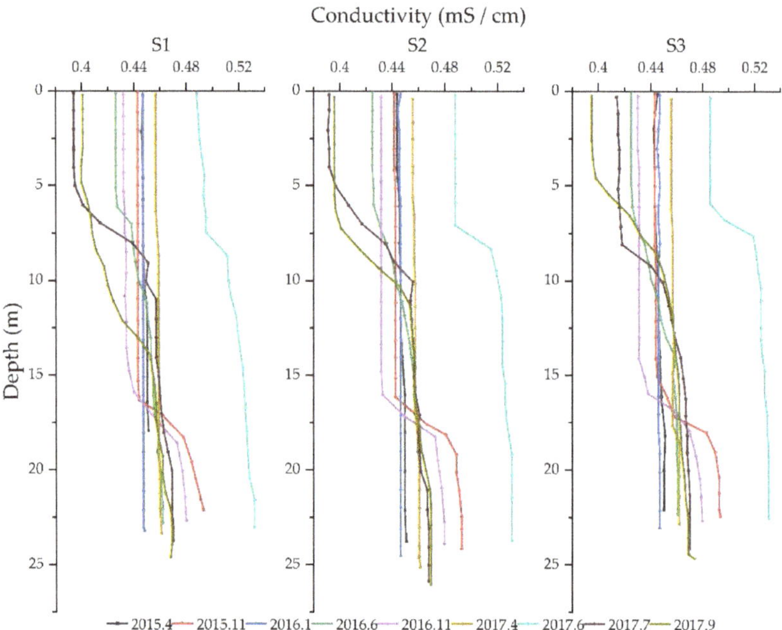

Figure 5. Vertical profile of conductivity in Lake Yangzong.

Similar to the temperature, DO and pH classification, the changes in conductivity could also be divided into four seasons. In April 2015, January 2016 and April 2017, except for the difference in value, the vertical and horizontal conductivity variation was insignificant. In spring, conductivity slightly increased with depth in the vertical direction, but with a very small amount. The changes at the three monitoring sites were similar. The conductivity in April 2017 (0.458 mS/cm) was higher than in April 2015 (0.447 mS/cm) and had no stratification. In summer, the stable stratification of conductivity began to form. At this stage, there was a sharp increase in the middle part of the vertical profile. In June 2017, the conductivity showed an abnormally high value (up to 0.532 mS/cm). The horizontal distribution difference was small in the same month. Except in July 2017, the surface conductivity of the northern part was higher than the southern and central parts. Among them, the highest conductivity was recorded in June 2017. In autumn, the conductivity was higher than that of summer. Vertically, the conductivity increased sharply within the range of 16 m to 20 m. Horizontally, there was no significant difference among the three monitoring sites.

3.2.4. Chlorophyll-*a* (Chl-*a*)

Compared with other parameters, the variation in Chl-*a* demonstrated distinct characteristics, and the concentration of Chl-*a* was higher in autumn and winter (Figure 6).

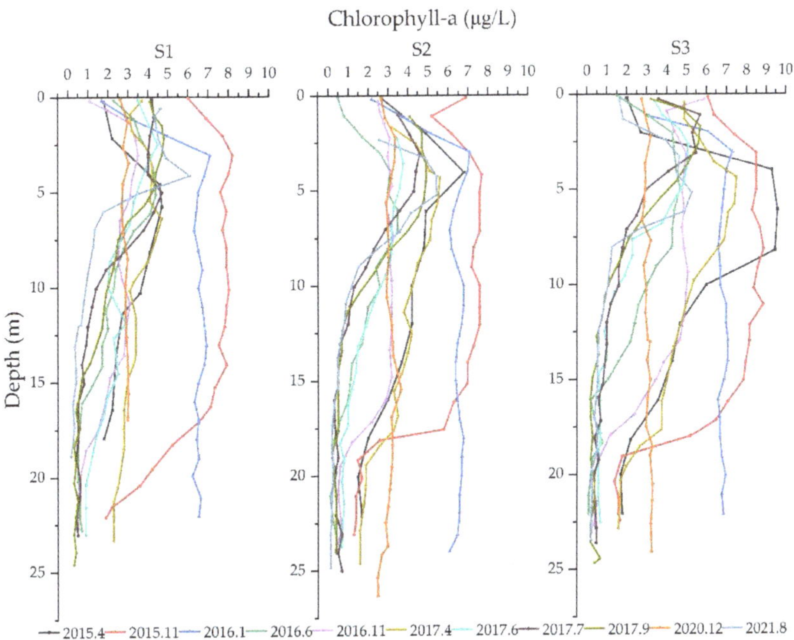

Figure 6. Vertical profile of Chlorophyll-a in Lake Yangzong.

Significant changes could be found in different seasons. There were obvious differences in both Chl-*a* value and variation trends in winter and early spring. In spring, the difference in the vertical and horizontal directions was obvious. The concentration of Chl-*a* was significantly high in the northern part of the lake in April 2015, and there was a clear peak within the 4 m to 8 m water depth. In the central and northern parts of the lake, there were lower values between the 4 m and 6 m water depths, respectively. In April 2017, the concentration of Chl-*a* first increased and then decreased with an increase in depth. The concentration in the central and northern parts increased sharply within 4 m of the epilimnion, while in the southern part of the lake, the Chl-*a* concentration increased less within 6 m of the epilimnion and did not change much in other depth ranges. The Chl-*a* concentration in April 2015 was higher than that of April 2017. In spring, the changes in the lake were disordered, and the concentration of Chl-*a* was higher in the northern part of the lake. In summer, a stable peak formed in the vertical profile. In July 2017, a peak was recorded from 0 m to 5 m at a value of ~5 µg/L. In June 2016 and 2017, the peak and depth decreased more than in July 2017. The stable stratification structure formed at all monitoring sites and all months during this stage, except for the variation in June 2017 in the southern part. In autumn, the Chl-*a* concentration in November 2015 was extremely high and clustered in the range of 2 m to 18 m water depth. The stratification in this stage was obvious, and the variation trend of all monitoring sites was consistent. In winter, the concentration was higher than in early spring. In January 2016, the concentration of Chl-*a* increased within the top 4 m and remained similarly constant at ~6 µg/L below 4 m. In December 2020, Chl-*a* demonstrated little change at each depth.

3.2.5. Phycocyanin

Phycocyanin and Chl-*a* have the same function; they can be used to estimate the amount of plankton and the trophic state of the water lakes. The difference between them is that Chl-*a* exists in almost all eukaryotes, while phycocyanin mainly exists in the cyanobacteria. The seasonal distribution of phycocyanin in Lake Yangzong had its own characteristics (Figure 7).

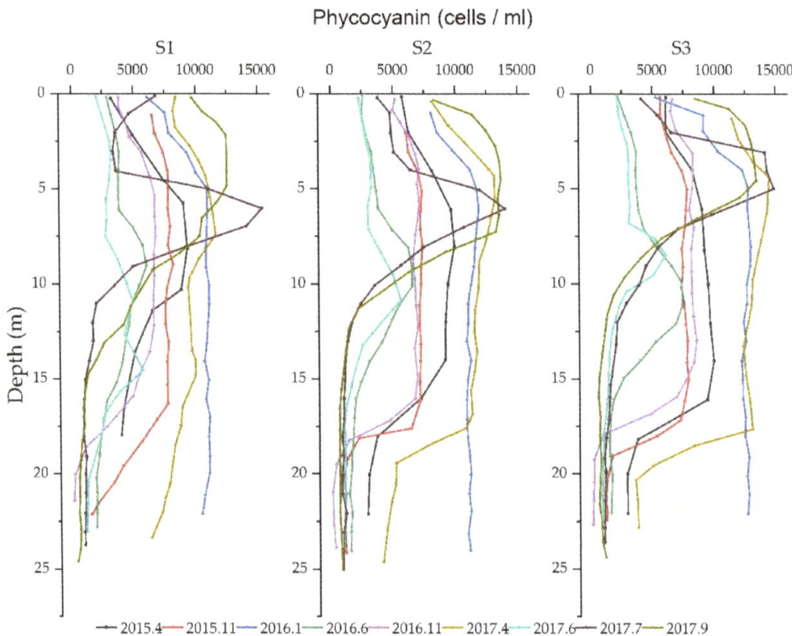

Figure 7. Vertical profile of phycocyanin in Lake Yangzong.

Using the four seasonal classifications, the variation trend of phycocyanin was different from that of Chl-*a*. The phycocyanin concentration gradually decreased from spring, was lowest in summer and then gradually increased to its highest in winter. The concentration of phycocyanin was higher in spring than in autumn.

In winter and early spring, the phycocyanin concentration was at its highest level and was largely distributed at water depths of 4 m to 17 m. In January 2016, the phycocyanin concentration first increased, then stabilized. In April 2015, the phycocyanin concentration gradually increased from south to north, with little change in the horizontal direction in the remaining months. The concentration in April 2017 was higher than in April 2015. In summer, stratification stably developed, and compared to spring, the range was narrower. In June 2016 and June 2017, the phycocyanin concentration peaked between a water depth of 10 m to 12 m in the central and northern part of the lake, and stratification was obvious. In July 2017, the phycocyanin concentration increased sharply at water depths ranging from 3 m to 8 m and reached a peak at about 6 m water depth and then suddenly decreased. In September 2017, the phycocyanin concentration decreased sharply after reaching a peak at 5 m to 7 m water depth and was highest at water depths ranging from 0 m to 3 m compared to the other months. In autumn, the concentration increased and remained at a middle level. The stratification maintained a similar trend at different monitoring sites. In November 2015 and 2016, the phycocyanin concentration in the entire lake was relatively high and stable at water depths ranging from 4 m to 16 m water depth, with small vertical changes.

3.3. Seasonal Variation Characteristics of CRQI

The relative amount of cyanobacteria in Lake Yangzong was relatively high in spring, summer and winter (Figure 8).

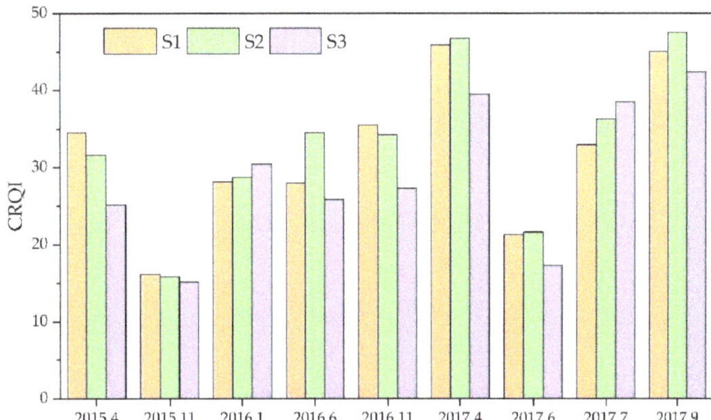

Figure 8. Cyanophyte relative quantity index from 2015–2017 in Lake Yangzong. Note: The ordinate value in the figure was calculated using the formula CRQI = [PC]/[Chl-*a*], and its unit is cells/μg.

Figure 8 shows that from April 2015 to April 2017, the CRQI decreased first and then increased. In September 2017, the CRQI was the highest among the three monitoring loci, and the maximum value was recorded in the central part of the lake. Comparing April 2015 to April 2017, an increase in CRQI could be observed, especially in the central part of the lake. It also showed that in 2017, the number of cyanobacteria was higher than that of 2016, except in June 2017. The minimum value of the CRQI monitored was recorded in November 2015, with few differences among the three monitoring sites. Compared with November 2015, the CRQI increased in November 2016, most notably in the southern and central parts of the lake.

3.4. TN and TP Contents and Their Correlation with Other Indexes

The TN content in Lake Yangzong demonstrated a certain change in different seasons. The monitoring data showed that the TN content was lower in the summer and autumn of 2016, 2017 and 2018, and higher in winter and spring (Figure 9), which was related to less water in winter and spring. The TN content in August 2021 increased compared to 2018 (about 28%). The TP content decreased from 2016 to 2018 (Figure 10). The TP content in May 2019 did not change significantly from May 2018, but in August 2021 (about 0.04 mg/L), it was significantly higher than that of August 2018 (0.03 mg/L).

During May 2018 and May 2019, the TN and TP contents at the Hypolimnion of the lake were significantly higher than at the epilimnion. In May 2018 and 2019, TN and TP contents of epilimnion were relatively uniform, but in May 2019, TN and TP contents of hypolimnion were higher in the middle of the lake area. In December 2020, the TN and TP contents in the epilimnion and hypolimnion were extremely high. The TN content at the epilimnion was higher in the south and higher at the hypolimnion in the east. The TP content at the epilimnion was relatively uniform in the southeast and higher in the hypolimnion. In August 2021, the TN and TP contents at the epilimnion were relatively uniform (0.78 mg/L < TN < 1.21 mg/L, 0.02 mg/L < TP < 0.05 mg/L). The TN content at the hypolimnion was significantly higher in the north than in the south, and the TP at the bottom was higher in the northeast (about 0.14 mg/L).

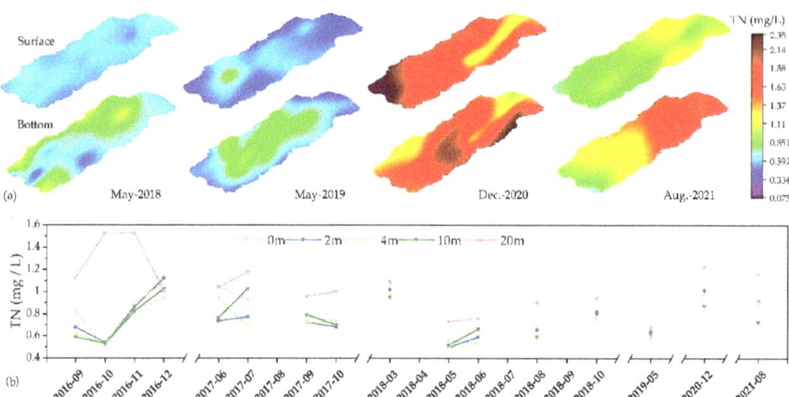

Figure 9. Change in TN content in Lake Yangzong from 2017 to 2021. (**a**) indicates the distribution of TN content in the whole lake at the surface and bottom. (**b**) represents the mean value of TN content at 5 different depths.

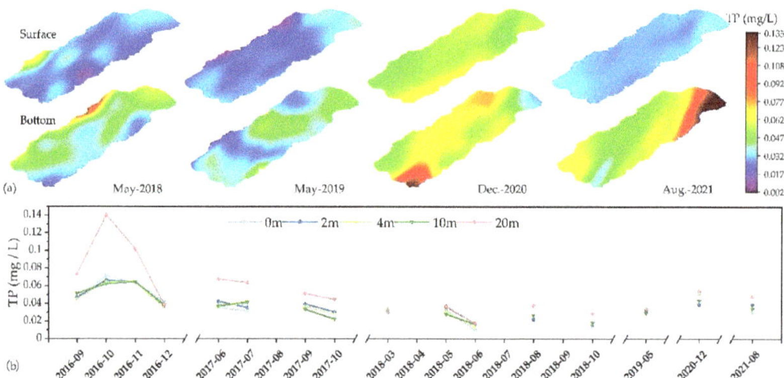

Figure 10. Changes in TP content in Lake Yangzong from 2017 to 2021. (**a**) indicates the distribution of TP content in the whole lake at the surface and bottom. (**b**) represents the mean value of TP content at 5 different depths.

In the vertical distribution, the TP and TN contents were significantly higher at 20 m depth than at 0 m, 2 m, 4 m and 10 m depths.

Correlation analysis of various indicators of Lake Yangzong, including water quality parameters, TN and TP contents in October and September 2016, June, July and September 2017, December 2020 and August 2021 (Figure 11) revealed a high correlation coefficient between DO and Chl-*a* contents ($p = 0.83$), and the correlation between temperature and DO, pH and Chl-*a* was also relatively high ($p \geq 0.54$). However, the correlation between TN and TP was relatively low ($p = 0.31$).

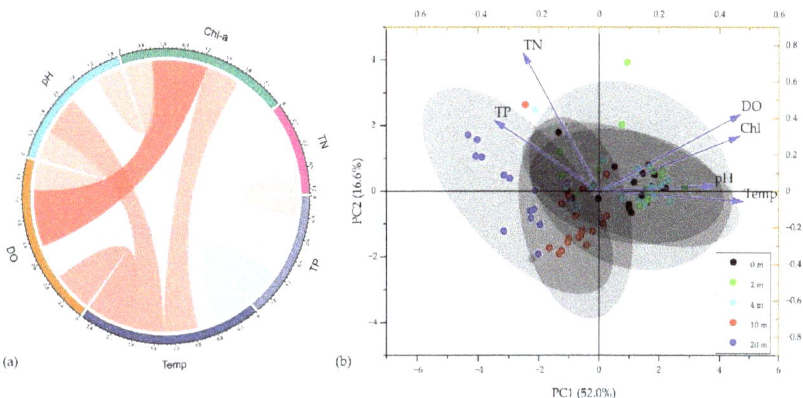

Figure 11. Correlation (**a**) and principal component (**b**) analysis of various water quality parameters.

There were also differences in the TN and TP compositions at different sites and seasons (Figures 12 and 13).

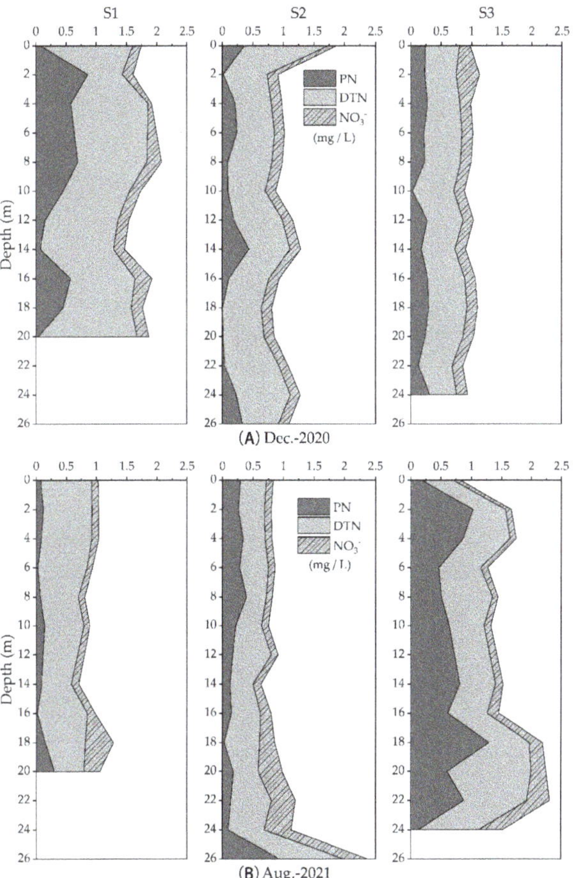

Figure 12. Nitrogen composition accumulation diagram in December 2020 and August 2021.

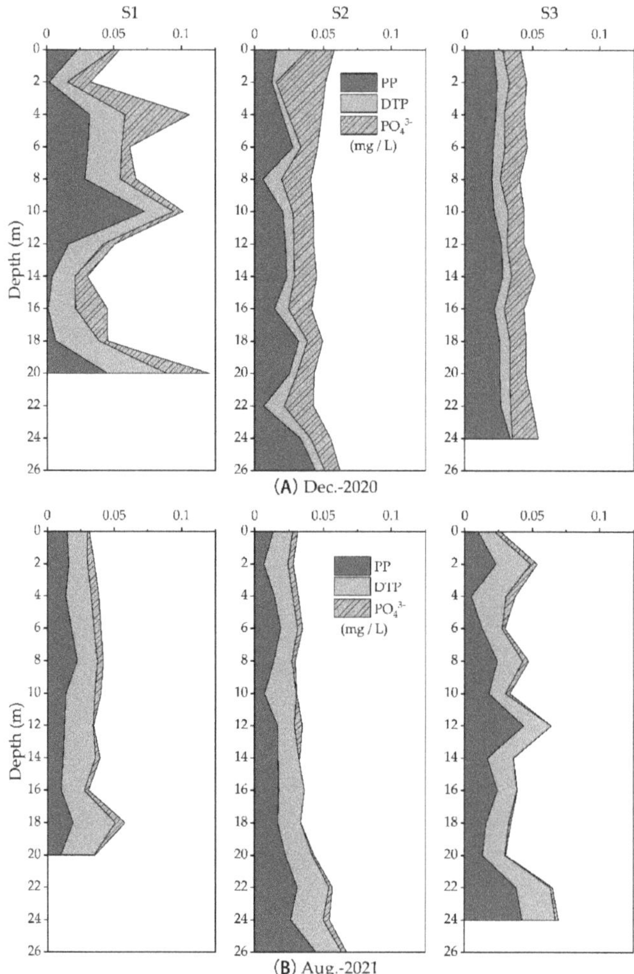

Figure 13. Phosphorus composition accumulation diagram in December 2020 and August 2021.

The content of particulate nitrogen (PN) and dissolved total nitrogen (DTN) at S1 was significantly higher in December 2020 than in August 2021, and the opposite was found at S3. In December 2020, the nitrate nitrogen (NO_3^-) content changed very little with water depth, and in August 2021, the NO_3^- content gradually increased with depth.

At S1, in December 2020, the particulate phosphorus (PP) content was significantly higher than in August 2021 and fluctuated greatly with water depth. There was little difference between the two seasons in dissolved total phosphorus (DTP). The content of PP at S3 in August 2021 also fluctuated with water depth. The orthophosphate (PO_4^{3-}) content of the whole lake in December 2020 was significantly higher than in August 2021.

4. Analysis Discussion

4.1. Mixing Type of Lake Yangzong

The seasonal temperature stratification and mixing characteristics of Lake Yangzong could be divided into six types [29]. Since the lake was located in temperate, subtropical mountainous areas, was affected by ocean climate cycled once a year and had a minimum water temperature $\geq 4\ ^\circ$C, it was classified as a warm monomictic lake. Lake Yangzong maintained this form from spring to autumn, during which the thermocline switched between present and absent, forming a dynamic cyclical process from stratification to mixing.

4.2. Water Quality Parameters

Water temperature is an important factor in determining the primary productivity of lakes, affecting their physical properties, chemical reaction processes and biological activity. Variations in water temperature and the formation and disappearance of thermocline significantly influenced the levels of the chemical parameters [9,30]. The environmental parameters follow the thermal structure and stratification in response to the climatological condition. According to the seasonal variation in water temperature, April, June, July, September and November could be classified as the stratification period, and January and December as the mixed period.

DO is the molecular oxygen dissolved in the water from the air. It is influenced by temperature, algae growth, biochemical reaction, etc. During the thermal stratification period, algal photosynthetic and atmosphere reaeration efficiency in epilimnion was high. With increasing water depth, photosynthesis weakened. Respiration of the upper aquatic organisms consumed the oxygen produced by photosynthesis, causing a lack of oxygen in the deeper waters. In addition, DO deletion in the hypolimnion is due to geochemical DO consumption during decomposition and stable stratification prevent mixing. Therefore, in this period, DO decreased from the surface to the bottom, especially in summer, DO in the middle layer of the lake decreased sharply and tended to be ~0 (0.34 mg/L) at the bottom layer. According to a study by Kalff Jacob et al. [10], the solubility of DO in freshwater mainly depended on the water temperature. Under constant air pressure, a lower water temperature led to a higher concentration of DO. Therefore, compared with December 2020 and January 2016, the winter season, January had a higher DO (7.925 m/L) at lower temperatures (Figures 2 and 3).

The concentration of CO_2 in water usually affects pH change, limiting the amount of phytoplankton at the surface and the decomposition of organic matter at the bottom. The photosynthesis of algae at the epilimnion consumed a large amount of CO_2, causing a reduction in radical acid ions and an increase in pH value. The bottom layer displayed low denitrification, which ultimately reduced the pH. Temperature plays an important role in the growth of algae and the decomposition of organic matter. Therefore, pH level was mostly controlled by photosynthetic activity. The pH value was kept at a high level in the upper profile and a low level in the lower profile. The variation in pH value in Lake Yangzong was in line with the trend. In the mixing period at a lower temperature, the photosynthesis of planktonic algae was weak, the physical and chemical reaction in the lake was not significant, the acidity and alkalinity of the lake changed slightly in the vertical direction and the water was homogeneous in this period.

The vegetation coverage in the catchment of Lake Yangzong was low and showed a steep slope. In the rainy seasons, a large amount of solid matter and industrial and agricultural pollutants were transported into the lake by rivers and surface runoff, which contained large amounts of nutrients. After entering the lake, a portion of the nutrients is consumed by aquatic organisms, with the rest sinking into the bottom of the lake. In addition, the nutrients in the surface sediments migrate upward and are released into the water when the bottom conditions, especially the redox status, change. As a result, the conductivity at the epilimnion was lower, and that of the middle and lower layers increased with depth. From November, Lake Yangzong entered the dry season until the middle of May the following year. During this period, the precipitation was very low, leading to greater evaporation of the lake water than the amount of precipitation, increasing the concentration of salt substances in the lake and increasing the conductivity [31]. The conductivity remained at the same level in the vertical profile. In the study of Deng et al. [32], it was found that the quantity of some cyanobacteria had a good correlation with the electrical conductivity, which was related to the K^+, Cl^- and NO_2^- plasma brought by agricultural emissions. Large amounts of nutrients and pollutants due to domestic sewage discharge from residential areas, wastewater from factories and fertilizers entered the lake through seepage, increasing its conductivity. This also led to a higher CRQI value in April 2017 than in April 2015 (Figure 8). The pH value and the conductivity were abnormally high in June 2017 (Figures 4 and 5) and might be attributed to the second phase of the arsenic pollution treatment project launched in June 2017. Through the Ferric Salt Coagulation method, a large amount of $FeCl_3$ was injected into Lake Yangzong [33], which might have led to an abnormally low phycocyanin concentration and CRQI.

The concentration of Chl-*a* and phycocyanin is used as indicators of phytoplankton biomass. The growth of algae was affected by temperature, light, nutrients and other factors. Epilimnion of Lake Yangzong has large hydrodynamic force and is not suitable for algal growth, so the Chl-*a* content is higher at 2.5–9 m. Algae growth positively correlated with temperature and DO in Lake Yangzong ($p \geq 0.60$). During the thermal stratification of Lake Yangzong, the middle and upper water body temperature was suitable, and the DO was sufficient. At the same time, large amounts of precipitation in Lake Yangzong during rainy seasons with large amounts of pollutants provided sufficient nutrients for the algae to bloom. In the mixing period, the conductivity varied little at different water depths, indicating a homogeneous distribution of nutrients and algae at all depths.

4.3. Trophic Status and the Eutrophication

Eutrophication of water bodies is an aging phenomenon of water bodies. TN and TP contents are important indicators of eutrophication levels of lakes. However, the correlation between TN and TP in Lake Yangzong was generally low ($p = 0.31$).

The average content of TN in Lake Yangzong above 20 m depth was 0.79 mg/L, and that of TP was 0.04 mg/L, both of which were classified as Grade III water quality according to GB 3838-2002. However, the water nutrition below 20 m belonged to Grade IV (TN = 1.05 mg/L, TP = 0.06 mg/L). A sharp increase in TN and TP contents at the bottom of the lake indicated a significant release of nitrogen and phosphorus from sediments. The TN and TP contents at the hypolimnion were higher in the north than in other places in August 2021, and as the northern part was deeper than the southern part, the surface sediments released TN and TP much easier under anaerobic conditions. In the dry season, the lake water mainly comes from the rivers entering the lake. Therefore, in December 2020, the TN and TP contents at point S1 of Lake Yangzong were significantly high.

The TN and TP content in the water column showed no obvious stratification with changes in water depth, which was different from other parameters. The TN and TP content in August 2021 increased significantly compared to August 2018 (Figures 9b and 10b), indicating the degree of nutrition in Lake Yangzong was still increasing.

5. Conclusions

Analysis of the vertical and horizontal spatial distribution characteristics of water quality parameters and nutrition clearly showed that Lake Yangzong undergoes complex seasonal changes.

Lake Yangzong was identified as a typical warm monomictic lake. Its water quality parameters showed obvious changes following stratification, excluding winter, indicating that the water quality parameters were strongly influenced by temperature variations. As the lake water temperature change almost followed the change in air temperature, the lake water quality parameters were also highly influenced by changes in air temperature.

In spring and autumn, the CRQI index was higher, indicating a higher risk of cyanobacterial bloom. Though the contents of TN and TP in Lake Yangzong were not high (the TN contents, especially, were still lower than 2.0 mg/L, the threshold value for algae blooming), it returned to higher values in December 2020 (TN = 1.3 mg/L, TP = 0.06 mg/L), causing a rise in the inter-annual variation. These findings suggest that Lake Yangzong is facing a serious algae blooming threat. The contents of different forms of nitrogen and phosphorus have increased at the bottom of the lake, showing that the nitrogen and phosphorus released from sediments were strengthened. Therefore, it is necessary to intensify lake water quality monitoring and control human activities and endogenous release to prevent further deterioration of the water of Lake Yangzong.

Author Contributions: Conceptualization—H.Z.; Original draft preparation—W.X.; Supervision, Writing—review and editing, H.Z.; Conceptualization, Supervision, Resources, Writing—review and editing, Foundations acquisition—H.Z.; Investigation, Data Curation, L.D., W.X., X.W., H.L., D.L. and Y.Z. All authors have read and agreed to the published version of the manuscript.

Funding: This work was supported by the Special Project for Social Development of Yunnan Province (202103AC100001), Natural Science Foundation of Yunnan Province (2018FH 001-047) and NSFC (41807447).

Data Availability Statement: The data that support the findings of this study are available from the corresponding author upon reasonable request.

Acknowledgments: We thank all the graduate students who participated in the field works and laboratory analyses.

Conflicts of Interest: The authors declare no conflict of interest.

References

1. Wu, H.; Zhang, H.C.; Li, Y.L.; Chang, F.Q.; Duan, L.Z.; Zhang, X.N.; Peng, W.; Liu, Q.; Liu, F.W.; Zhang, Y. Plateau lake ecological response to environmental change during the last 60 years: A case study from freshwater Lake Yangzong, SW China. *J. Soils Sediments* **2021**, *21*, 1550–1562. [CrossRef]
2. Kalinkina, N.; Tekanova, E.; Korosov, A.; Zobkov, M.; Ryzhakov, A. What is the extent of water brownification in lake onego, russia?—Sciencedirect. *J. Great Lakes Res.* **2020**, *46*, 850–861. [CrossRef]
3. Ozguven, A.; Yetis, A.D. Assessment of Spatiotemporal Water Quality Variations, Impact Analysis and Trophic Status of Big Soda Lake Van, Turkey. *Water Air Soil Pollut.* **2020**, *231*, 260. [CrossRef]
4. Iqbal, M.M.; Shoaib, M.; Farid, H.U.; Lee, J.L. Assessment of water quality profile using numerical modeling approach in major climate classes of Asia. *Int. J. Environ. Res. Public Health* **2018**, *15*, 2258. [CrossRef] [PubMed]
5. Iqbal, M.M.; Li, L.; Hussain, S.; Lee, J.L.; Mumtaz, F.; Elbeltagi, A.; Waqas, M.S.; Dilawar, A. Analysis of Seasonal Variations in Surface Water Quality over Wet and Dry Regions. *Water* **2022**, *14*, 1058. [CrossRef]
6. Iqbal, M.M.; Hussain, S.; Cheema, M.J.M.; Lyul, J. Seasonal effect of agricultural pollutants on coastline environment: A case study of the southern estuarine water ecosystem of the boseong county Korea. *Pak. J. Agri. Sci* **2022**, *59*, 117–124.
7. Farid, H.U.; Ahmad, I.; Anjum, M.N.; Khan, Z.M.; Iqbal, M.M.; Shakoor, A.; Mubeen, M. Assessing seasonal and long-term changes in groundwater quality due to over-abstraction using geostatistical techniques. *Environ. Earth Sci.* **2019**, *78*, 386. [CrossRef]
8. Baterdene, A.; Nagao, S.; Zorigt, B.; Ochir, A.; Fukushi, K.; Davaasuren, D.; Gankhurel, B.; Munkhsuld, E.; Tsetgee, S.; Yunden, A. Seasonal Variation and Vertical Distribution of Inorganic Nutrients in a Small Artificial Lake, Lake Bulan, in Mongolia. *Water* **2022**, *14*, 1916. [CrossRef]
9. Ellah, R. Physical properties of inland lakes and their interaction with global warming: A case study of lake nasser, egypt. *Egypt. J. Aquat. Res.* **2020**, *46*, 103–115. [CrossRef]
10. Kalff, J. *Limnology-Inland Water Ecosystems*; Higher Education Press: Beijing, China, 2011.
11. Githukia, C.; Onyango, D.; Lusweti, D.; Ramkat, R.; Kowenje, C.; Miruka, J.; Lung'ayia, H.; Orina, P. An Analysis of Knowledge, Attitudes and Practices of Communities in Lake Victoria, Kenya on Microcystin Toxicity. *Open J. Ecol.* **2022**, *12*, 198–210. [CrossRef]
12. Xie, Y.H.; Li, C.Y.; Yang, Z.L. Research on plankton community structure of Yangzonghai Lake. *Water Resour. Prot.* **2015**, *4*, 47–51. (In Chinese)
13. Zhang, H.C.; Chang, F.Q.; Duan, L.Z.; Li, H.Y.; Zhang, Y.Y.; Meng, H.W.; Wen, X.Y.; Wu, H.; Lu, Z.M.; Bi, R.X.; et al. Water quality characteristics and variations of Lake Dian. *Adv. Earth Sci.* **2017**, *32*, 651–659. (In Chinese)
14. Zhang, Y.; Chang, F.Q.; Zhang, X.N.; Li, D.L.; Liu, Q.; Liu, F.W.; Zhang, H.C. Release of Endogenous Nutrients Drives the Transformation of Nitrogen and Phosphorous in the Shallow Plateau of Lake Jian in Southwestern China. *Water* **2022**, *14*, 2624. [CrossRef]
15. Cai, M.; Zhang, H.C.; Chang, F.Q.; Li, T.; Hu, J.J.; Duan, L.Z.; Zhang, Y. Nitrogen and Phosphorus Content Changes of the Lake Water Samples from the Typical Yunnan Plateau Lakes. *Resour. Environ. Yangtze Basin* **2019**, *28*, 237–245. (In Chinese)
16. Zhang, Y.; Zhang, H.C.; Liu, Q.; Duan, L.Z.; Zhou, Q.C. Total nitrogen and community turnover determine phosphorus use efficiency of phytoplankton along nutrient gradients in plateau lakes. *J. Environ. Sci.* **2023**, *124*, 699–711. [CrossRef]
17. Wu, H.; Zhang, H.C.; Chang, F.Q.; Duan, L.Z.; Zhang, X.N.; Peng, W.; Liu, Q.; Zhang, Y.; Liu, F.W. Isotopic constraints on sources of organic matter and environmental change in Lake Yangzong, Southwest China. *J. Asian Earth Sci.* **2021**, *217*, 104845. [CrossRef]
18. Li, X.M.; Li, S.Y. Dynamic characteristics of nitrogen, phosphorus, chlorophyll a and eutrophication trend in Yangzonghai Lake. *Chin. J. Ecol.* **2014**, *30*, 43–46. (In Chinese)

19. Yang, C.L.; Li, S.Y.; Liu, R.B.; Li, Z.Y.; Liu, K. An Analysis on External Loading of Nitrogen and Phosphorus in Lake Yangzonghai. *Shanghai Environ. Sci.* **2014**, *033*, 47–52. (In Chinese)
20. Yan, H.; Xiang, Q.Q.; Wang, P.; Zhang, J.Y.; Lian, L.H.; Chen, Z.Y.; Li, C.J.; Chen, L.Q. Trophodynamics and health risk assessment of toxic trace metals in the food web of a plateau freshwater lake. *J. Hazard. Mater.* **2022**, *439*, 129690. [CrossRef]
21. Zhang, Y.; Chang, F.Q.; Liu, Q.; Li, H.Y.; Duan, L.Z.; Li, D.L.; Chen, S.X.; Zhang, H.C. An Eco-Environmental Risk Assessment of Heavy Metal Contamination in Surface Sediments of Lake Yangzong, Southwestern China SSRN (Preprint). 16 February 2022. Available online: https://ssrn.com/abstract=4022388 (accessed on 16 February 2022).
22. Bi, P.J.; Liu, C.; Li, S.Z. Variation of water quality of Yangzonghai Lake affected by arsenic pollution. *Water Resour. Prot.* **2014**, *30*, 84–89. (In Chinese)
23. Wang, Z.H.; He, B.; Pan, X.J.; Zhang, K.G.; Jiang, G.B. Levels, trends and risk assessment of arsenic pollution in Yangzonghai Lake, Yunnan Province, China. *Sci. China Chem.* **2010**, *53*, 1809–1817. [CrossRef]
24. Yuan, L.N.; Yang, C.L.; Li, X.M.; Li, S.Y.; Sheng, S.L.; Li, Z.Y.; Liu, R.B.; Liu, K. Effect of daily thermal stratification on dissolved oxygen, pH, total phosphorus concentration, phytoplankton and algae density of a deep plateau lake: A case study of Lake Yangzonghai, Yunnan Province. *J. Lake Sci.* **2014**, *26*, 161–168. (In Chinese)
25. Yang, C.L.; Li, S.Y.; Yuan, L.N.; Yang, L.X.; Liu, W.H. The Eutrophication Process in a Plateau Deepwater Lake: Response to the Changes of Anthropogenic Disturbances in the Watershed. In Proceedings of the 2012 International Conference on Biomedical Engineering and Biotechnology, Macau, Macao, 28–30 May 2012; pp. 1815–1821. [CrossRef]
26. Zhu, Y. Study on Environment ental Background of Yangzonghai Lake Basin. *Environ. Sci. Surv.* **2008**, *5*, 75–78. (In Chinese)
27. Zhang, Y.; Chang, F.Q.; Liu, Q.; Li, H.Y.; Duan, L.Z.; Li, D.L.; Chen, S.X.; Zhang, H.C. Contamination and eco-risk assessment of toxic trace elements in lakebed surface sediments of Lake Yangzong, southwestern China. *Sci. Total Environ.* **2022**, *843*, 157031. [CrossRef]
28. Yang, K.; Yu, Z.; Luo, Y. Analysis on driving factors of lake surface water temperature for major lakes in Yunnan-Guizhou Plateau. *Water Res.* **2020**, *184*, 116018. [CrossRef]
29. Sheng, J. *Lake Sediments and Environmental Changes*; Science Press: Beijing, China, 2010. (In Chinese)
30. Hou, P.F.; Chang, F.Q.; Duan, L.Z.; Zhang, Y.; Zhang, H.C. Seasonal Variation and Spatial Heterogeneity of Water Quality Parameters in Lake Chenghai in Southwestern China. *Water* **2022**, *14*, 1640. [CrossRef]
31. Deng, X.L.; Kong, G.F.; Yang, S.Q.; Dong, S.M. Study on the relationship between water quality and inflow of Yangzong lake. *J. China Hydrol.* **2008**, *28*, 43–45. (In Chinese)
32. Deng, J.M.; Qin, B.Q.; Sarvala, J.; Salmaso, N.; Zhu, G.W.; Ventaelä, A.; Zhang, Y.L.; Gao, G.; Nurminen, L.; Kirkkala, T.; et al. Phytoplankton assemblages respond differently to climate warming and eutrophication: A case study from Pyhäjärvi and Taihu. *J. Great Lakes Res.* **2016**, *42*, 386–396. [CrossRef]
33. Liu, R.B.; Yang, C.L.; Li, S.Y.; Sun, P.S.; Shen, S.L.; Li, Z.Y.; Liu, K. Arsenic mobility in the arsenic-contaminated Yangzonghai Lake in China. *Ecotoxicol. Environ. Saf.* **2014**, *107*, 321–327. [CrossRef]

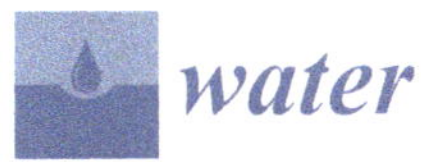

Article

Seasonal Variation in the Water Quality and Eutrophication of Lake Xingyun in Southwestern China

Yanbo Zeng [1], Fengqin Chang [1], Xinyu Wen [2,*], Lizeng Duan [1], Yang Zhang [1], Qi Liu [1] and Hucai Zhang [1,*]

[1] Institute for Ecological Research and Pollution Control of Plateau Lakes, School of Ecology and Environmental Science, Yunnan University, Kunming 650504, China
[2] School of Geography and Engineering of Land Resources, Yuxi Normal University, Yuxi 653100, China
* Correspondence: xywen0801@163.com (X.W.); zhanghc@ynu.edu.cn (H.Z.)

Abstract: It is crucial to understand the spatial-temporal variation of water quality for the water safety and eutrophication migration in plateau lakes. To identify the variation property and the main causes of eutrophication and continuous water quality deterioration, the water quality, including the water temperature (WT), dissolved oxygen (DO), pH, Chl-a, turbidity, total nitrogen (TN) and total phosphorus (TP), of Lake Xingyun was monitored from 2016 to 2021, and their spatial and temporal distribution characteristics were analyzed. The results show that there is no obvious thermal stratification in the vertical direction; pH and DO decrease with depth, which is caused by both physical and biochemical processes, especially at the bottom of Lake Xingyun, which has an anaerobic environment. The chlorophyll content was higher during the high-flow periods and varied significantly in the vertical direction; the spatial variation of water quality in Lake Xingyun was more obvious in the low-flow period and alkaline throughout the year. The average content of total phosphorus (TP) ranged between 0.33 and 0.53 mg/L during the high-flow periods and between 0.22 and 0.51 mg/L during the low-flow periods, while the average content of total nitrogen (TN) ranged between 1.92 and 2.62 mg/L and 1.36 and 2.53 mg/L during the high- and low-flow periods, respectively. The analysis of the inflow samples shows that exogenous nitrogen and phosphorus is the main pollution source affecting the nitrogen and phosphorus content of Lake Xingyun. The trophic level index (TLI) shows that Lake Xingyun is in eutrophication all year round, and even in areas less affected by the exogenous nutrient, there are still conditions for cyanobacterial blooms. This study shed new light on the water quality, eutrophication status and changes in Lake Xingyun, providing suggestions for controlling lake pollution and eutrophication mitigation.

Keywords: Lake Xingyun; water quality; spatial variation; temporal variation

Citation: Zeng, Y.; Chang, F.; Wen, X.; Duan, L.; Zhang, Y.; Liu, Q.; Zhang, H. Seasonal Variation in the Water Quality and Eutrophication of Lake Xingyun in Southwestern China. *Water* 2022, 14, 3677. https://doi.org/10.3390/w14223677

Academic Editors: George Arhonditsis and Shengrui Wang

Received: 8 September 2022
Accepted: 10 November 2022
Published: 14 November 2022

Publisher's Note: MDPI stays neutral with regard to jurisdictional claims in published maps and institutional affiliations.

1. Introduction

Natural lakes are waters maintained at a certain scale and duration, whereby specific biological species and combinations are formed under a specific geological and geographical background. During the evolution of natural lakes, from formation to maturity and final extinction, there are numerous interactions between tectonic movements, climate change, sea levels and human activities [1], which appear as prominent regional characteristics. As a unique natural complex, the plateau lake features a steep lakeshore and a deep lake basin, with fewer rivers flowing out of the lake and along its residence period.

Once pollution or eutrophication occur in the lake, it is a big and time-consuming challenge to effectively treat and mitigate the pollution within a short time. Unfortunately, owing to the effects of land use [2], climate change [3], geological structure [4], surface runoff [5], sewage discharge [6] and other natural processes and human activities, lake pollution and eutrophication problems have become one of the most significant ecological and environmental problems affecting humankind, severely restricting the development of the local economy. Thus, recognizing and understanding the temporal and spatial changes in water quality at the watershed scale is important for evaluating water quality, aquatic ecosystem rehabilitation and water management [7,8].

WT is an essential factor affecting various physical and chemical processes and dynamic phenomena of a lake. It defines the lake type, biological community structure and aquatic ecosystem productivity. A lake's WT is mainly influenced by seasons, which are synchronous and highly correlated with the regional temperature [9,10]. Xu et al. have monitored Lake Yangzong for a long time, and the analysis results show that Lake Yangzong is a warm, monophasic lake, and the temperature has a significant impact on the water quality parameters. In spring and autumn, cyanobacteria are in danger of blooming [11]. Studies have shown that global warming since the 20th century has led to an increased spread of harmful algal blooms worldwide, which is closely related to the global temperature rise [12,13]. Chlorophyll-a (Chl-a) is one of the leading indicators of phytoplankton abundance in inland waters and can be used to measure the nutrient status of lake water [14,15] because its content can indicate the number of algae. Algae distribution is controlled by temperature, pH, DO, light intensity, organic carbon content, wind direction and nutrients such as TN and TP, which are related to its biology and population ecology. According to the accumulated monitoring data of the Yuxi Environmental Monitoring Station over several years, in 2000, cyanobacteria started to bloom. They began to appear at a few bays of Lake Xingyun, followed by periodic outbreaks from April to November every year. Although the mechanism of the water bloom is still not fully understood, it was found that the variation in algae had a close relationship with Chl-a and algae cell density [16], which are the most intuitive factors reflecting the number of algae. They are often used for determinations of the alert level for cyanobacteria bloom. However, Chl-a concentration is more accessible to measure than algal cell density and can predict algal biomass.

Many studies on the spatial and temporal variations in surface water quality in reservoirs or lakes have been published [8,16–21]. At the same time, Najafzadeh et al. used a Data-Driven Model (DDM) to assess the Water Quality Index (WQI) of rivers [22], and they found that the WQI of the Karun River was categorized as "relatively poor" quality. Furthermore, based on reliability analysis, the Karun River had only a 19% chance of having a better quality index; the DDM model exhibited better performance in estimating the WQI classification. Chen et al. used principal component analysis/factor analysis (PCA/FA) to reveal potential sources of contamination in the Danjiangkou Reservoir Basin, and the results showed that high-pollution regions are polluted by chemicals and industries, while low-pollution regions are polluted by sewage from agriculture and livestock [23]. However, from the perspective of regional conditions, the accumulation, systematisms and interpretation of data for plateau lakes in Yunnan, China, still need to be improved.

Lake Xingyun is a typical plateau lake located in Yunnan and plays an irreplaceable role in regulating the local climate, providing water sources and beautifying the environment. It is a place where people have lived for thousands of years. However, in recent years, the influence of human activities, the discharge of large amounts of industrial wastewater and domestic sewage and the unreasonable crops planting structure have not only led to hydrological fluctuations but also to more serious consequences such as eutrophication [11,24], which is severely impacting the local ecological environment and water security and restricting the development of the local economy and people's quality of life. Over the past few decades, studies on Lake Xingyun mainly concentrate on the analysis of phytoplankton biomass and its population structure [25], sedimentation [26], tributary loadings [27], climate change and watershed land-use evolution [28]. There is no comprehensive characterization of the dynamic changes in water quality in Lake Xingyun, and few correlations between eutrophication and water quality indicators have been identified. Therefore, it is necessary to accurately and comprehensively study the trophic level of the lake, especially the nitrogen and phosphorus levels, the pollutant sources and the variation in water quality parameters. This study uses our continuous water quality monitoring data from 2016 to 2021 to:

(1)   Investigate the spatial and temporal distribution characteristics of water quality parameters;

(2) Evaluate Lake XingYun's water quality and nutrient content;

(3) Investigate the potential source of nutrient dynamics and the relationship between nutrient status and water quality indicators.

The purpose of the investigation was to deeply understand the fate of physiochemical contaminants in the lake and to put forward corresponding countermeasures and suggestions.

2. Materials and Methods

2.1. Study Area

Lake Xingyun is a lake in Yunnan Province, southwest China (Figure 1), located in the eastern part of the Jiangchuan Basin in Yunnan Province. Lake Xingyun was mainly formed in the Miocene-Pliocene, has a history of about 3.4 million years and has entered the middle and late stages of lake marsh development. The geography scope of Lake Xingyun is 24°17′20″–24°23′03″ N, 102°45′20″–102°48′20″ E. The lake surface is located at a mean sea level of 1722 m. From north to south, it has a length of 10.5 km and is 3.8 km wide from east to west. The maximum water depth is 10 m, with an average depth of 7 m, a total lake area of 34.7 km^2 and a total drainage area of 386 km^2. The lake is maintained by rainfall and runoff in the watershed, with more than 14 rivers flowing into the lake [29].

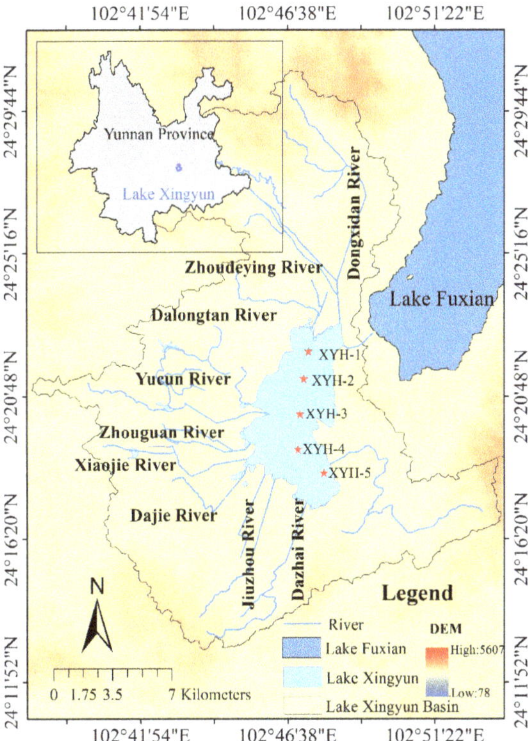

Figure 1. The location of Lake Xingyun and sampling sites.

Quaternary alluvium makes up the majority of the lake's catchment area; the reason for this is that it was deposited during the period when the lake was at a high altitude [30]. Presently, it is a semi-closed, shallow water plateau lake with a subsidence center near the east side of its basin. As a result, the east side of the lake basin is steeper than the west side, which is less steep, with many hot springs around it. The main rivers flowing into the lake are the eastern and western rivers, including the Luosipu River, Yucun River, Dazhuang River, Haiku River and more than ten seasonal rivers. Lake Xingyun used to be a river in the upper reaches of Lake Fuxian, which was connected with Lake Fuxian through a man-made river called Gehe river. Later, after the "Lake Xingyun–Lake Fuxian Outflow Diversion" project, the Xiaojie River was manually excavated and became the only outlet of Lake Xingyun. According to local county records of Jiangchuan, in the prime time of the lake basin, it was connected to Lake Fuxian and Lake Qilu, named the ancient Lake Fuxian. In the Quaternary period, when the north–south faulting activity occurred, the earth's crust subsided along the fault, dividing the Fuxian Lake and eventually forming Lake Xingyun.

The lake is located in the subtropical semi-humid plateau monsoon climatic zone and has dry and wet seasons. About 81.4% of the precipitation is concentrated from May to October, forming the rainy season, and the average annual precipitation is about 717.5 mm (Figure 2). Based on the meteorological data from 2016 to 2021, the average air temperature was about 16.8 °C for many years, with warm climates from May to September and cold climates from October to April. The main direction of the wind is southwest, the annual average wind speed is about 2.2 m/s and the maximum wind speed value is 33 m/s. The soil in the basin is dominated by red earth, combined with purple and paddy soil. The primary vegetation in the area is the secondary coniferous forest, such as the Yunnan pine forest and Huashan pine.

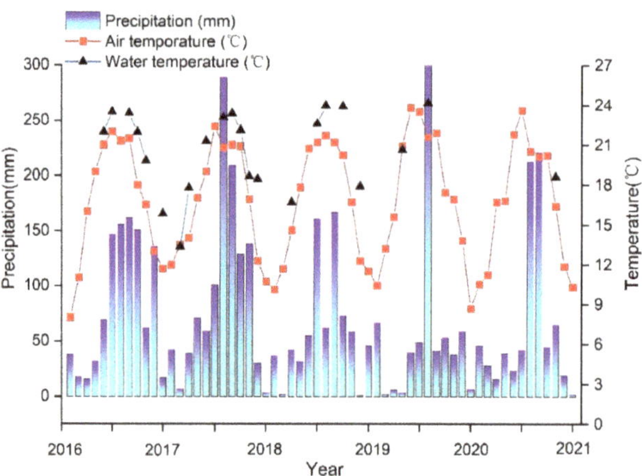

Figure 2. Monthly average air temperature and precipitation of Lake Xingyun from 2016 to 2020 (data provided by the Jiangchuan meteorological station).

2.2. Study Methods

Given the seasonal variations in the vertical and spatial distribution of the physical and chemical properties of the lake, field surveys and monitoring were performed in Lake Xingyun. The instrument used for measurement was the YSI 6600 V multi-parameter water quality monitor (Xylem, Inc., Rye Brook, NY, USA). The water quality parameters of WT, pH, DO, conductivity, turbidity and depth at selected points were measured after calibrating the instrument. During the fieldwork, the probe was placed vertically in the lake, and data were collected at an average interval of about 1 m from the top to the bottom of the water column. From 2016 to 2021, water samples were collected at five sites (XYL-1, XYL-2, XYL-3, XYL-4 and XYL-5) in Lake Xingyun, sealed in a pre-cleaned 1 L polyethylene plastic bottle after sampling and kept in cold storage. TN and TP in the water were measured using a spectrophotometer in the laboratory.

2.3. Data Analysis

The water quality evaluation method was based on the national "environmental quality standard of surface water" (GB3838-2002) for classification [25], and the water quality inferior to V was classified as VI. The environmental quality level standard of each evaluation project was the same as the national standard. In the evaluation process, the evaluation of the trophic level mainly adopted the comprehensive trophic level index *TLI* (Σ) (a number used to indicate the health of lakes in New Zealand) [31] scoring standard to score each sampling point (with two decimals), which is given as:

$$TLI\ (\Sigma) = \sum_{j=1}^{m} W_j \cdot TLI(j) \tag{1}$$

where *TLI(j)* is used to represent the single trophic level index for parameter j, and W_j represents the correlative weighted score of the trophic level index and is expressed by the following equation:

$$W_j = \frac{R_{1j}^2}{\sum_{j=1}^{m} R_{1j}^2} \tag{2}$$

where R_{1j} represents the relative coefficient of Chl-a and parameter j, and m represents the number of parameters. Table 1 displays the findings of 26 Chinese major lakes and reservoirs, which were used to determine the correlation coefficients of Chl-a and other parameters [32].

Table 1. The correlation coefficient for Chl-a to other parameters of Chinese lakes and reservoirs.

Parameters	Chl-a	TP	TN	SD
$R1j$	1	0.84	0.82	−0.83
R_{1j}^2	1	0.7056	0.6724	0.6889
Wj	0.2663	0.1879	0.1790	0.1834

The equations for each eutrophication index are as follows [33]:

$$TLI(Chl\text{-}a) = 10\ (2.5 + 1.086 \ln \text{Chl-a}) \tag{3}$$

$$TLI(TP) = 10\ (9.463 + 1.624 \ln \text{TP}) \tag{4}$$

$$TLI(TN) = 10\ (5.453 + 1.694 \ln \text{TN}) \tag{5}$$

$$TLI(SD) = 10\ (5.118 - 1.94 \ln \text{SD}) \tag{6}$$

The unit of Chl-a is μg L^{-1}, the unit of SD is m, and the other units are all mg L^{-1}.

The Carson index method was used to grade each monitoring point (Table 2). Water quality parameters (temperature, Chl-a, DO, pH, turbidity) were determined. The TP and TN data used in this study were measured using the ultraviolet spectrophotometer UV-1750. The data were processed using software such as Grapher and Arcgis for geospatial analysis.

Table 2. The standard of grading and classification.

Score Value	0–35	36–45	46–55	56–65	66–75	76–85	86–95	96–100
Level	Dystrophic	Lower-Mesotrophic	Mesotrophic	Upper-Mesotrophic	Eutrophic	Hypertrophic	Severe Eutrophic	Abnormal Eutrophic

As the annual precipitation in the lake area is mainly concentrated from May to October, its water level is affected primarily by rain. The water level varied by about 1 m before and after the rainy season. The data used in this study were based on a combination of monitoring data from the year 2016 to 2021. We divided Lake Xingyun into rainy and dry seasons. The water quality characteristics in the vertical and spatial distribution of Lake Xingyun were analyzed and combined with the nitrogen and phosphorus test data. We also investigated the seasonal change and influencing factors of water quality parameters in Lake Xingyun before and after the rainy season.

3. Results

3.1. Analysis of Temporal and Spatial Changes

3.1.1. Analysis of Temporal and Spatial Changes in WT

As a shallow lake, no thermocline exists in Lake Xingyun (Figure 3), and the WT was mainly affected by air temperature changes. The WT below 1 m from the surface showed a significant downward trend and varied significantly in different seasons, showing a monthly increase in spring and reaching a maximum in summer. The pattern was in line with the cyclical pattern of temperature variation. Water temperature is an essential factor affecting various physical and chemical processes and dynamic phenomena of a lake; it defines the lake type, biological community structure and aquatic ecosystem productivity.

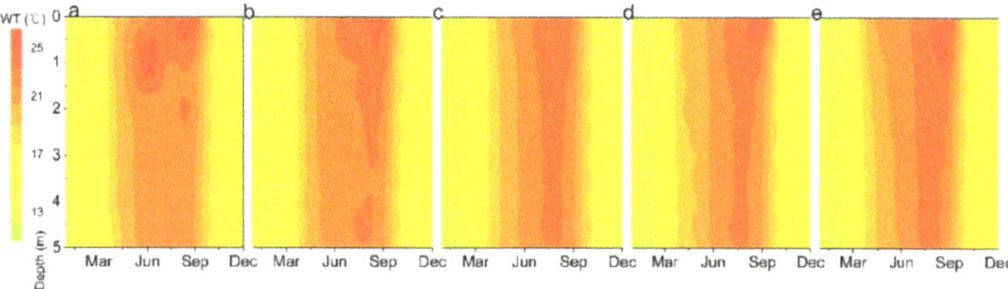

Figure 3. Spatiotemporal variation in temperature in Lake Xingyun, measured during 2017. (**a**) XYH-1. (**b**) XYH-2. (**c**) XYH-3. (**d**) XYH-4. (**e**) XYH-5.

3.1.2. Analysis of Temporal and Spatial Changes in pH

The pH change curve of Figure 4 shows that the pH value of Lake Xingyun was overall alkaline and changed slightly in the vertical direction, with all the measured results at five observatory sites ranging from 7.5 to 9.6. Most of them showed a downward trend with increasing depth, probably due to planktonic algal respiration, as on the top, they performed photosynthesis and absorbed CO_2 from the surrounding water, which raises the pH. The algae in the bottom layer, which get less sunlight, exhale CO_2, thus reducing the pH value of the surrounding water. Compared with the dry season, the pH of Lake Xingyun is higher during the wet season. This enhanced the hydrodynamics of the lake, leading to a significant difference in its pH during the wet season. The mean pH values of Lake Xingyun were as follows: north > central > south. Seasonally, the highest pH values were in autumn, followed by summer, and they were relatively low in winter and spring, with significant temporal differences and little overall spatial variation.

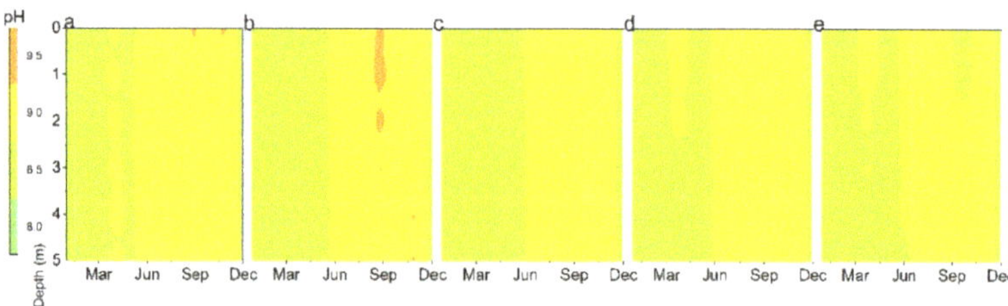

Figure 4. Spatiotemporal variation in pH in Lake Xingyun, measured during 2017. (**a**) XYH-1. (**b**) XYH-2. (**c**) XYH-3. (**d**) XYH-4. (**e**) XYH-5.

3.1.3. Analysis of Temporal and Spatial Changes of DO

The DO variation shows that, in the vertical direction, its change trend was similar to that of pH, which generally showed a downward trend with an increased depth (Figure 5). This might result from photosynthesis by planktonic algae, which releases O_2 on the surface and increases the O_2 content in the water. When the bottom planktonic algae respire, they consume O_2 in the surrounding water, reducing the O_2 content. Compared with the dry season, the DO in Lake Xingyun is generally higher during the wet season. This may be due to the increase in algae caused by the rise in WT in the wet season and the growth of dissolved O_2 content in the water due to photosynthesis. In addition, due to the increase in precipitation, an increase in the hydrodynamic intensity allows the surface water to supply DO supplement quickly, and at the same time, with a rise in the seasonal water level, sunlight at the bottom decreases, the respiration of the underlying plants increases and the O_2 consumption increases, leading to a rise in DO in the surface water of Lake Xingyun. In contrast, the DO in the bottom decreased, showing a quite different variation situation.

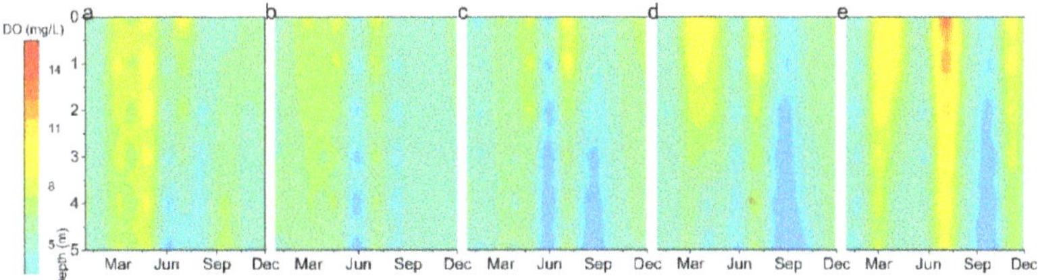

Figure 5. Spatiotemporal variation in DO in Lake Xingyun. measured during 2017. (**a**) XYH-1. (**b**) XYH-2. (**c**) XYH-3. (**d**) XYH-4. (**e**) XYH-5.

Dissolved O_2 in Lake Xingyun varied significantly in the vertical direction. The DO at the bottom of the lake was extremely low in summer. In spring, wind disturbance is stronger, solar radiation is weaker, the density difference of the water is smaller and the water is mixed evenly, so the DO in Lake Xingyun was more evenly distributed in the vertical direction. In terms of concentration, a higher DO concentration was observed in summer, which was also the season in which the vertical gradient of DO was the largest and the planktonic algae grew in large quantities. Due to the abundant light and heat on the surface layer of the lake, algae and floating plants gather in large quantities on the surface layer to carry out photosynthesis and produce O_2, making the surface layer of the lake rich in O_2. Due to the high density of algae on the surface layer, the O_2 in the atmosphere has a difficult time penetrating to the bottom of the water, making the bottom layer of the lake severely hypoxic. Thus, the vertical DO gradient was larger during this time. The southwest wind prevails in Lake Xingyun all year round, making large amounts of planktonic algae gather in the lake's northern part to grow and develop under the wind effects. As they release O_2 during photosynthesis, the DO content in this area was also high. On the seasonal time scale, the differences in dissolved O_2 in Lake Xingyun were significant, with some concentration differences among months, resulting in the obvious spatial heterogeneity of DO in Lake Xingyun.

3.1.4. Analysis of Temporal and Spatial Changes of Chlorophyll

According to the chlorophyll curve in Figure 6, the chlorophyll content in the wet season is higher than that in the dry season, which could be due to the rising summer temperature, causing the algae to grow steadily. The detected Chl-a data showed a large difference between the highest and lowest values, with significant vertical variation, showing a decreasing trend from the surface layer to the bottom layer. The surface layer of the water had suitable light and heat conditions, especially in summer, and the algae were concentrated in the upper layer of the water. In terms of regional distribution, the vertical variation varied among regions, with a small vertical concentration gradient in the dry season, a larger vertical gradient in the north and central parts of the lake, a smaller one in the south during the rainy season and significant spatial and temporal variation in Chl-a in the whole lake.

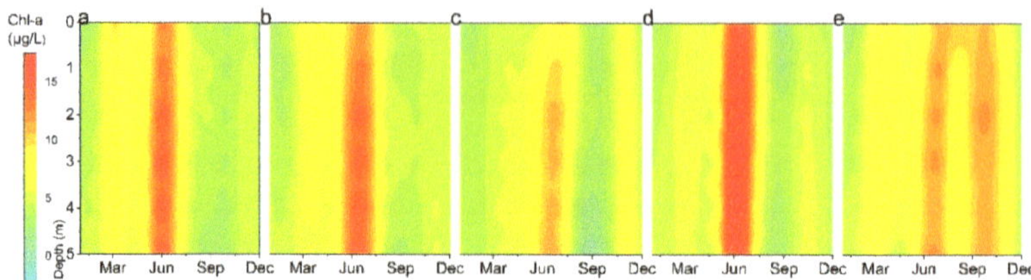

Figure 6. Spatiotemporal variation in chlorophyll in Lake Xingyun, measured during 2017. (**a**) XYH-1. (**b**) XYH-2. (**c**) XYH-3. (**d**) XYH-4. (**e**) XYH-5.

3.1.5. Analysis of Temporal and Spatial Changes in Turbidity

The monitoring results (Figure 7) show that the turbidity of Lake Xingyun ranged from 5 to 50 NTU. It also varied with seasons. The turbidity of the surface water was higher in the rainy season, which may partly be due to the increase in rainfall during the wet season, which increased the amount of sand that flowed into the lake, leading to the rise in the lake's NTU. In terms of spatial scale, the turbidity of the east coast was higher than that of the west coast in the dry season. There were obvious seasonal differences in the vertical variation in turbidity in Lake Xingyun. Algae proliferated during summer and autumn, and the transmission rate of light in the water was lower, so the turbidity of the water layer was also higher where the algae were denser. When the water was disturbed during the rainy season, the insoluble substances settled in the bottom mud returned to the water and increased the water turbidity. The turbidity of Lake Xingyun also had significant spatial and temporal differences.

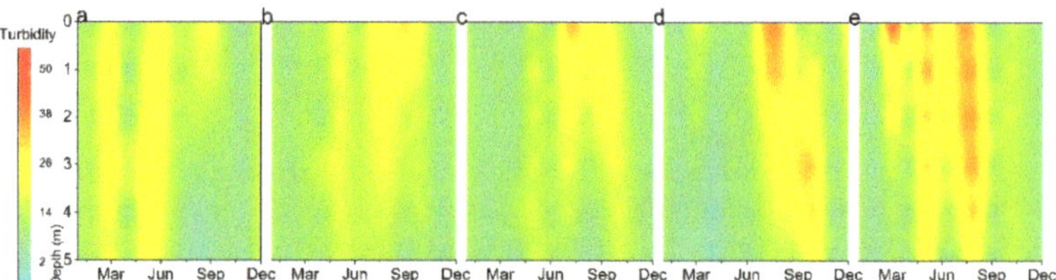

Figure 7. Spatiotemporal variation in turbidity in Lake Xingyun, measured during 2017. (**a**) XYH-1. (**b**) XYH-2. (**c**) XYH-3. (**d**) XYH-4. (**e**) XYH-5.

3.1.6. Inter-Annual Changes in Water Quality

The water quality parameter values were measured at five monitoring sites (Figure 8). The monitoring data showed that the interannual variation trend of water quality parameters at the five monitoring sites was the same, and the variation trend of each monitoring site was also the same in the same year. The peak value of the Chl-a concentration in the five areas occurred in summer (July to August), while the lowest value occurred approximately in November. The Chl-a concentration varied significantly between seasons, with higher concentrations in summer than in winter. During rising summer temperatures, and when rainfalls were relatively abundant, there was a stable growth of algae. When the temperature dropped gradually in October, the metabolism of algae was suspended, and they entered a dormant state, inhibiting the growth and reproduction of algae. The change in the DO content was related to the intensity of algae photosynthesis and lake wave action, consistent with previous studies' results. The pH value varied between 8 and 9.6, the average pH value is higher in autumn and the water of Lake Xingyun is alkaline. April to July was generally the peak period for Chl-a concentrations.

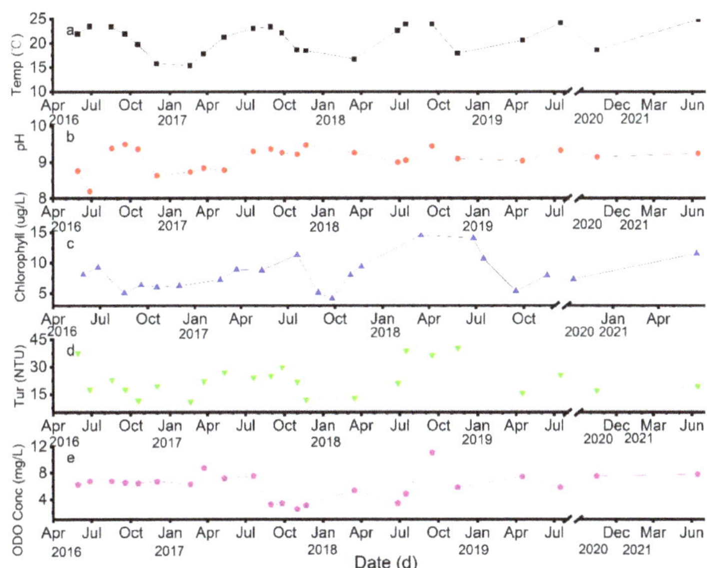

Figure 8. Water quality trend lines of Lake Xingyun from 2016 to 2021. (**a**) TW. (**b**) pH. (**c**) Chl-a. (**d**) Turbidity. (**e**) DO.

3.2. Water Quality Evaluation

The main reason for the eutrophication process of lakes is that primary producers (plants and algae with photosynthesis) use the nutrients available in the lake to produce organic matter. The three primary sources of nutrients were lake input, organic matter circulation [34] and sediment release. N and P are the main controlling factors of lake plant biomass.

Tables 3 and 4 show that the water quality of Lake Xingyun was in the V class all year round. In terms of the nutrient level, it was almost in the eutrophication and medium eutrophication stages.

Table 3. The evaluation table of water quality in Lake Xingyun in 2016.

Monitoring Site	Depth (m)	2016.5					2016.10				
		TN (mg/L)	TP (mg/L)	Average	Trophic Level	Category	TN (mg/L)	TP (mg/L)	Average	Trophic Level	Category
XYH-1	0	2.68	0.60	72.9	Eutrophic	V	2.09	0.45	66.07	Eutrophic	V
	1	2.33	0.46	70.8	Eutrophic	V	2.53	0.46	67.11	Eutrophic	V
	2	2.13	0.34	67.8	Eutrophic	V	2.78	0.46	67.34	Eutrophic	V
	3	2.34	0.30	66.9	Eutrophic	V	2.82	0.46	67.19	Eutrophic	V
	4	2.90	0.34	67.1	Eutrophic	V	2.43	0.46	66.96	Eutrophic	V
	5	1.87	0.30	65.2	Eutrophic	V	2.15	0.44	66.19	Eutrophic	V
	6	1.93	0.29	64.5	Eutrophic	V	2.60	0.46	66.30	Eutrophic	V
	7	1.90	0.31	64.2	Eutrophic	V	2.80	0.47	66.07	Eutrophic	V
	8	2.43	0.38	66.8	Eutrophic	V	/	/	/	/	/
	9	2.67	0.42	67.6	Eutrophic	V	/	/	/	/	/
	Average	2.62	0.37	67.4	Eutrophic	V	2.53	0.46	66.65	Eutrophic	V
XYH-2	0	1.98	0.28	65.22	Eutrophic	V	/	/	/	/	/
	1	1.84	0.29	65.79	Eutrophic	V	/	/	/	/	/
	2	2.11	0.29	64.37	Upper-mesotrophic	V	/	/	/	/	/
	3	1.66	0.29	65.06	Eutrophic	V	/	/	/	/	/
	4	1.80	0.29	64.99	Upper-mesotrophic	V	/	/	/	/	/
	5	1.96	0.28	64.36	Eutrophic	V	/	/	/	/	/
	6	3.44	0.29	63.83	Upper-mesotrophic	V	/	/	/	/	/
	7	2.14	0.29	63.49	Upper-mesotrophic	V	/	/	/	/	/
	8	2.49	0.29	63.58	Upper-mesotrophic	V	/	/	/	/	/
	9	2.65	0.38	65.19	Eutrophic	V	/	/	/	/	/
	Average	2.2	0.30	64.59	Eutrophic	V	/	/	/	/	/

Table 3. *Cont.*

Monitoring Site	Depth (m)	2016.5					2016.10				
		TN (mg/L)	TP (mg/L)	Average	Trophic Level	Category	TN (mg/L)	TP (mg/L)	Average	Trophic Level	Category
XYH-3	0	2.06	0.26	64.24	Upper-mesotrophic	V	2.24	0.42	63.98	Upper-mesotrophic	V
	1	1.95	0.29	65.62	Eutrophic	V	1.94	0.43	64.48	Upper-mesotrophic	V
	2	1.91	0.28	63.44	Upper-mesotrophic	V	1.91	0.44	65.46	Eutrophic	V
	3	1.77	0.28	63.44	Upper-mesotrophic	V	2.32	0.42	65.46	Eutrophic	V
	4	1.83	0.29	63.74	Upper-mesotrophic	V	2.61	0.43	65.42	Eutrophic	V
	5	1.98	0.30	63.87	Upper-mesotrophic	V	2.35	0.44	65.46	Eutrophic	V
	6	1.91	0.28	64.00	Upper-mesotrophic	V	1.71	0.43	65.22	Eutrophic	V
	7	2.10	0.35	64.22	Upper-mesotrophic	V	1.76	0.44	65.28	Eutrophic	V
	8	2.18	0.33	63.75	Upper-mesotrophic	V	1.66	0.44	65.65	Eutrophic	V
	Average	1.97	0.30	64.04	Upper-mesotrophic	V	2.06	0.43	65.16	Eutrophic	V
XYH-4	0	1.41	0.24	62.30	Upper-mesotrophic	IV	/	/	/	/	/
	1	1.84	0.27	63.38	Upper-mesotrophic	V	/	/	/	/	/
	2	2.00	0.26	63.22	Upper-mesotrophic	V	/	/	/	/	/
	3	1.90	0.28	64.29	Upper-mesotrophic	V	/	/	/	/	/
	4	2.74	0.31	65.14	Eutrophic	V	/	/	/	/	/
	5	2.09	0.31	65.48	Eutrophic	V	/	/	/	/	/
	6	2.46	0.35	65.97	Eutrophic	V	/	/	/	/	/
	7	2.55	0.35	65.55	Eutrophic	V	/	/	/	/	/
	Average	2.12	0.30	64.42	Upper-mesotrophic	V	/	/	/	/	/
XYH-5	0	1.93	0.27	65.05	Eutrophic	V	1.94	0.45	65.30	Eutrophic	V
	1	1.86	0.27	65.51	Eutrophic	V	1.84	0.47	65.58	Eutrophic	V
	2	1.74	0.26	64.63	Upper-mesotrophic	V	2.90	0.46	65.74	Eutrophic	V
	3	1.95	0.28	64.77	Upper-mesotrophic	V	2.82	0.46	65.54	Eutrophic	V
	4	2.28	0.31	66.26	Eutrophic	V	2.04	0.54	66.85	Eutrophic	V
	5	2.38	0.36	67.48	Eutrophic	V	2.02	0.48	66.11	Eutrophic	V
	Average	2.02	0.29	65.62	Upper-mesotrophic	V	2.26	0.48	65.85	Eutrophic	V

Table 4. The evaluation table of water quality in Lake Xingyun in 2017.

Monitoring Site	Depth (m)	2017.5					2017.10				
		TN (mg/L)	TP (mg/L)	Average	Trophic Level	Category	TN (mg/L)	TP (mg/L)	Average	Trophic Level	Category
XYH-1	0	3.2	0.31	65.07	Upper-mesotrophic	V	1.93	0.32	61.68	Upper-mesotrophic	V
	1	2.61	0.29	65.50	Upper-mesotrophic	V	1.44	0.32	63.10	Upper-mesotrophic	V
	2	2.83	0.27	65.22	Upper-mesotrophic	V	1.31	0.31	62.98	Upper-mesotrophic	V
	3	2.63	0.26	64.78	Upper-mesotrophic	V	1.31	0.31	63.06	Upper-mesotrophic	V
	4	2.19	0.27	65.34	Upper-mesotrophic	V	1.21	0.43	64.97	Upper-mesotrophic	V
	5	2.32	0.27	65.05	Upper-mesotrophic	V	1.21	0.33	63.18	Upper-mesotrophic	V
	6	2.02	0.28	65.37	Upper-mesotrophic	V	1.35	0.33	63.06	Upper-mesotrophic	V
	7	2.39	0.34	66.42	Eutrophic	V	1.24	0.33	62.89	Upper-mesotrophic	V
	8	2.61	0.36	66.43	Eutrophic	V	1.31	0.34	63.15	Upper-mesotrophic	V
	9	2.59	0.35	66.52	Eutrophic	V	1.28	0.42	64.39	Upper-mesotrophic	V
	Average	2.54	0.30	65.57	Upper-mesotrophic	V	1.36	0.34	63.25	Upper-mesotrophic	V
XYH-3	0	2.75	0.31	64.77	Upper-mesotrophic	V	1.76	0.33	63.00	Upper-mesotrophic	V
	1	2.44	0.32	65.73	Upper-mesotrophic	V	1.77	0.33	63.72	Upper-mesotrophic	V
	2	2.05	0.30	65.38	Upper-mesotrophic	V	1.66	0.33	64.33	Upper-mesotrophic	V
	3	2.43	0.30	65.46	Upper-mesotrophic	V	1.71	0.29	63.75	Upper-mesotrophic	V
	4	2.06	0.27	65.23	Upper-mesotrophic	V	1.67	0.28	63.20	Upper-mesotrophic	V
	5	2.41	0.29	65.94	Upper-mesotrophic	V	1.45	0.31	63.21	Upper-mesotrophic	V
	6	2.08	0.35	67.35	Eutrophic	V	1.44	0.31	63.08	Upper-mesotrophic	V
	7	2.03	0.45	68.68	Eutrophic	V	1.06	0.33	63.43	Upper-mesotrophic	V
	8	/	/	/	/	/	1.16	0.36	63.70	Upper-mesotrophic	V
	Average	2.28	0.33	66.07	Eutrophic	V	1.52	0.32	63.49	Upper-mesotrophic	V
XYH-5	0	1.87	0.34	65.69	Upper-mesotrophic	V	1.64	0.34	68.62	Eutrophic	V
	1	1.45	0.30	65.11	Upper-mesotrophic	V	1.73	0.38	69.77	Eutrophic	V
	2	1.78	0.31	65.17	Upper-mesotrophic	V	1.62	0.33	68.62	Eutrophic	V
	3	1.94	0.31	65.02	Upper-mesotrophic	V	1.61	0.35	68.43	Eutrophic	V
	4	2.27	0.32	65.74	Upper-mesotrophic	V	1.25	0.31	66.63	Eutrophic	V
	5	2.24	0.33	65.61	Upper-mesotrophic	V	1.31	0.31	66.17	Eutrophic	V
	Average	1.93	0.32	65.39	Upper-mesotrophic	V	1.53	0.34	68.04	Eutrophic	V

Based on the trend of recent years (Figure 9), the TN content ranged from 1.25 to 2.75 mg/L, and that of TP ranged from 0.26 to 1.6 mg/L. High TN concentration values in Lake Xingyun were usually observed during spring and summer, with a significant increase in spring, a slight decrease and subsequent increase in late spring or early summer and a decrease in autumn. These could be due to Lake Xingyun being in the dry period in spring, with a low flow velocity, coupled with the low utilization of N by plants, the weak denitrification at low temperatures, most of the N being stored in the water and the input of industrial and agricultural wastewater, leading to abundant N nutrients in the lake and higher TN in spring compared to other seasons. The change in TP in all seasons of the year showed an obvious periodicity, characterized by a rise in spring, a peak in summer and a decrease in winter, similar to the WT and TN cycles. The high TP in Lake Xingyun in summer may be due to the strong metabolism of aquatic animals and more excretion in summer, while cyanobacteria proliferate greatly in summer, and the algae are rich in phosphorus. On the other hand, it may also be due to the suspension of the bottom mud under the wind and waves, and the phosphorus in the bottom mud released in a dissolved form, making the TP content high.

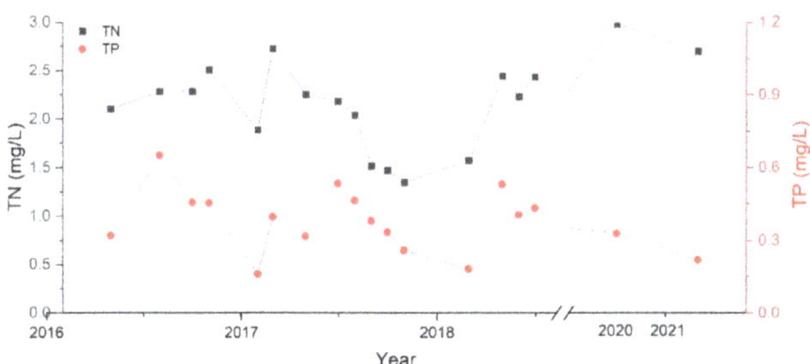

Figure 9. Average TN and TP concentrations in Lake Xingyun from 2016 to 2021.

Water quality along the depth gradient in 2017 was analyzed using heatmaps (Figure 10), showing a poor water quality at the top and bottom of the lake. The water quality of Lake Xingyun demonstrated a large seasonal variation. Eutrophication levels were higher in the wet season. The distribution of TP and TN along the depth gradient is illustrated on the heatmap (Figure 11). The highest concentrations of TP and TN were observed at the surface and near the bottom of the lake. Seasonally, the distribution patterns of TP and TN were varied significantly.

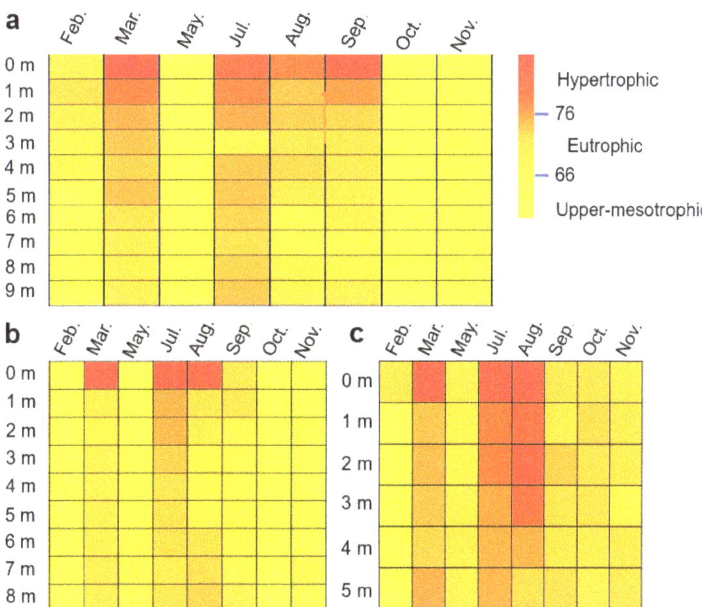

Figure 10. Heat map of the temporal and spatial distribution of water quality in Lake Xingyun in 2017. (**a**) XYH-1, (**b**) XYH-3, (**c**) XYH-5.

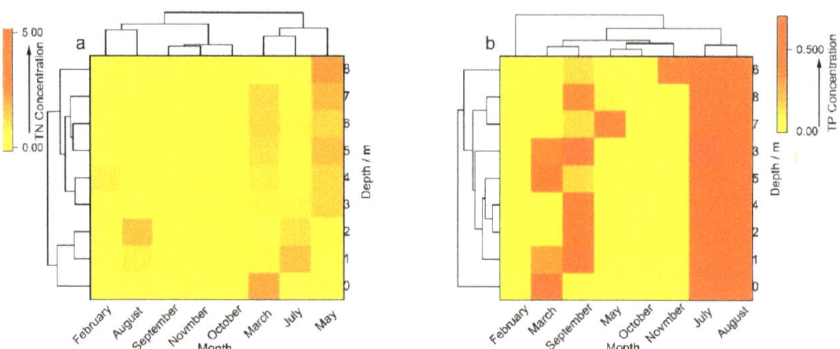

Figure 11. Heatmap showing the temporal and spatial distribution patterns of TN and TP in Lake Xingyun. (**a**) TN. (**b**) TP.

The TN and TP data of six rivers (river 1 to river 6: Dajie river, Xiaojie river, Zhouguan river, Yucun river, Dalongtan river and Zhoudeying river, respectively) entering Lake Xingyun in 2017 were analyzed (Figure 12). The TN and TP values of inflow rivers were higher than those of the lake. Therefore, the primary source of nutrients in Lake Xingyun may be related to the external input. We hypothesize that the water quality of Lake Xingyun may be greatly affected by the nitrogen and phosphorus nutrients from rainwater washing the surrounding farmland during the wet season, reflecting an enormous external load.

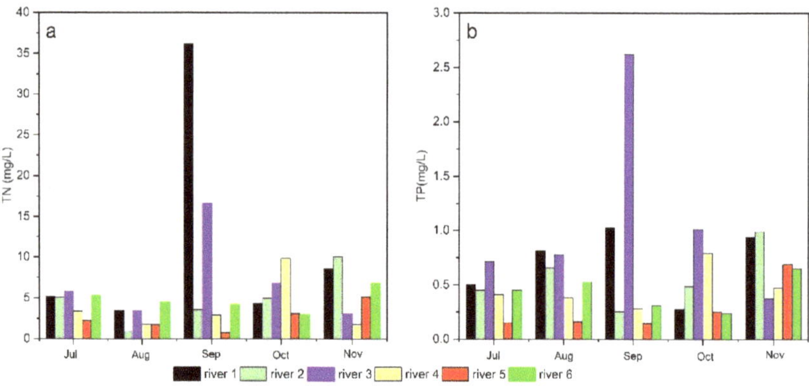

Figure 12. Average TN and TP concentrations in Lake Xingyun inflow rivers in 2017. (**a**) TN. (**b**) TP.

There was little variation in the TP content at different depths in the wet season, which could be related to the relatively high DO concentration in cold weather. The oxidation-reduction potential of surface sediments remained at a high value, conducive to the formation of iron oxide or hydroxide [35], with a high content of ferric iron absorbing phosphorus released by sediments. During the dry season, the TP content increased with increased water depth. This may be because the WT in the dry season was relatively high, there was decreased oxidation-reduction potential in surface sediments and there was decreased adsorption capacity of phosphorus released by deposits. The release of phosphorus, adsorbed initially on iron oxide or hydroxide, increased the TP content in the deep-water area [36]. In addition, low oxidation-reduction potentials and hypoxia conditions in surface sediments stimulated the release of polyphosphates by microorganisms [37]. In some lakes, polyphosphates may account for a significant proportion of the total phosphorus in sediments. The TN content is similar to the TP content in seasons and spaces, and the TN content is higher in the wet season. The release of nitrogen from sediment is affected by microbial activity, the degradation of organic matter and the internal phosphorus load in sediment [27,38,39]. Compared with phosphorus, nitrogen cycling in lake sediments is mainly regulated by microbial activity [40]. During the rainy season, the TN content increases significantly with depth, which may be due to the solid microbial activity in the warm season and the decomposition of organic matter in sediments by microorganisms, resulting in the release of ammonium into the water and an increase in the nitrogen concentration in deep water. Therefore, there were also internal sources of pollution in Lake Xingyun.

In regard to the treatment of Lake Xingyun, based on the obtained scores of TP and TN, first of all, we should reduce the use of pesticides and fertilizers in the surrounding farmlands and promote the plantation of green vegetables. Second, wetland sewage treatment systems should be constructed at the surrounding rivers entering the lake to reduce the impact of domestic sewage and livestock pollution. Lastly, the desalting of Lake Xingyun should be made a priority.

3.3. Comprehensive Analysis

Correlation analysis showed that DO was significantly and positively correlated with TN (F = 0.62, *p* = 0.0026), TN showed a positive correlation (F = 0.57, *p* = 0.0074) with TP and temperature was significantly and positively correlated with TP (F = 0.51, *p* = 0.019) (Figure 13a). A graphical biplot was developed from Principal Component Analysis (PCA) by integrating the indexes of TN and TP with five water quality data matrixes (Figure 13b). The results showed that PC1 was positively correlated with most water quality indicators (i.e., TN, TP and DO) but negatively correlated with chlorophyll. PC2 was positively correlated with temperature, pH and Cond.

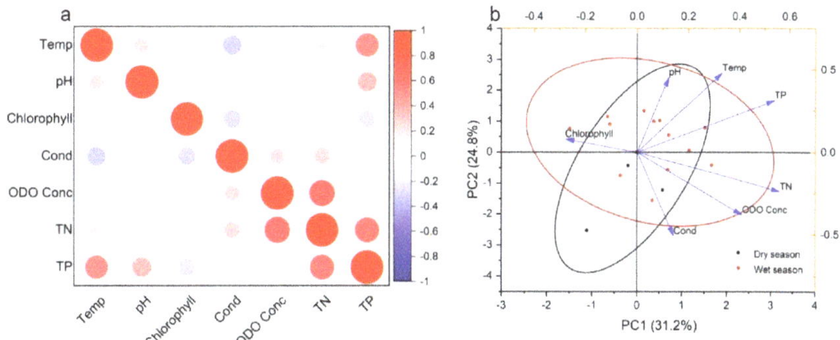

Figure 13. Correlations and principal component analysis of each water parameter in Lake Xingyun. (**a**) Correlations analysis. (**b**) principal component analysis.

4. Discussion

With the increase in water depth, WT, turbidity, pH and DO showed an overall decreasing trend. Factors associated with WT are affected by the hydrological regime, the orographical conditions of its basin [41] and the anthropogenic impact [42,43]. The decrease in pH and DO may be related to algae photosynthesis, pH is an important factor controlling the eutrophication of lakes. An analysis of the spatial and temporal variation characteristics of Lake Cajititlan's water quality showed significant spatial differences in biochemical oxygen demand (BOD) and pH values, with significant temporal differences in alkalinity, total chlorine, electrical conductivity, COD, total hardness, ammonia nitrogen, pH, total dissolved solids and temperature. The parameters with the most significant spatial variation were pH, NO_3^-, NO_2^-, conductivity, hardness of water and NO_2^-, while pH and temperature were the most influential parameters distinguishing seasons. The temporal behavior of these parameters was related to the transport route of seasonal pollutants [44]. The phosphorus cycle in sediments and the distribution of other micropaleontology, such as ostracism [45,46], also directly affect the forms of inorganic carbon sources in water bodies. Its formation is related to the local geological background, such as the climatic conditions, terrestrial heat flow, etc., the interactions between different nutrient elements, the algal propagation, DO and other factors in artificial eutrophication. High pH is one of the characteristics of Yunnan Plateau lakes. Studies on the relationship between pH and algae showed that cyanobacteria dominated in low acid lakes, while color bacteria dominated in acid lakes. Acidic lakes have more species than less acidic lakes [47], and cyanobacteria were utterly absent from habitats with a pH of less than 4 [48]. DO is inversely proportional to salinity and temperature but positively correlated with hydrodynamic intensity, and the DO level is related to algae's photosynthesis. The influencing factors are not only related to its natural geological background and climatic conditions but also to the sewage and industrial wastewater released by human activities and agricultural pollution [49]. During the wet season, the hydrodynamics of Lake Xingyun were enhanced as wind and waves, resulting in an increase in the water mixing and DO. In the dry season, the parameters have some spatial heterogeneity in the horizontal direction due to differences in the lake basin landscape, geographical location, wind and wave action. The seasonal variation in chlorophyll in Lake Xingyun may be related to algal succession and urges further experimental analysis and research.

The analysis results of nutrients showed high concentrations of nitrogen and phosphorus in Lake Xingyun, with TN concentrations ranging from 1.64 to 2.57 mg/L and TP concentrations ranging from 0.27 to 0.52 mg/L. The content of TP in the wet season was almost twice that in the dry season, showing that the nutrient discharge into the lake was pronounced in the wet season, and the exogenous pollution was serious. Fertilization residues in farmland are washed into the lake by rainwater; the main source of nitrogen in the catchment is vegetable cultivation. Most of the TN enters the lake through tributaries during the rainy season. This is the reason for the increase in the annual TN load in the lake over the past 30 years [50]. Analysis results show that, since 2000, the water pollution index has gradually increased and is positively correlated with farmland, fully indicating that the deterioration of water quality may be the result of non-point source pollution [51]. Luo et al. [27] discussed the nutrient concentrations in the main tributaries of Lake Xingyun from January 2010 to April 2018. Their results revealed that the multi-year average load for TN is 183.8 t/year, and the multi-year average load for TP is 23.3 t/year. The external loads of TN and TP increase significantly during high-flow periods. Among them, the TN and TP load account for 84.9% and 84.0% of the annual load during the rainy season (May–October), respectively. Similar findings have been observed in other lakes, Zhen et al. analyzed the monitoring data of Lake Poyang (Jiujiang, China) and found that the increasingly intensive chemical fertilizer use by local farmers has caused a relatively high concentration of total nitrogen and total phosphorus and chemical oxygen demand (COD) in Lake Poyang, leading to moderate to severe eutrophication, which has worsened in recent years [52]. Chen et al. collected data on 11 water quality indicators from the mainstream and major tributaries of the Danjiangkou Reservoir Basin in central China during 2012–2014 to analyze the water spatial-temporal variation and evaluate the water pollution situation. Their results showed that the study area was being affected by industrial and domestic sewage [23].

In addition, the release of the TN and TP from the sediment is also a potential cause of eutrophication in the lake [27,53–55]. Cheng et al. [39] showed that the concentration of TN in the surface sediment of Lake Xingyun is 5140 mg/kg, and that of TP is 4790 mg/kg, which are the highest concentrations in the nine great plateau lakes of Yunnan. Sakamoto et al. [56] estimated that the bottom sediments of Lake Xingyun store 25% of the TN input to the catchment. In addition to the internal load contained in the sediment, cyanobacteria and water hyacinth in the lake also produce internal pollution due to their own decomposition, and the biomass of both nutrient types has increased in recent years. It has been shown that internal loading is a key component of lake eutrophication [57,58]. The water quality evaluation of Lake Xingyun was in the V class all year round, with almost all points being heavy eutrophic. The nutrients in Lake Xingyun were mainly derived from the simultaneous effects of external input and internal pollution.

Despite the analysis of the spatiotemporal variation patterns of water quality, and the seasonal characteristics of the water quality and eutrophication in Lake Xingyun being discussed, long-term observations of hydrodynamic conditions, nutrients and phytoplankton in Lake Xingyun are lacking—for example, to identify its causes for changes in the phytoplankton community structure and the driving mechanism of bloom. At the same time, the impact of human activities on the water quality of Lake Xinyun has not been fully understood. Our study only collected samples from six rivers entering the lake for analysis, while the other seasonal rivers were not sampled, as the exogenous nitrogen and phosphorus pollution from them usually discharge large amounts of pollutants, affecting the nutrients in Lake Xingyun, Therefore, much work remains to be done to gain a more comprehensive and in-depth understanding of the ecological and environmental impacts of the Lake Xingyun watershed.

5. Conclusions

By using water quality monitoring data from 2016 to 2021 on Lake Xingyun, complex seasonal variations in water quality parameters were revealed. The results show that there is no obvious thermal stratification in the vertical direction; changes in pH and DO are caused by physical and biochemical processes, especially at the bottom of Lake Xingyun, which has an anaerobic environment. The pH of Lake Xingyun is alkaline throughout the year, with the increase in depth. WT, turbidity, pH and DO show an overall decreasing trend. In the dry season, the surface spatial heterogeneity of WT, turbidity, pH and Chl-a was greater than that of the bottom. Except for DO, small differences were observed in the spatial heterogeneity of pH and DO in the surface and bottom waters, but they were larger for turbidity and Chl-a. During the wet season, the hydrodynamics of Lake Xingyun were enhanced as wind and waves, resulting in an increase in the water mixing and DO. The content of TP in the wet season was almost twice that in the dry season. The nutrient loads from the catchment area are continuously high, providing excessive nutrients for the growth of planktonic algae, and even in areas where the exogenous nutrient input is low, there are still conditions for cyanobacterial blooming. In addition, the intra-lake load is also an important cause of lake eutrophication; sediments, especially under hypoxic conditions, are potentially major sources of nutrients that may stimulate lake eutrophication, indicating that the nutrient-rich surface sediment was a significant source of N and P that might be released under hypoxic circumstances at the sediment–water interface or resuspended by wind at the lake's surface. In addition to the internal load contained in the sediment, cyanobacteria and water hyacinth in the lake also produce internal pollution due to their own decomposition. Eutrophication leads to the loss of biodiversity, the disruption of food chains and the reduction in ecosystem services. Algal blooming and toxic by-products have negative impacts on fisheries and economic development, leading to declining ecosystem services.

The effective control will need long-term and arduous work; however, the first step should be to gradually control and reduce the area of farmland near the lake, reduce the use of pesticides and fertilizers and promote green vegetable plantations. Secondly, wetland sewage treatment systems should be constructed at the surrounding rivers that enter the lake to reduce the impact of domestic sewage and livestock pollution, regular dredging and management should be performed to reduce the lake's internal pollution hazards and another alternate strategy for reducing eutrophication may be to directly remove the accumulated algae from surface waters, thus achieving the sustainable development and utilization of lake resources.

Author Contributions: H.Z. and Y.Z. (Yanbo Zeng) designed the study. F.C., X.W., Q.L., L.D. and Y.Z. (Yang Zhang) collected and analyzed the data used. Y.Z. (Yanbo Zeng) wrote the first draft of the article. H.Z. contributed to the final version of the manuscript. All authors have read and agreed to the published version of the manuscript.

Funding: This research was funded by the Yunnan Provincial Government Scientist workshop, the Special Project for Social Development of Yunnan Province (Grant No. 202103AC100001), the Natural Science Foundation of Yunnan Province (2018FH 001-047) and the Yunnan Provincial Government Scientist workshop and the China Scholarship Council (No. 202007030022).

Institutional Review Board Statement: Not applicable.

Informed Consent Statement: Not applicable.

Data Availability Statement: Not applicable.

Acknowledgments: We thank all the students who participated in the fieldwork and laboratory analyses.

Conflicts of Interest: The author declares no conflict of interest.

References

1. Beets, D.J.; van der Spek, A.J.F. The Holocene evolution of the barrier and the back-barrier basins of Belgium and the Netherlands as a function of late Weichselian morphology, relative sea-level rise and sediment supply. *Geol. En Mijnb. Neth. J. Geosci.* **2000**, *79*, 3–16. [CrossRef]
2. Duncan, R. Regulating agricultural land use to manage water quality: The challenges for science and policy in enforcing limits on non-point source pollution in New Zealand. *Land Use Policy* **2014**, *41*, 378–387. [CrossRef]
3. Whitehead, P.G.; Wilby, R.L.; Battarbee, R.W.; Kernan, M.; Wade, A.J. A review of the potential impacts of climate change on surface water quality. *Hydrol. Sci. J.* **2009**, *54*, 101–123. [CrossRef]
4. Bilgin, A.; Konanç, M.U. Evaluation of surface water quality and heavy metal pollution of Coruh River Basin (Turkey) by multivariate statistical methods. *Environ. Earth Sci.* **2016**, *75*, 1029. [CrossRef]
5. Abbaspour, K.C.; Yang, J.; Maximov, I.; Siber, R.; Bogner, K.; Mieleitner, J.; Zobrist, J.; Srinivasan, R. Modelling hydrology and water quality in the pre-alpine/alpine Thur watershed using SWAT. *J. Hydrol.* **2007**, *333*, 413–430. [CrossRef]
6. Zhen, S.; Zhu, W. Analysis of isotope tracing of domestic sewage sources in Taihu Lake—A case study of Meiliang Bay and Gonghu Bay. *Ecol. Indic.* **2016**, *66*, 113–120. [CrossRef]
7. Bu, H.; Tan, X.; Li, S.; Zhang, Q. Temporal and spatial variations of water quality in the Jinshui River of the South Qinling Mts., China. *Ecotoxicol. Environ. Saf.* **2010**, *73*, 907–913. [CrossRef]
8. Iscen, C.F.; Emiroglu, Ö.; Ilhan, S.; Arslan, N.; Yilmaz, V.; Akiska, S. Application of multivariate statistical techniques in the assessment of surface water quality in Uluabat Lake, Turkey. *Environ. Monit. Assess.* **2007**, *144*, 269–276. [CrossRef]
9. Van Vliet, M.T.H.; Ludwig, F.; Zwolsman, J.J.G.; Weedon, G.P.; Kabat, P. Global river temperatures and sensitivity to atmospheric warming and changes in river flow. *Water Resour. Res.* **2011**, *47*, W02544. [CrossRef]
10. O'Reilly, C.M.; Alin, S.R.; Plisnier, P.-D.; Cohen, A.S.; McKee, B.A. Climate change decreases aquatic ecosystem productivity of Lake Tanganyika, Africa. *Nature* **2003**, *424*, 766–768. [CrossRef]
11. Xu, W.; Duan, L.; Wen, X.; Li, H.; Li, D.; Zhang, Y.; Zhang, H. Effects of Seasonal Variation on Water Quality Parameters and Eutrophication in Lake Yangzong. *Water* **2022**, *14*, 2732. [CrossRef]
12. Huang, C.; Li, Y.; Yang, H.; Sun, D.; Yu, Z.; Zhang, Z.; Chen, X.; Xu, L. Detection of algal bloom and factors influencing its formation in Taihu Lake from 2000 to 2011 by MODIS. *Environ. Earth Sci.* **2013**, *71*, 3705–3714. [CrossRef]
13. Qin, B.; Zhu, G.; Gao, G.; Zhang, Y.; Li, W.; Paerl, H.W.; Carmichael, W.W. A Drinking Water Crisis in Lake Taihu, China: Linkage to Climatic Variability and Lake Management. *Environ. Manag.* **2009**, *45*, 105–112. [CrossRef] [PubMed]
14. Wang, Y.; Ma, H.; Sheng, D.; Wang, D. Assessing the Interactions between Chlorophyll *a* and Environmental Variables Using Copula Method. *J. Hydrol. Eng.* **2012**, *17*, 495–506. [CrossRef]
15. Modabberi, A.; Noori, R.; Madani, K.; Ehsani, A.H.; Mehr, A.D.; Hooshyaripor, F.; Kløve, B. Caspian Sea is eutrophying: The alarming message of satellite data. *Environ. Res. Lett.* **2020**, *15*, 124047. [CrossRef]
16. Wu, G.; Xu, Z. Prediction of algal blooming using EFDC model: Case study in the Daoxiang Lake. *Ecol. Model.* **2011**, *222*, 1245–1252. [CrossRef]
17. Liu, W.-C.; Yu, H.-L.; Chung, C.-E. Assessment of Water Quality in a Subtropical Alpine Lake Using Multivariate Statistical Techniques and Geostatistical Mapping: A Case Study. *Int. J. Environ. Res. Public Health* **2011**, *8*, 1126–1140. [CrossRef] [PubMed]
18. Palma, P.; Alvarenga, P.; Palma, V.L.; Fernandes, R.M.; Soares, A.; Barbosa, I. Assessment of anthropogenic sources of water pollution using multivariate statistical techniques: A case study of the Alqueva's reservoir, Portugal. *Environ. Monit. Assess.* **2009**, *165*, 539–552. [CrossRef]
19. Joshi, J.; Abhyankar, A.A. Evaluation of Spatial and Temporal Variations in Water Quality of River Ganga. *Int. J. Ecol. Dev.* **2019**, *34*, 69–81.
20. Sun, X.; Zhang, H.; Zhong, M.; Wang, Z.; Liang, X.; Huang, T.; Huang, H. Analyses on the Temporal and Spatial Characteristics of Water Quality in a Seagoing River Using Multivariate Statistical Techniques: A Case Study in the Duliujian River, China. *Int. J. Environ. Res. Public Health* **2019**, *16*, 1020. [CrossRef]
21. Yerel, S. Assessment of Surface Water Quality using Multivariate Statistical Analysis Techniques: A Case Study from Tahtali Dam, Turkey. *Asian J. Chem.* **2009**, *21*, 4054–4062.

22. Najafzadeh, M.; Homaei, F.; Farhadi, H. Reliability assessment of water quality index based on guidelines of national sanitation foundation in natural streams: Integration of remote sensing and data-driven models. *Artif. Intell. Rev.* **2021**, *54*, 4619–4651. [CrossRef]

23. Chen, P.; Li, L.; Zhang, H. Spatio-Temporal Variations and Source Apportionment of Water Pollution in Danjiangkou Reservoir Basin, Central China. *Water* **2015**, *7*, 2591–2611. [CrossRef]

24. Zhang, Y.; Chang, F.; Zhang, X.; Li, D.; Liu, Q.; Liu, F.; Zhang, H. Release of Endogenous Nutrients Drives the Transformation of Nitrogen and Phosphorous in the Shallow Plateau of Lake Jian in Southwestern China. *Water* **2022**, *14*, 2624. [CrossRef]

25. *GB3838-2002*; Environmental Quality Standard for Surface Water. State Environmental Protection Administration (SEPA): Beijing, China, 2002; pp. 1–8.

26. Zhou, X.F. Characterization and sources of sedimentary organic matter in Xingyun Lake, Jiangchuan, Yunnan, China. *Environ. Earth Sci.* **2016**, *75*, 1–11. [CrossRef]

27. Luo, L.; Zhang, H.; Luo, C.; McBridge, C.; Muraoka, K.; Zhou, H.; Hou, C.; Liu, F.; Li, H. Tributary Loadings and Their Impacts on Water Quality of Lake Xingyun, a Plateau Lake in Southwest China. *Water* **2022**, *14*, 1281. [CrossRef]

28. Chen, X.; Huang, X.; Wu, D.; Chen, J.; Zhang, J.; Zhou, A.; Dodson, J.; Zawadzki, A.; Jacobsen, G.; Yu, J.; et al. Late Holocene land use evolution and vegetation response to climate change in the watershed of Xingyun Lake, SW China. *CATENA* **2021**, *211*, 105973. [CrossRef]

29. Wu, D.; Chen, X.; Lv, F.; Brenner, M.; Curtis, J.; Zhou, A.; Chen, J.; Abbott, M.; Yu, J.; Chen, F. Decoupled early Holocene summer temperature and monsoon precipitation in southwest China. *Quat. Sci. Rev.* **2018**, *193*, 54–67. [CrossRef]

30. Hillman, A.L.; Abbott, M.B.; Finkenbinder, M.S.; Yu, J. An 8600 year lacustrine record of summer monsoon variability from Yunnan, China. *Quat. Sci. Rev.* **2017**, *174*, 120–132. [CrossRef]

31. Zhao, L.; Li, Y.; Zou, R.; He, B.; Zhu, X.; Liu, Y.; Wang, J.; Zhu, Y. A three-dimensional water quality modeling approach for exploring the eutrophication responses to load reduction scenarios in Lake Yilong (China). *Environ. Pollut.* **2013**, *177*, 13–21. [CrossRef]

32. Jin, X. *Lacustrine Environment in China*; Ocean Press: Beijing, China, 1995.

33. Xu, M.; Yu, L.; Zhao, Y.; Li, M. The Simulation of Shallow Reservoir Eutrophication Based on MIKE21: A Case Study of Douhe Reservoir in North China. *Procedia Environ. Sci.* **2012**, *13*, 1975–1988. [CrossRef]

34. Doi, H. Spatial patterns of autochthonous and allochthonous resources in aquatic food webs. *Popul. Ecol.* **2008**, *51*, 57–64. [CrossRef]

35. Markovic, S.; Liang, A.; Watson, S.B.; Guo, J.; Mugalingam, S.; Arhonditsis, G.; Morley, A.; Dittrich, M. Biogeochemical mechanisms controlling phosphorus diagenesis and internal loading in a remediated hard water eutrophic embayment. *Chem. Geol.* **2019**, *514*, 122–137. [CrossRef]

36. Yang, C.; Yang, P.; Geng, J.; Yin, H.; Chen, K. Sediment internal nutrient loading in the most polluted area of a shallow eutrophic lake (Lake Chaohu, China) and its contribution to lake eutrophication. *Environ. Pollut.* **2020**, *262*, 114292. [CrossRef]

37. Xie, L.; Xie, P.; Tang, H. Enhancement of dissolved phosphorus release from sediment to lake water by Microcystis blooms—An enclosure experiment in a hyper-eutrophic, subtropical Chinese lake. *Environ. Pollut.* **2003**, *122*, 391–399. [CrossRef]

38. Zheng, Y.; Jiang, X.; Hou, L.; Liu, M.; Lin, X.; Gao, J.; Li, X.; Yin, G.; Yu, C.; Wang, R. Shifts in the community structure and activity of anaerobic ammonium oxidation bacteria along an estuarine salinity gradient. *J. Geophys. Res. Biogeosci.* **2016**, *121*, 1632–1645. [CrossRef]

39. Cheng, N.; Liu, L.; Hou, Z.; Wu, J.; Wang, Q. Pollution characteristics and risk assessment of surface sediments in nine plateau lakes of Yunnan Province. In Proceedings of the 4th International Conference on Energy Engineering and Environmental Protection (EEEP), Xiamen, China, 19–21 November 2019.

40. Liu, C.; Du, Y.; Yin, H.; Fan, C.; Chen, K.; Zhong, J.; Gu, X. Exchanges of nitrogen and phosphorus across the sediment-water interface influenced by the external suspended particulate matter and the residual matter after dredging. *Environ. Pollut.* **2018**, *246*, 207–216. [CrossRef]

41. Pekárová, P.; Halmová, D.; Miklánek, P.; Onderka, M.; Pekár, J.; Škoda, P. Is the Water Temperature of the Danube River at Bratislava, Slovakia, Rising? *J. Hydrometeorol.* **2008**, *9*, 1115–1122. [CrossRef]

42. Borman, M.M.; Larson, L.L. A case study of river temperature response to agricultural land use and environmental thermal patterns. *J. Soil Water Conserv.* **2003**, *58*, 8–12.

43. Huguet, F.; Parey, S.; Dacunha-Castelle, D.; Malek, F. Is there a trend in extremely high river temperature for the next decades? A case study for France. *Nat. Hazards Earth Syst. Sci.* **2008**, *8*, 67–79. [CrossRef]

44. Gradilla-Hernandez, M.S.; de Anda, J.; Garcia-Gonzalez, A.; Meza-Rodríguez, D.; Yebra Montes, C.; Perfecto-Avalos, Y. Multivariate water quality analysis of Lake Cajititlan, Mexico. *Environ. Monit. Assess.* **2020**, *192*, 5. [CrossRef] [PubMed]

45. Flynn, K.J.; Clark, D.R.; Mitra, A.; Fabian, H.; Hansen, P.J.; Glibert, P.M.; Wheeler, G.L.; Stoecker, D.K.; Blackford, J.C.; Brownlee, C. Ocean acidification with (de)eutrophication will alter future phytoplankton growth and succession. *Proc. R. Soc. B-Biol. Sci.* **2015**, *282*, 20142604. [CrossRef] [PubMed]

46. Ramin, N.; Bates, M.H. The effects of pH on the aluminum, iron and calcium phosphate fraction of lake sediments. *Water Res.* **1979**, *13*, 813–815.

47. Brezonik, P.L.; Crisman, T.L.; Schulze, R.L. Planktonic communities in florida softwater lakes of varying ph. *Can. J. Fish. Aquat. Sci.* **1984**, *41*, 46–56. [CrossRef]

48. Brock, T.D. Lower pH limit for the existence of blue-green algae: Evolutionary and ecological implications. *Science* **1973**, *179*, 480–483. [CrossRef] [PubMed]
49. Tutmez, B.; Hatipoglu, Z.; Kaymak, U. Modelling electrical conductivity of groundwater using an adaptive neuro-fuzzy inference system. *Comput. Geosci.* **2006**, *32*, 421–433. [CrossRef]
50. Jin, X.L.; Jin, S.; Liu, H.; Li, Y. Research on amount of total phosphorus entering into Fuxian Lake and Xingyun Lake through dry ad wet deposition caused by surrounding phosphorus chemical factories. *Environ. Sci. Surv.* **2010**, *29*, 39–42. (In Chinese)
51. Zhang, X.; Shao, Y. LUCC impact on water quality change in Xingyun Lake basin. *E3S Web Conf.* **2020**, *198*, 04025. [CrossRef]
52. Zhen, L.; Li, F.; Huang, H.; Dilly, O.; Liu, J.; Wei, Y.; Yang, L.; Cao, X. Households' willingness to reduce pollution threats in the Poyang Lake region, southern China. *J. Geochem. Explor.* **2011**, *110*, 15–22. [CrossRef]
53. Noori, R.; Ansari, E.; Bhattarai, R.; Tang, Q.; Aradpour, S.; Maghrebi, M.; Haghighi, A.T.; Bengtsson, L.; Kløve, B. Complex dynamics of water quality mixing in a warm mono-mictic reservoir. *Sci. Total Environ.* **2021**, *777*, 146097. [CrossRef]
54. Zhou, B.; Fu, X.; Wu, B.; He, J.; Vogt, R.; Yu, D.; Yue, F.; Chai, M. Phosphorus Release from Sediments in a Raw Water Reservoir with Reduced Allochthonous Input. *Water* **2021**, *13*, 1983. [CrossRef]
55. Shou, C.-Y.; Tian, Y.; Zhou, B.; Fu, X.-J.; Zhu, Y.-J.; Yue, F.-J. The Effect of Rainfall on Aquatic Nitrogen and Phosphorus in a Semi-Humid Area Catchment, Northern China. *Int. J. Environ. Res. Public Health* **2022**, *19*, 10962. [CrossRef] [PubMed]
56. Sakamoto, M.S.M.; Maruo, M.; Murase, J.; Song, X.; Zhang, Z. Distribution and dynamics of nitrogen and phosphorus in the Fuxian and Xingyun lake system in the Yunnan Plateau, China. *Yunnan Geogr. Environ. Res.* **2002**, *14*, 1–9.
57. Qin, B.; Paerl, H.W.; Brookes, J.D.; Liu, J.; Jeppesen, E.; Zhu, G.; Zhang, Y.; Xu, H.; Shi, K.; Deng, J. Why Lake Taihu continues to be plagued with cyanobacterial blooms through 10 years (2007–2017) efforts. *Sci. Bull.* **2019**, *64*, 354–356. [CrossRef]
58. Sondergaard, M.; Kristensen, P.; Jeppesen, E. Phosphorus Release from Resuspended Sediment in the Shallow and Wind-Exposed Lake Arreso, Denmark. *Hydrobiologia* **1992**, *228*, 91–99. [CrossRef]

Article

Seasonal Variations in Water Quality and Algal Blooming in Hypereutrophic Lake Qilu of Southwestern China

Donglin Li [1], Fengqin Chang [2,*], Xinyu Wen [3], Lizeng Duan [2] and Hucai Zhang [2,*]

1   Institute for International Rivers and Eco-Security, Yunnan University, Kunming 650504, China
2   Institute for Ecological Research and Pollution Control of Plateau Lakes,
    School of Ecology and Environmental Science, Yunnan University, Kunming 650504, China
3   School of Geography and Engineering of Land Resources, Yuxi Normal University, Yuxi 653100, China
*   Correspondence: changfq@ynu.edu.cn (F.C.); zhanghc@ynu.edu.cn (H.Z.)

Abstract: Understanding the spatiotemporal distributions and variation characteristics of water quality parameters is crucial for ecosystem restoration and management of lakes, in particular, Lake Qilu (QL), a typical plateau shallow lake on the Yunnan-Guizhou Plateau, southwestern China. To identify the main causes of harmful algal blooming and continuous water quality decline, the total phosphorus (TP), total nitrogen (TN), water temperature (WT), dissolved oxygen (DO), chlorophyll-a (Chl-a), pH, and turbidity in hypereutrophic Lake Qilu from January 2017 to December 2021 were analyzed. The results showed a complex pattern in spatiotemporal distribution and variation. WT showed no significant change in the vertical profile. DO and pH value variations were caused by both physical and biochemical processes, especially at the bottom of Lake QL with an anaerobic environment. The Trophic State Index (TSI) assessment results showed that Lake QL is a eutrophic (70.14% of all samples, 50 < TSI < 70) to a hypereutrophic lake (29.86%, 70 < TSI) with poor water quality (WQI < 25). TP and WT were the main factors controlling harmful algal blooms (HABs) based on the statistical analysis of Principal Component Analysis (PCA), Random Forest Model (RFM), and Correlation Analysis (CA). In lake QL, TP loading reduction and water level increase might be the key strategies for treating HABs in the future. Based on our results, reducing TP loading may be more effective than reducing TN to prevent HABs in the highly eutrophicated Lake Qilu.

Keywords: Lake Qilu; seasonal variation; water temperature; dissolved oxygen; chlorophyll-a; pH; turbidity

Citation: Li, D.; Chang, F.; Wen, X.; Duan, L.; Zhang, H. Seasonal Variations in Water Quality and Algal Blooming in Hypereutrophic Lake Qilu of Southwestern China. *Water* **2022**, *14*, 2611. https://doi.org/10.3390/w14172611

Academic Editor: George Arhonditsis

Received: 18 July 2022
Accepted: 22 August 2022
Published: 25 August 2022

Publisher's Note: MDPI stays neutral with regard to jurisdictional claims in published maps and institutional affiliations.

1. Introduction

Global eutrophication and harmful algal blooms (HABs) in water bodies, for instance, rivers, lakes, ponds, streams, etc., are expected to intensify with climate change and anthropogenic pressure [1–3] and may seriously affect our living environments and standard [4,5]. Lakes are an important natural and strategic water resource, and play a critical role in the integrated processes of social and economic development [6]. Lakes are affected by changes in the atmosphere, lithosphere, biosphere, and terrestrial hydrosphere, as they are sensitive to variations in natural processes. Thus, it is important to study climate, environmental and biological changes, and evolution [7]. However, over the last decades, human activities have led to nutrient over-enrichment, especially in nitrogen (N) and phosphorus (P), due to urban, agricultural, and industrial expansion, accelerating the rates of primary production or eutrophication [1,8,9]. Many lake and reservoir ecosystems have been degraded, with their ecological functions being, or have been lost, which not only endangers the healthy development of lake and reservoir ecosystems but also seriously threatens the productivity and lives of people near these lake basins [10].

Both the natural environment and human activities have affected the physical and chemical parameters of lakes. The effects of anthropogenic activities on lake water quality could exceed those of the natural environment, which not only accelerates the evolutional processes of lakes but also disbalances the ecosystem, leading to water pollution and eutrophication [4,11]. HABs are the most serious results that originated from the above effects [9].

Nutrients influence algal biomass and community structure [12–14]. An aquatic ecosystem heavily loaded with nutrients can display N limitation, P limitation, and co-limitation the limiting nutrient could change both seasonally and spatially [15]. Dodds et al., [12] found that N limitation (13%), P limitation (18%), and N + P limitation (44%) occur based on the data from 158 limitation bioassays. The limitation of nutrients on phytoplankton is mainly the concentration and relative ratio of nutrients, but there are still controversies in lakes with different nutrient levels [14,16–18]. Besides, water temperature (WT) is often considered one of the most important determinants of growth and metabolism in freshwater algae without nutrient limitations [19,20]. However, previous studies have found that the density of HABs (e.g., *P. agardhii*) does not have a linear relationship with WT [21]. In other words, the reproduction of HABs may not be in the summer with the highest temperature [20,21]. In the background of global warming, the frequency and duration of HABs are expected to increase in eutrophic ecosystems [1,19]. Algal blooms also significantly alter the physicochemical properties of water (non-stabilized pH changes, organic matter enrichment, dissolved oxygen (DO) depletion/super-saturation), which can deteriorate the ecosystem [22]. Moreover, the mass development of HABs increases turbidity and restricts light diffusion in affected ecosystems. HABs also cause oxygen depletion by respiration and bacterial decomposition of dense blooms, which can result in the death and loss of aquatic animals and hydrophytes [23,24].

In recent years, more and more attention has been paid to performing research on lakes in the Yunnan-Guizhou Plateau [13,25,26]. Lake Qilu (QL) is one of nine major lakes in the Yunnan province, southwestern China, and its water quality has reached hypereutrophic levels, classified in the inferior Grade V standard (GB3838-2002). In the past decades, research on lake QL focused mainly on eutrophication, pollution characteristics, sedimentation, and watershed land-use changes [13,25–27], and less often on water quality. Based on the relation between HABs proliferation and eutrophication, the water quality parameter changes should be investigated [1,2,28], especially determining whether algal blooms are limited by N or P, or both N and P [16,17,29]. The interactions between water quality parameters (e.g., DO, pH value, and turbidity) and HABs are complex. Real-time monitoring of a lake's water quality could enhance our understanding of the health status of its ecosystem and allow correct assessment of the ecological condition of the lake [6,30].

Lake QL is not only the center of surface runoff of water collection in a basin but also the sink of many kinds of pollutants. As a result of anthropogenic activities, the water quality of Lake QL has been continuously declining for the past three decades [31]. The eutrophic conditions of the lake have been dominated by cyanobacteria species in dry and rainy seasons [13]. HABs with high Chl-a concentrations are the primary environmental problem at the moment, but the most important water quality parameters influencing HABs remain unclear. Therefore, the objectives of this study were to (1) understand the spatiotemporal distribution characteristics of each water quality parameter; (2) assess the water quality and the nutrition level of Lake QL, and; (3) identify the major factors controlling HABs and provide targeted advice to control the pollution in lake QL. This not only contributes to the timely implementation of remedial actions but also provides a theoretical basis for understanding the changing trends in hypertrophsic lakes under the joint influence of natural and human activities.

2. Sampling and Methods

2.1. Overview of the Study Area

Lake QL (24°4′36″ N–24°14′21″ N, 102°33′48″ E–102°52′36″ E; 1795.7 m a.s.l.) is located in Tonghai County, Yuxi City, Yunnan Province, southwestern China. It is bordered to the north by Jiangchuan county, to the west by the Yuxi River (the upper reach of the

Qujiang River), and to the east by the Longdong River, which is part of the Nanpanjiang river system. Lake QL has a rectangular outline oriented northeast to southwest. The lake's surface area is 36.95 km^2 when the surface water level is located at an altitude of 1795.7 m a.s.l. Its length from east to west is 10.4 km, with an average width from north to south of about 3.5 km [13]. The maximum water depth is 6.84 m, and the lake capacity is 1.47×10^8 km^3 (Figure 1). In recent decades (e.g., 1958, 1983, 2012, etc.), Lake QL has experienced many episodes of reduced area and falling water levels. Due to the anthropogenic discharge of water and expanding farmland, the area of Lake QL has dropped sharply from 33.53 km^2 in February 2012 to 21.8 km^2 in April 2013, and 23 km^2 in 2015 [13].

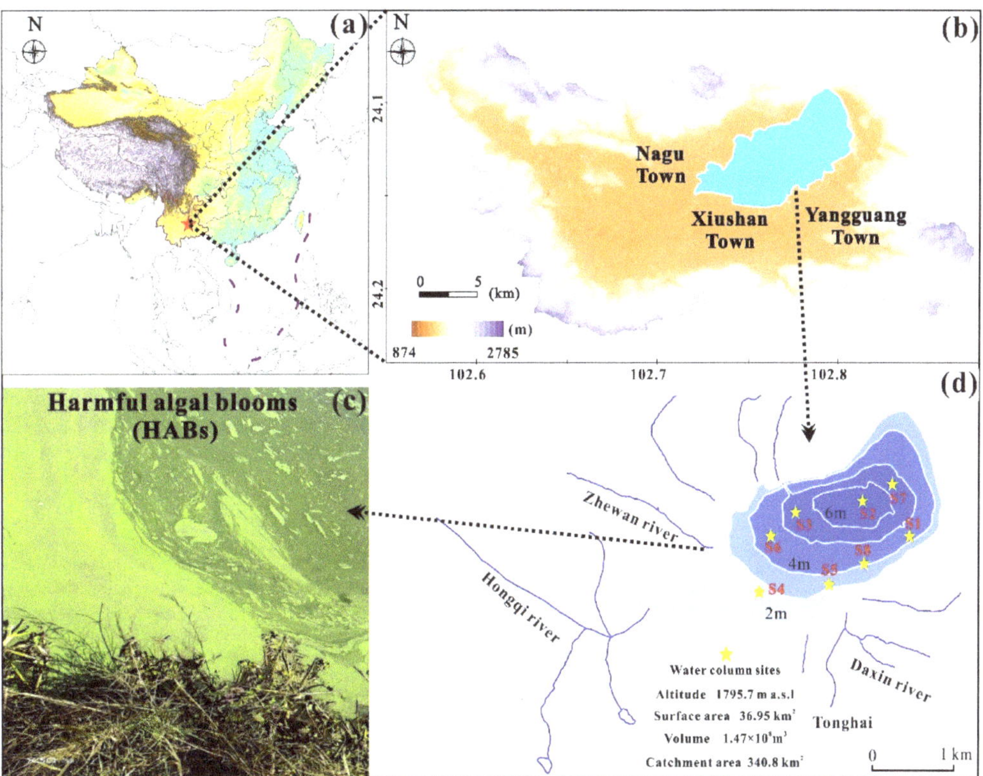

Figure 1. Sampling locations in Lake Qilu, southwestern China. (**a**) Location of lake Qilu; (**b**) Basin of Lake Qilu; (**c**) Main inflows in lake Qilu; and (**d**) the dense harmful algal blooms in lake Qilu.

The lake area, located in southwestern China (Figure 1a), is characterized by a subtropical monsoon climate, with an average annual precipitation of 881.0 mm and an annual average temperature of 15.6 °C. Lake QL is a closed shallow plateau lake with no obvious outlet and is discharged only through natural karst caves. Water input is derived mainly from rainfall and runoff. More than a dozen seasonal watercourses enter the lake, such as the Hongqi, Daxin, and Zhewan rivers. Lake QL is an important water resource for the city (Figure 1c). The basin of lake QL is the focus of social and economic development for the county and the basis for its survival and development. In Tonghai city, lake QL is colloquially referred to as the "Mother Lake" because it has various functions, including industrial and agricultural water usage, regulation and storage, flood control, shipping, tourism, aquaculture, etc. [31].

2.2. Measurements and Sampling

Based on the shape and hydrological condition of the lake, sampling points were set up at the center of the lake (S6), estuarine (S3, S4, S5), the deepest point (S2), and the area of the shore where there was no obvious freshwater input (S8, S1, S7). Sampling sites were marked using the Magellan GPS2000XL satellite-based global positioning system (Figure 1c). To determine the seasonal variation and spatial distribution of water quality, the lake was monitored and sampled for a complete year (from Jan. 2017 to Dec. 2017), after which the sampling time was randomly selected to understand the dynamic changes in water quality (Jan., Mar., Jun., Jul., and Aug. in 2018, while Apr., May, and Aug. in 2020, May., and Dec. in 2021). A complete year of data was used to identify seasonal variation characteristics, and other random sampling data were used to understand the lake's water quality dynamics. Through 22 times of monitoring and sampling at 8 points, we obtained approximately 880 datasets (nutrients to environmental factors). To investigate the water quality parameter seasonal variation pattern at different depths and times, we averaged the data at 8 points using intervals of two meters. A multi-parameter water quality monitor (YSI 6600V2) was used to measure water temperature, Chl-a, DO, pH value, turbidity, and other water quality parameters at an interval of one meter. It was reported that it could be ideal to monitor industrial sewage, sewage discharge estuaries, rivers, lakes, swamps, estuaries, coasts, and potable water [7]. The instrument was cross-calibrated before each test to ensure data reliability and accuracy. Water columns were collected at intervals of one meter at each site. The Tonghai County Meteorological Bureau provided the lake's meteorological data (air temperature and precipitation).

The total nitrogen (TN) and total phosphorus (TP) were measured by a UV-spectrophotometer (UV-2600). TN was determined using an alkaline potassium persulfate digestion-UV spectrophotometric method (GB11894-89) at wavelengths of 220 nm and 275 nm, and the ammonium molybdate spectrophotometric method (GB11893-89) at wavelengths of 700 nm for TP, established by the State Environmental Protection Administration of China (SEPAC) [29,32,33]. Program blank and standard samples were used in the test and analysis process to ensure the analysis's accuracy. All glassware and plastic containers were soaked in HNO_3 for 24 h before use and thoroughly washed with deionized water.

2.3. Water Quality and Trophic State Assessment

To access the water quality of Lake QL, the National Sanitation Foundation Water Quality Index (NSFWQI) was performed [34,35]. The NSFWQI was calculated using the following equation:

$$WQI = \sum_{i=1}^{n} QiWi, \tag{1}$$

Here, Qi and Wi represent the sub-index and weight of water quality parameters, while n refers to the number of water quality parameters. Based on recent work [35,36], the weights of six water quality parameters (0.21 for WT, 0.17 for pH, 0.17 for turbidity, 0.21 for DO, 0.13 for TP, and 0.13 for TN) were calculated. The NSFWQI classification of the water quality is shown in Table 1.

To access the trophic state of Lake QL, the Trophic State Index (TSI) is calculated independently from Secchi disk depth (SD), chlorophyll a (Chl-a), and total phosphorus concentration (TP) using the following equations (Equations (2)–(5) and separated water parameter concentrations [37].

$$TSI(SD) = 60 - 14.41 ln(Secchi\ disk\ depth) \tag{2}$$

$$TSI(Chl\text{-}a) = 9.81 ln(Chlorophyll\ a) + 30.6 \tag{3}$$

$$TSI(TP) = 14.42 ln(total\ phosphorus) + 4.15 \tag{4}$$

$$TSI = (TSI(SD) + TSI(Chl\text{-}a) + TSI(TP))/3 \tag{5}$$

Here, Chl-a represents chlorophyll-a concentration (µg/L), TP represents the total phosphorus concentration (µg/L), and SD represents the Secchi disk depth (meter). The TSI classification of the trophic state is shown in Table 1.

Table 1. Water quality and trophic classification.

Water Quality	Excellent	Good	Medium	Bad	Very bad
NSFWQI [1]	91–100	71–90	51–70	26–50	0–25
	Classification	Oligotrophic	Mesotrophic	Eutrophic	Hypereutrophic
Trophic status	Chl-a [2]	<0.95	0.95–7.3	7.3–56	>56
	SD [2]	>8	8–2	2–0.5	<0.5
	TP [2]	<6	6–24	24–96	>96
	TSI [2]	<30	30–50	30–70	>70
	Mean. TP [3]	<10	10	30–100	>100
	Max.Chl-a [3]	<8	30	25–75	>75

[1] NSFWQI [35], [2] Trophic status [37], [3] Trophic status index (TSI) [38].

2.4. Statistical Procedures

Statistical analyses of chemical analytical data were performed using the IBM SPSS 24.0 and 2019 Microsoft Excel software [13]. CorelDraw X8 (Corel, Ottawa, ON, Canada) was used to generate the map of sampling locations. Ordinary Kriging interpolations of water parameters were computed in Surfer 16 at a 99% confidence interval. Correlation analysis (CA) was conducted using the spearman method in OriginPro 2021 to detect the relationships between the Chl-a and each water quality parameter (pH, DO, TP, TN, etc.). The dimension reduction of the dataset was determined using Principal Component Analysis (PCA) by the "factoextra" and "FactoMineR" packages in R. Random Forest Model (RFM) was performed using the "randomForest" package to select the important factors affecting HABs (https://www.r-project.org/ (accessed on 15 July 2022)).

3. Results and Discussions

3.1. Water Temperature Changes

From January 2017 to May 2021, water temperature (WT) was inherently synchronized with seasonal change at each monitoring point (Figure 2b). The results showed that the WT of lake QL ranged from 13.45 °C to 26 °C during the whole monitoring period. Spatially, WT decreased gradually from the lake's surface to its bottom, especially during the summer. The average water temperature at 8 sample sites at 0 m, 2 m, and 4 m was 24.2 °C, 22.9 °C, and 22.1 °C, respectively. Seasonally, WT in the northern part of the lake was higher than in the south in January. WT in the center of the lake was higher than in other parts of the water body in July (Figure 2a). Due to the shallow depth of the lake, WT showed no significant changes in its vertical profile. Overall, these results revealed that water column exchange was weaker in summer than in other seasons. WT stratification occurred in many deep-water lakes and reservoirs and generally showed a stable thermocline in spring, and the thermocline thickness and temperature gradient increased continuously under the influence of water thermal conditions, transparency, light, and lake shape [39]. The seasonal variations of air temperature and characteristics of the lake basin were stratified according to seasons [28]. Water quality parameters such as DO, pH, Chl-a, and turbidity were affected by temperature stratification and also changed in vertical distribution.

Changes in WT were consistent with those of air temperature and precipitation (Figure 2c). It shows that during the monitoring period, WT changed synchronously with air temperature, with no apparent lag effect, and that WT was slightly higher than air temperature (except in February and April). The Tonghai county meteorological station was approximately 150 m higher than the current lake surface. The warming of the basin and the difference in altitude were the main reasons why the WT of Lake QL was higher than that recorded at the Tonghai meteorological station. Another possible reason is that the water temperature was measured during the day and the air temperature was the average value for the whole day. Precipitation was concentrated between June and October of each year. Simultaneously, precipitation resulted in a lower water level in June and a higher one in September.

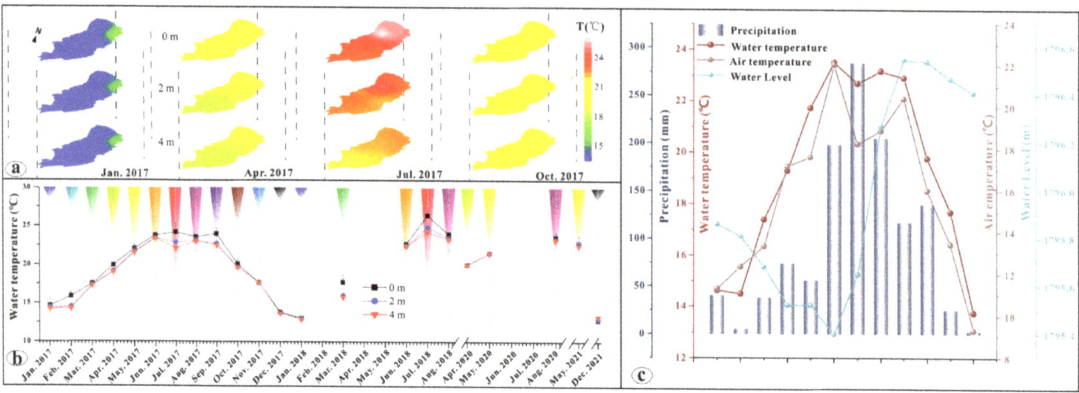

Figure 2. Seasonal and interannual variations in water temperature in lake QL. (**a**) Seasonal variations in water temperature; (**b**) Interannual variations in water temperature; (**c**) Monthly changes in precipitation, air temperature, water temperature, and water levels in Lake QL.

Lake QL is a representative shallow lake on the Yunnan-Guizhou Plateau. WT in the lake exhibited strong seasonal variability. Figure 2a shows the absence of thermocline at lake QL, which differed significantly from other deep-water bodies such as Lakes Fuxian [39] and Yangzong [40]. Increasing air temperatures in May of 2017 resulted in rising WT (Figure 2b) due to thermal energy exchange between surface water and air, causing the surface WT to be higher than those at the bottom of the lake. Furthermore, this condition caused decreasing water density in surface water and hindered caloric exchange with bottom water, especially in summer. In winter, due to the disappearance of the thermocline and shallow lake characteristics, the exchange capacity of the water body was enhanced, with the surface and bottom water exhibiting the same properties. Since there was no temperature stratification phenomenon in the lake water, the heat transformation was relatively rapid, making WT sensitive to temperature changes but without any lag (Figure 2c), which differs from the changing characteristics of deep-water lakes and reservoirs on the Yunnan-Guizhou Plateau [40]. As a heat reservoir, lake QL plays a vital role in regulating the basin's climate and maintaining salubrious temperatures in the surrounding area. WT is often considered one of the most important determinants of growth and metabolism in freshwater algae without nutrient limitations [1]. The effects of WT on eutrophication and HABs are discussed in Section 3.7.

3.2. Dissolved Oxygen (DO) Characteristics and Variability

Dissolved oxygen (DO) concentration is essential in defining water health status in aquatic ecosystems and certainly affects the well-being of zooplankton. Our results indicated that the DO concentration in lake QL ranged from 0.67 to 14.96 mg/L during the monitoring time. Figure 3a shows that the DO concentration of the lake's water decreased from the surface to the bottom in all seasons in terms of vertical variation. The DO concentration at the bottom of the water was significantly lower than at the surface from July to October (Figure 3b), and the lowest DO concentration was 0.67 mg/L in July, showing a relatively anaerobic environment at the bottom of the lake. Spatially, the DO concentration showed strong heterogeneity at the different lakeshores (Figure 3a). DO concentration was controlled by physical and biochemical processes (i.e., wind and wave disturbance, shipping, and other factors). These results were related to the lake's seasonal and annual changes in DO concentration. Fluctuations in DO concentration could also be caused by planktonic photosynthesis and the respiration of microorganisms. Notably, because of the shallow water depth (~1 m) and abundant nutrients, a large number of aquatic plants (dominant by *Potamogeton pectinatus*, a pollution-tolerant species, and *Phragmites communis*, etc.) lived in the western part of lake QL [41], resulting in a relatively lower DO concentration. In summer, because of microorganism respiration, the DO concentration at the bottom water approached anaerobic levels (Figure 3). Thus, based on these results, more attention should be paid to the reductive dissolution of heavy metals and P, which causes water pollution under anoxic conditions [42–44].

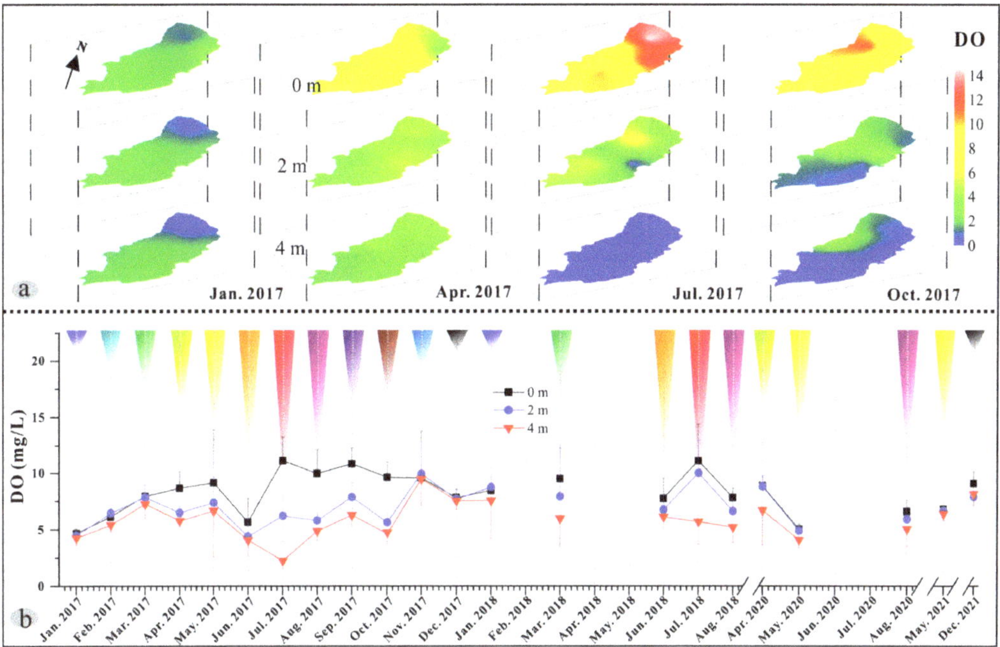

Figure 3. Seasonal and annual variations in dissolved oxygen in Lake QL. (**a**) Seasonal variations in dissolved oxygen; (**b**) Interannual variations in dissolved oxygen.

3.3. Variations in Water Turbidity

Turbidity describes the degree to which suspended matter blocks light passing through water and is expressed in terms of the mean number of particles [26]. Our monitoring data revealed that the turbidity of Lake QL ranged from 0.32 to 22.9 NTU, changed seasonally, increased from January to June, and decreased from June to December 2017, with the highest values observed in June (Figure 4). With increasing depth, turbidity also increased, and the highest turbidity levels were measured at a depth of 4 m in April. From the spatial perspective, turbidity on the eastern lakeshore was consistently greater than on the western shore, indicating that the dense HABs may have been caused due to a decline in turbidity [6]. Another important factor controlling turbidity is sediment suspension due to disturbance, shipping, and other factors, resulting in a high turbidity value at the bottom of April 2017.

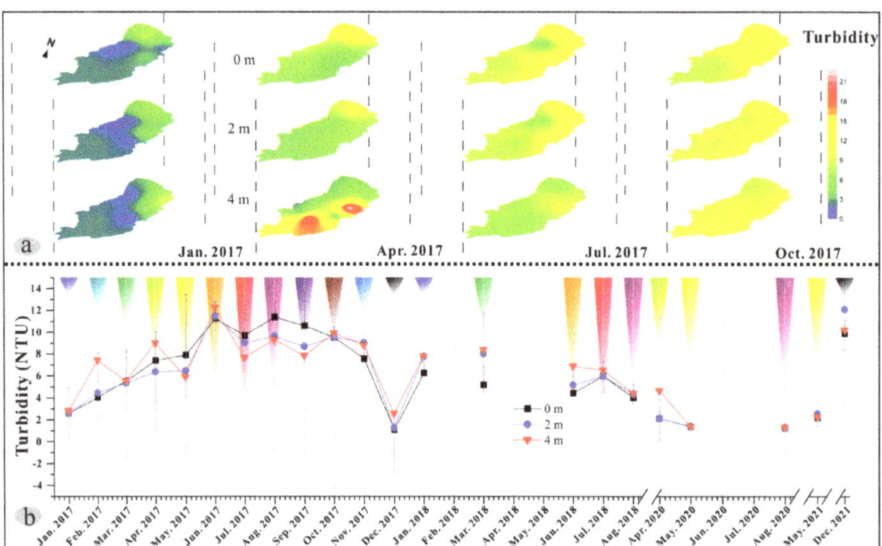

Figure 4. Seasonal variations in turbidity in Lake QL. (**a**) Temporal and spatial variation characteristics in turbidity; (**b**) Interannual variations in turbidity.

3.4. pH Characteristics and Variability

pH value has an important impact on controlling the P cycle sediment and eutrophication [17,45]. H^+, CO_3^{2-}, and HCO_3^- ions in water have a dominant influence on pH value. Our monitoring data showed that the pH of lake QL ranged from 7.45 to 9.15 and changed seasonally. The pH values from the lake surface to the bottom decreased on a seasonal basis, and the values were relatively high on the eastern lakeshore (Figure 5). Vertically, the pH value of the surface was significantly higher than the lake bottom from July to October.

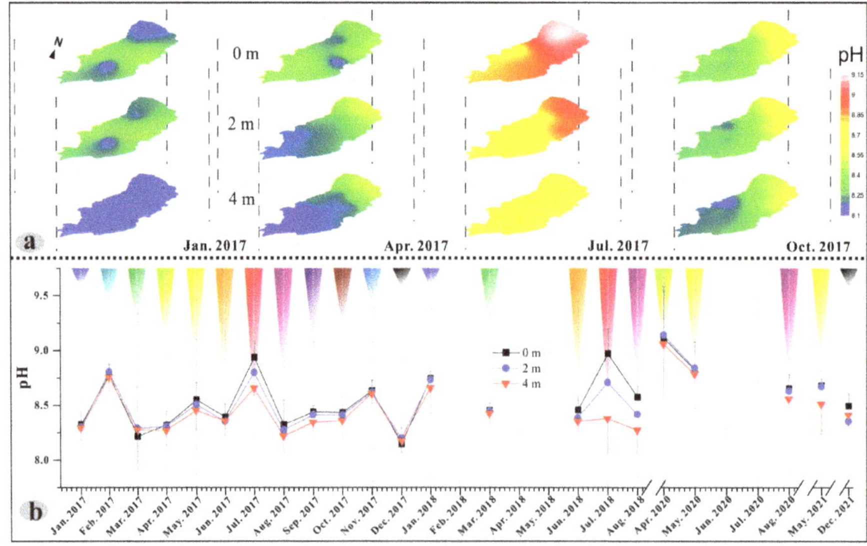

Figure 5. Seasonal and annual pH variations in Lake QL. (**a**) Temporal and spatial variation characteristics in pH; (**b**) Interannual variations in pH.

The pH of Lake QL changed because of phytoplankton photosynthesis and the exchange of carbon dioxide (CO_2) in the air. Liu et al., [46] reported that HCO_3^- significantly increased algal growth compared with CO_3^{2-} when the pH values ranged from 8.0 to 9.5. Lake QL is a typical naturally alkaline lake with pH values ranging from 8.14 to 9.15 (Figure 5), similar to most natural lakes and reservoirs on the Yunnan-Guizhou Plateau [40]. pH value decreased with water depth in lake QL (Figure 5a), possibly due to the decomposition of microorganisms at the sediment-water interface producing large quantities of acidic substances (e.g., H_2S), which were easily detected from July to October 2017 (Figure 5b). In addition, with the increase of WT, the dissolution rate of CO_2 decreased, leading to an increase in pH in the water column. A high pH value was observed during summer (Figure 5a). pH is an important factor regulating P cycle sediment and lake eutrophication. Thus, an increase in pH reduced the P absorption capacity of ferric iron (oxy)hydroxides, thereby increasing the potential for sediment P release [47].

3.5. Chlorophyll-a (Chl-a) Characteristics and Variability

Chlorophyll-a (Chl-a) is an essential component of phytoplankton organisms and a valuable index for evaluating phytoplankton biomass and standing stock. Our monitoring results showed that the concentration of Chl-a in Lake QL ranged from 1.7 µg/L to 119.9 µg/L. Based on the assessment of maximum Chl-a concentration (75 µg/L), the trophic status of Lake QL was classified as hypereutrophic (Table 1). Simultaneously, the Chl-a changes differently at various water depths (Figure 6a), its concentration increased continuously from 2017 to 2021, indicating a constantly decreasing water quality (Figure 6b).

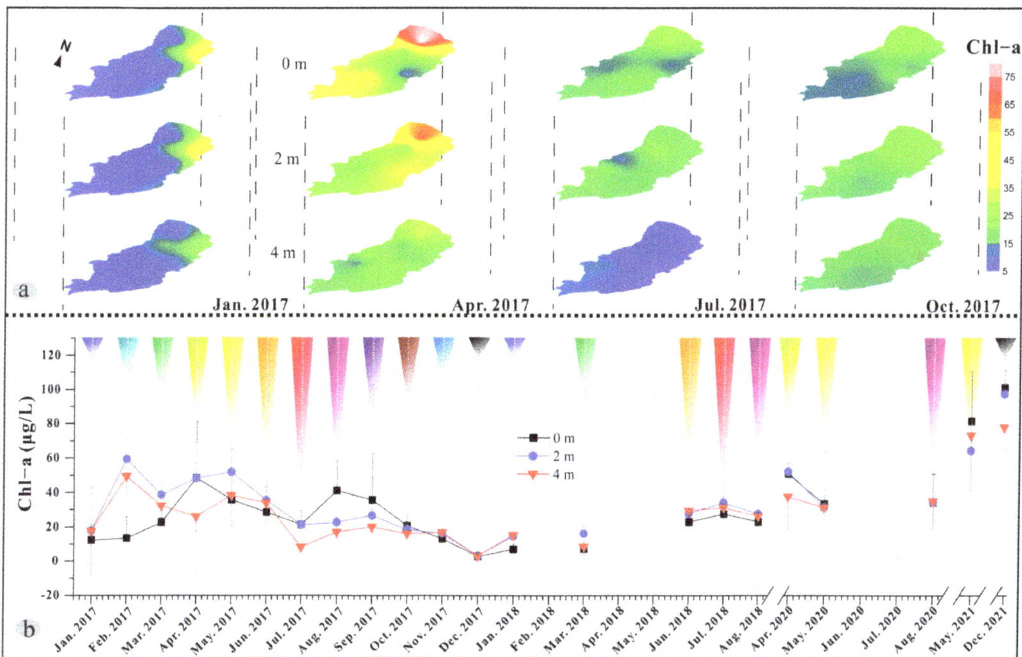

Figure 6. Seasonal and annual variations in Chl-a in Lake Qilu. (**a**) Temporal and spatial variation characteristics in Chl-a; (**b**) Interannual variations in Chl-a.

From the perspective of spatial distribution, we observed a significant variation in seasonal Chl-a concentration, which was significantly higher on the eastern lakeshore than that of the west from January to April, corresponding to dense HABs in the west. In terms of vertical distribution, the Chl-a concentration at the bottom was significantly lower than that of the surface from July to October, indicating that relatively dense HABs appeared on the surface rather than at the bottom (Figure 6a). From the perspective of annual change, we observed that the concentration of Chl-a was higher than in other months from February 2017 to May 2017. The time of HABs was inconsistent with other eutrophic lakes (e.g., Lake Taihu [43] and Lake Dianchi [48]).

3.6. Assessment of Trophic State and Water Quality

The monitoring dataset showed that the TN and TP concentrations ranged from 1.6 to 3.57 (2.19 ± 0.37) mg/L and 0.02 to 0.4 (0.12 ± 0.04) mg/L, respectively. They were inferior to Class V (GB3838-2002) for a long time [25]. Notably, TN and TP concentrations continuously increased from 2017 to 2021, showing an increasing trophic status in Lake QL (Figure 7). Compared with the so-called nine large lakes (>30 km^2), the concentrations of TN and TP were higher than Lake Chenghai (0.75 ± 0.12, 0.05 ± 0.01), Erhai (0.57 ± 0.51, 0.05 ± 0.05), Yangzong (0.69 ± 0.11, 0.02 ± 0.00), Lugu (0.31 ± 0.11, 0.01 ± 0.00), and Fuxian (0.20 ± 0.32, 0.02 ± 0.00) [13]. Based on the TP and TN concentration threshold (TP, 0.05 mg/L; TN, 1.4 mg/L) in the trophic state reported by Liang et al., [16], Lake QL was classified as a hypereutrophic lake. According to a review by Tanvir et al., [38], the trophic state of the lake was hypereutrophic when the mean TP and maximum Chl-a concentration were higher than 100 μg/L and 75 μg/L, respectively, (Table 2). Based on the work of Carlson and Simpson [37], the percentages of all samples were eutrophic for Chl-a (80.21%), TP (34.72%), and the depth of Secchi disk transparency (30.21%). The hypereutrophic percentage was 6.60%, 65.28%, and 69.44%, respectively. There were 70.14% and 29.86% of comprehensive TSI that showed a eutrophic and hypereutrophic level in lake Qilu, respectively. All assessments of trophic state results indicated that Lake QL was a eutrophic to the hypereutrophic lake. Simultaneously, NSFWQI assessment of all numbers showed that the water quality of Lake QL was very bad (0 ≤ WQI ≤ 25) [49]. Another interesting finding is that TP concentration continuously increased from January to May of 2017 and decreased after May but was relatively steady for TN, which led to a change in TN/TP ratios.

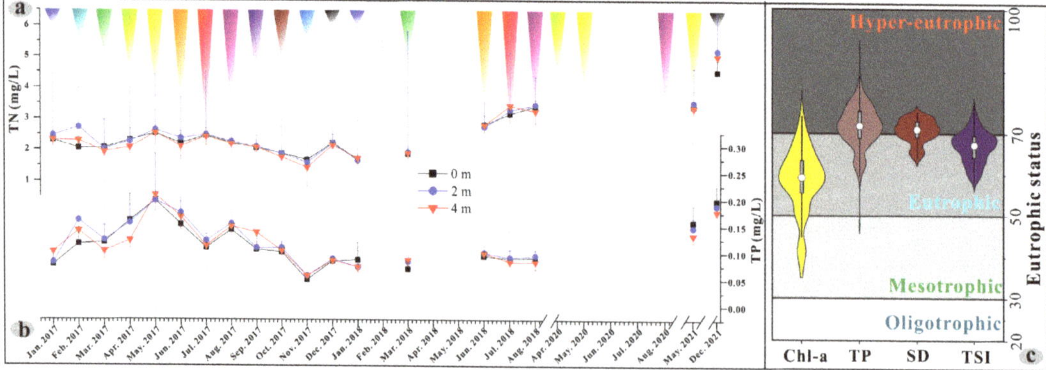

Figure 7. Seasonal and annual variations of TN and TP in Lake Qilu. (**a**) Seasonal and annual variations of TN; (**b**) Seasonal and annual variations of TP; and (**c**) the eutrophic status of Chl-a, TN, and TP.

Table 2. The ranges of TP and TN, and the trophic state and proportion.

TSI	TP (μg/L)	TN (mg/L)	Chl-a (μg/L)	SD (meter)	TP (μg/L)	TSI	WQI
Oligotrophic	0–50 1.40%	0–1.4 0.00%	<0.95 0.00%	>8 0.00%	<6 0.00%	<30 0.00%	0–25 100.00%
Mesotrophic	50–100 31.20%	1.4–2 31.90%	0.95–7.3 13.19%	2-8 0.00%	6–24 0.35%	30–50 0.00%	26–50 0.00%
Eutrophic	>100 68.40%	>2 68.10%	7.3–56 80.21%	0.5-2 34.72%	24–96 30.21%	50–70 70.14%	50–75 0.00%
Hypereutrophic			>56 6.60%	<0.5 65.28%	>96 69.44%	>70 29.86%	75–100 0.00%

Nutrient inputs are the primary cause of eutrophication in a lake [13,25]. It is situated at the lowest point in a basin and the accumulation point of all pollutants in a watershed. Social and economic activities may exacerbate the fragility of the lacustrine ecosystem [31]. On the one hand, more than ten seasonal watercourses carried anthropogenic effluents into the lake, including the Hongqi, Zhewan, and Daxin rivers. Every month, TN and TP concentrations were higher than the thresholds of Class V (GB3838-2002), especially in May, possibly related to precipitation transporting many rocks, soil weathering products, and agricultural pollutants from the basin into the lake at the beginning of the rainy season (Figure 7).

On the other hand, the basin land-use type was mainly farmland that produces large quantities of vegetables along these rivers. In addition, there are now many industrial enterprises in the basin, especially on the lake's south shore at Xiushan and Yangguang, and to the north near Nagu (Figure 1b). At the beginning of the rainy season, large quantities of chemical fertilizers, domestic sewage, and weathering products from soil and rock were transported into the lake, making it the principal reason for the apparent imbalance in the lake's ecosystem. Previous studies also showed that TP and TN were the major pollutants in Lake QL [27,31,50]. These results, therefore, indicate that the decrease in both N and P should be taken into account for HABs' movement and treatment in future critical lake managing strategies [51].

3.7. Control Factors of Harmful Algal Blooms

The input of excessive nutrients and climate change caused eutrophication and increased harmful algal blooms (HABs) in Lake QL (Figure 8). However, despite the many water quality parameters, the primary factor controlling algal productivity remains unknown. Herein, we combined PCA, RFM, and CA to address such issues. PCA results showed that 53.6% of the variables, including PC1 and PC2 accounted for 33.2% and 20.4% of all datasets, respectively (Figure 8c). PCA results found no significant difference in water quality parameters at different depths, indicating a strong water mixing process. Chl-a, TP, and TN were highly and positively correlated with PC1, indicating a eutrophication status. DO, pH, and other water parameters were significantly correlated with PC2, suggesting "physical parameters" (Figure 8c). PCA results showed that Lake QL's trophic status and conditions were the primary driving factors for HABs. Based on the results from RFM, six variables explained 60.31% of the variation in Chl-a ($p < 0.001$), with TP, WT, and turbidity the first three important variables for Chl-a (Figure 8a). CA confirmed that Chl-a concentration was significantly correlated with TP ($r = 0.60$, $p < 0.001$) and TN ($r = 0.49$, $p < 0.05$), revealing the ongoing climate warming and eutrophication that could accelerate and promote the intensity and frequency of HABs (Figure 8b). Previous studies also showed that TP and TN were the major pollutants in lake QL [27,31,50].

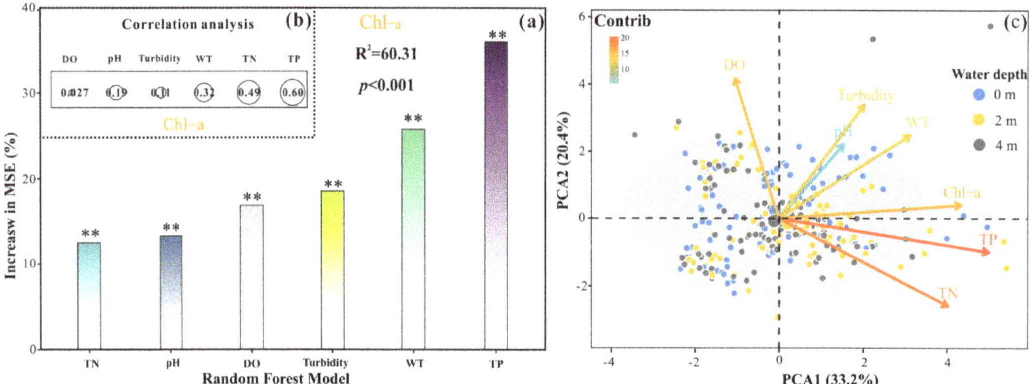

Figure 8. Comprehensive analysis of each water parameter in Lake Qilu. (**a**) Random Forest Model of variables prediction for Chl-a; (**b**) The correlations between Chl-a and each parameter based on correlation analysis; and (**c**) the Principal Component Analysis of all water quality parameters. **: The significance of variables ($p < 0.01$).

The WT and Chl-a of lake QL sharply increased after the end of winter (January), accelerating the timing of HABs compared with other lakes (e.g., Lake Dianchi [6] and Taihu [47]) (Figure 2). Walls et al., [20] reported that cyanobacteria biomass increased with warming and was greatest at a WT ranging from 9 °C to 19 °C and declining at temperatures above those optimal for growth ($\geq$19 °C). The effects of WT on HABs suggest that we should move the timing of HAB control forward. An increase in turbidity was found in January in Lake QL, which was from HABs [26]. DO has an important effect on lake ecosystems and is an important index of ecological balance. DO can more accurately reflect the metabolism of an aquatic ecosystem than other variables [52]. A high DO concentration enhanced pollutant degradation in water so that the water column could be purified more quickly. In contrast, the pollutant degradation rate was sluggish in low DO concentration environments. Figure 3 shows that DO concentration exhibited seasonal variation, especially on the vertical scale. Based on CA, no significant correlation between DO and Chl-a was found in lake QL, indicating no significant effect of DO on HABs. However, we observed that HABs decreased DO concentrations due to the respiration and decomposition of cyanobacteria, leading to a decline in pH because of the increase in CO_2. Simultaneously, the increase in HABs resulted in a decrease in transparency and photocatalytic efficiency due to light interruption.

Ecologists and governments have had a long and strong interest in determining whether N and/or P limited HABs in lakes [16,17,29,51]. Generally, TP could limit the primary production in freshwater lakes partly because the fixation of atmospheric nitrogen could compensate for N limitations to support phytoplankton growth [53]. TP was more closely related to Chl-a than TN under oligo-mesotrophic and eutrophic conditions, and both TP and TN were important under hypereutrophic conditions [16]. TP was more correlated to Chl-a than TN in lake QL (Figure 9a–c). Further, the TN/TP ratio was reported to be a good index to predict nutrient limitation in the lake [16]. An interesting result was that Chl-a concentrations were negatively correlated with TN/TP ratios. In other words, Chl-a concentration increased with a decrease in TN/TP (Figure 9d) due to an increase in TP but only a weak change in TN (Figure 9d). The increase in TP could be related to the decrease in water level during the dry season, especially from January to May (Figure 9c), suggesting that Lake QL was mainly limited by TP for HABs. Overall, these results suggest that more attention should be paid to TP and not to ignore TN because we found that TN promoted the efficiency of P use by phytoplankton directly or by impacting diversity [13].

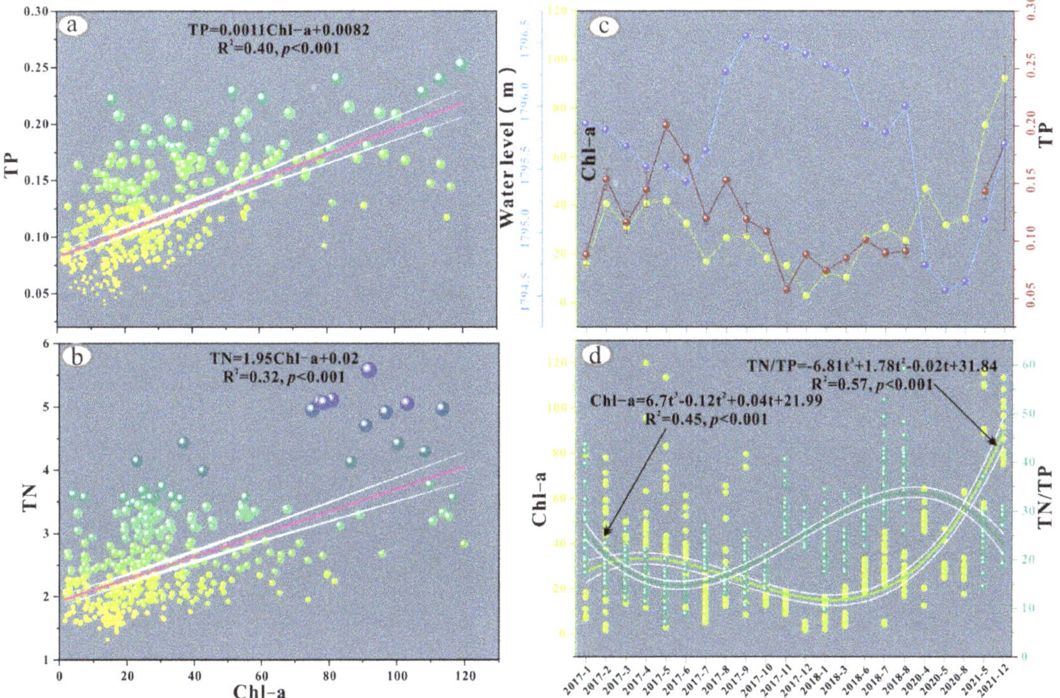

Figure 9. The effects of nutrients on HABs. (**a**) Linear regression between TP and Chl-a; (**b**) changes in water level, TP, and Chl-a during the monitoring time; (**c**) linear regression between TN and Chl-a; and (**d**) Chl-a concentrations increased with the decreasing of TN/TP ratios.

3.8. Environmental Implication

The trophic status indexes showed that Lake QL was a eutrophic to the hypereutrophic lake from 2017 to 2021, and the trend shows an ongoing increase. The reasons for the eutrophication of QL were the use of large quantities of chemical fertilizers and the discharge of domestic sewage, which both led to large quantities of nutrient accumulation into the lake. In summer, the high nutrients were probably released from bottom sediments, due to their strengthened heating and periodic deoxidation. Therefore, adjusting the agricultural production type and improving domestic sewage treatment capacity could be effective ways to control the eutrophication of QL. TP may have a more restrictive effect on algal blooming than TN, implying that we should reduce TP concentrations by inputting clean water to raise water levels and dilute water bodies to increase TN/TP ratios.

4. Conclusions

Based on our monitoring dataset of water parameters in Lake QL, complex seasonal changes in water quality parameters were identified. The lake water was well mixed during the monitoring period, resulting in no significant thermal stratification. Variations in DO and pH were caused by the physical (disturbing by shipping and wind) and biochemical processes (respiration organic matter degradation) associated with an anaerobic environment at the bottom of the water during summer.

Lake QL is a eutrophic to a hypereutrophic lake with poor water quality, which might continue to aggravate due to extreme climate conditions and intense human activities in the future. The ratios of TN/TP were negatively correlated with Chl-a, revealing that TP mainly limited phytoplankton growth more than TN. Thus, these findings enhance our understanding of the effects of water parameters and eutrophication on HABs, which suggests that reducing TP loading, increasing the water level, and advancing the timing of HABs treatment could be critical strategies for Lake Qilu management.

Author Contributions: Conceptualization, methodology, foundations acquisition, writing—review and editing, H.Z.; formal analysis, writing—original draft preparation, L.D. and F.C.; data curation, D.L. and X.W. All authors have read and agreed to the published version of the manuscript.

Funding: This research was funded by the Special Project for Social Development of Yunnan Province (Grant No. 202103AC100001).

Data Availability Statement: The original contributions presented in the study are included in the article. Further inquiries can be directed to the corresponding author.

Acknowledgments: Thanks to the teachers and students who participated in the field data collection.

Conflicts of Interest: The authors declare no conflict of interest.

Abbreviations

CA	Correlation analysis
Chl-a	Chlorophyll-a
DO	Dissolved oxygen
HABs	Harmful algal blooms
NSFWQI	National Sanitation Foundation Water Quality Index
PCA	Principal component analysis
QL	Qilu
RFM	Random Forest Model
SD	Secchi disk depth
SEPAC	State Environmental Protection Administration of China
TN	Total nitrogen
TN/TP	Total nitrogen/total phosphorus
TP	Total phosphorus
TSI	Trophic State Index
WT	Water temperature

References

1. Paerl, H.W.; Paul, V.J. Climate change: Links to global expansion of harmful cyanobacteria. *Water Res.* **2012**, *46*, 1349–1363. [CrossRef]
2. Przytulska, A.; Bartosiewicz, M.; Vincent, W.F. Increased risk of cyanobacterial blooms in northern high-latitude lakes through climate warming and phosphorus enrichment. *Freshw. Biol.* **2017**, *62*, 1986–1996. [CrossRef]
3. Ranjbar, M.H.; Hamilton, D.P.; Etemad-Shahidi, A.; Helfer, F. Impacts of atmospheric stilling and climate warming on cyanobacterial blooms: An individual-based modelling approach. *Water Res.* **2022**, *221*, 118814. [CrossRef] [PubMed]
4. Vinçon-Leite, B.; Casenave, C. Modelling eutrophication in lake ecosystems: A review. *Sci. Total Environ.* **2019**, *651*, 2985–3001. [CrossRef]
5. Zhang, Y.; Liang, J.; Zeng, G.; Tang, W.; Lu, Y.; Luo, Y.; Xing, W.; Tang, N.; Ye, S.; Li, X.; et al. How climate change and eutrophication interact with microplastic pollution and sediment resuspension in shallow lakes: A review. *Sci. Total Environ.* **2020**, *705*, 135979. [CrossRef] [PubMed]
6. Zhang, H.; Chang, F.; Duan, L.; Li, H.; Zhang, Y.; Meng, H.; Wen, X.; Wu, H.; Lu, Z.; Rongxin, B.; et al. Water Quality Characteristics and Variations of Lake Dian. *Adv. Earth Sci.* **2017**, *32*, 651–659, (In Chinese with English abstract). [CrossRef]
7. Yang, K.; Yu, Z.Y.; Luo, Y.; Zhou, X.L.; Shang, C.X. Spatial-Temporal Variation of Lake Surface Water Temperature and Its Driving Factors in Yunnan-Guizhou Plateau. *Water Resour. Res.* **2019**, *55*, 4688–4703. [CrossRef]
8. King, W.M.; Curless, S.E.; Hood, J.M. River phosphorus cycling during high flow may constrain Lake Erie cyanobacteria blooms. *Water Res.* **2022**, *222*, 118845. [CrossRef]
9. Diaz, R.J.; Rosenberg, R. Spreading Dead Zones and Consequences for Marine Ecosystems. *Science* **2008**, *321*, 926–929. [CrossRef]
10. Cheng, H.; Li, M.; Zhao, C.; Yang, K.; Li, K.; Peng, M.; Yang, Z.; Liu, F.; Liu, Y.; Bai, R.; et al. Concentrations of toxic metals and ecological risk assessment for sediments of major freshwater lakes in China. *J. Geochem. Explor.* **2015**, *157*, 15–26. [CrossRef]
11. Zeng, L.; Yan, C.; Guo, J.; Zhen, Z.; Zhao, Y.; Wang, D. Influence of algal blooms decay on arsenic dynamics at the sediment-water interface of a shallow lake. *Chemosphere* **2018**, *219*, 1014–1023. [CrossRef] [PubMed]

12. Dodds, W.K.; Smith, V.H.; Lohman, K. Nitrogen and phosphorus relationships to benthic algal biomass in temperate streams. *Can. J. Fish. Aquat. Sci.* **2002**, *59*, 865–874. [CrossRef]
13. Zhang, Y.; Zhang, H.; Liu, Q.; Duan, L.; Zhou, Q. Total nitrogen and community turnover determine phosphorus use efficiency of phytoplankton along nutrient gradients in plateau lakes. *J. Environ. Sci.* **2023**, *124*, 699–711. [CrossRef]
14. Conley, D.J.; Paerl, H.W.; Howarth, R.W.; Boesch, D.F.; Seitzinger, S.P.; Havens, K.E.; Lancelot, C.; Likens, G.E. Controlling eutrophication: Nitrogen and phosphorus. *Science* **2009**, *323*, 1014–1015. [CrossRef]
15. Conley, D.J. Biogeochemical nutrient cycles and nutrient management strategies. *Hydrobiologia* **1999**, *410*, 87–96. [CrossRef]
16. Liang, Z.; Soranno, P.A.; Wagner, T. The role of phosphorus and nitrogen on chlorophyll a: Evidence from hundreds of lakes. *Water Res.* **2020**, *185*, 116236. [CrossRef]
17. Qin, B.; Zhou, J.; Elser, J.J.; Gardner, W.S.; Deng, J.; Brookes, J.D. Water Depth Underpins the Relative Roles and Fates of Nitrogen and Phosphorus in Lakes. *Environ. Sci. Technol.* **2020**, *54*, 3191–3198. [CrossRef]
18. Schindler, D.W.; Hecky, R.E. Eutrophication: More Nitrogen Data Needed. *Science* **2009**, *324*, 721–722. [CrossRef]
19. Paerl, H.W.; Huisman, J. Climate change: A catalyst for global expansion of harmful cyanobacterial blooms. *Environ. Microbiol. Rep.* **2009**, *1*, 27–37. [CrossRef]
20. Walls, J.T.; Wyatt, K.H.; Doll, J.C.; Rubenstein, E.M.; Rober, A.R. Hot and toxic: Temperature regulates microcystin release from cyanobacteria. *Sci. Total Environ.* **2018**, *610–611*, 786–795. [CrossRef]
21. Park, H.-K.; Lee, H.-J.; Heo, J.; Yun, J.-H.; Kim, Y.-J.; Kim, H.-M.; Hong, D.-G.; Lee, I.-J. Deciphering the key factors determining spatio-temporal heterogeneity of cyanobacterial bloom dynamics in the Nakdong River with consecutive large weirs. *Sci. Total Environ.* **2021**, *755*, 143079. [CrossRef] [PubMed]
22. Kim, H.-J.; Nam, G.-S.; Jang, J.-S.; Won, C.-H.; Kim, H.-W. Cold Plasma Treatment for Efficient Control over Algal Bloom Products in Surface Water. *Water* **2019**, *11*, 1513. [CrossRef]
23. Jeppesen, E.; Sondergaard, M.; Meerhoff, M.; Lauridsen, T.L.; Jensen, J.P. Shallow lake restoration by nutrient loading reduction—Some recent findings and challenges ahead. *Hydrobiologia* **2007**, *584*, 239–252. [CrossRef]
24. Karlson, B.; Andersen, P.; Arneborg, L.; Cembella, A.; Eikrem, W.; John, U.; West, J.J.; Klemm, K.; Kobos, J.; Lehtinen, S.; et al. Harmful algal blooms and their effects in coastal seas of Northern Europe. *Harmful Algae* **2021**, *102*, 101989. [CrossRef]
25. Dong, Y.; Zhao, L.; Chen, Y.; Yu, Y.; Zhao, R.; Yang, G. Succession of Nine Plateau Lakes and Regulation of Ecological Safety in Yunnan Province. *Ecol. Econ.* **2015**, *31*, 185–191. (In Chinese with English Abstract). [CrossRef]
26. Yu, Z.Y.; Yang, K.; Luo, Y.; Yang, Y.L. Secchi depth inversion and its temporal and spatial variation analysis-A case study of nine plateau lakes in Yunnan Province of China. *Int. J. Appl. Earth Obs. Geoinf.* **2021**, *100*, 102344. [CrossRef]
27. Hillman, A.L.; O'Quinn, R.F.; Abbott, M.B.; Bain, D.J. A Holocene history of the Indian monsoon from Qilu Lake, southwestern China. *Quat. Sci. Rev.* **2020**, *227*, 106051. [CrossRef]
28. Ahn, C.Y.; Oh, H.M.; Park, Y.S. Evaluation of Environmental Factors on Cyanobacterial Bloom in Eutrophic Reservoir Using Artificial Neural Networks. *J. Phycol.* **2011**, *47*, 495–504. [CrossRef]
29. Su, X.; Steinman, A.D.; Oudsema, M.; Hassett, M.; Xie, L. The influence of nutrients limitation on phytoplankton growth and microcystins production in Spring Lake, USA. *Chemosphere* **2019**, *234*, 34–42. [CrossRef]
30. Iqbal, M.M.; Li, L.L.; Hussain, S.; Lee, J.L.; Mumtaz, F.; Elbeltagi, A.; Waqas, M.S.; Dilawar, A. Analysis of Seasonal Variations in Surface Water Quality over Wet and Dry Regions. *Water* **2022**, *14*, 1058. [CrossRef]
31. Dong, Y.; Liu, Y.; Li, Y.; Zhang, R. Analysis on Factors of Ecological Vulnerability of Qilu Lake in Yunnan Province. *Environ. Sci. Surv.* **2011**, *30*, 24–29. (In Chinese with English abstract). [CrossRef]
32. Li, Y.; Wang, L.; Yan, Z.; Chao, C.; Yu, H.; Yu, D.; Liu, C. Effectiveness of dredging on internal phosphorus loading in a typical aquacultural lake. *Sci. Total Environ.* **2020**, *744*, 140883. [CrossRef] [PubMed]
33. Shen, Y.S.; Huang, Y.Y.; Hu, J.; Li, P.P.; Zhang, C.; Li, L.; Xu, P.; Zhang, J.Y.; Chen, X.C. The nitrogen reduction in eutrophic water column driven by Microcystis blooms. *J. Hazard. Mater.* **2020**, *385*, 121578. [CrossRef] [PubMed]
34. Iqbal, M.M.; Shoaib, M.; Farid, H.U.; Lee, J.L. Assessment of Water Quality Profile Using Numerical Modeling Approach in Major Climate Classes of Asia. *Int. J. Environ. Res. Public Health* **2018**, *15*, 2258. [CrossRef] [PubMed]
35. Acuña-Alonso, C.; Álvarez, X.; Lorenzo, O.; Cancela, Á.; Valero, E.; Sánchez, Á. Assessment of water quality in eutrophized water bodies through the application of indexes and toxicity. *Sci. Total Environ.* **2020**, *728*, 138775. [CrossRef] [PubMed]
36. Valera, C.A.; Pissarra, T.C.T.; Martins, M.V.; do Valle, R.F.; Oliveira, C.F.; Moura, J.P.; Fernandes, L.F.S.; Pacheco, F.A.L. The Buffer Capacity of Riparian Vegetation to Control Water Quality in Anthropogenic Catchments from a Legally Protected Area: A Critical View over the Brazilian New Forest Code. *Water* **2019**, *11*, 549. [CrossRef]
37. Carlson, R.E.; Simpson, J. A Coordinator's Guide to Volunteer Lake Monitoring Methods. *N. Am. Lake Manag. Soc.* **1996**, *99*, 96.
38. Tanvir, R.U.; Hu, Z.; Zhang, Y.; Lu, J. Cyanobacterial community succession and associated cyanotoxin production in hypereutrophic and eutrophic freshwaters. *Environ. Pollut.* **2021**, *290*, 118056. [CrossRef]
39. Chen, J.; Lyu, Y.; Zhao, Z.; Liu, H.; Zhao, H.; Li, Z. Using the multidimensional synthesis methods with non-parameter test, multiple time scales analysis to assess water quality trend and its characteristics over the past 25 years in the Fuxian Lake, China. *Sci. Total Environ.* **2019**, *655*, 242–254. [CrossRef]
40. Wu, H.; Zhang, H.; Li, Y.; Chang, F.; Duan, L.; Zhang, X.; Peng, W.; Liu, Q.; Liu, F.; Zhang, Y. Plateau lake ecological response to environmental change during the last 60 years: A case study from freshwater Lake Yangzong, SW China. *J. Soils Sed.* **2021**, *21*, 1550–1562. [CrossRef]

41. Shen, Y.; Wang, H.; Liu, X. Aquatic flora and assemblge characteristic of submerged macrophytics in five lakes of the Yunnan province. *Resour. Environ. Yangtze Basin.* **2010**, *19*, 9. (In Chinese with English Abstract).
42. Chen, J.; Wang, S.; Zhang, S.; Yang, X.; Huang, Z.; Wang, C.; Wei, Q.; Zhang, G.; Xiao, J.; Jiang, F.; et al. Arsenic pollution and its treatment in Yangzonghai lake in China: In situ remediation. *Ecotoxicol. Environ. Saf.* **2015**, *122*, 178–185. [CrossRef] [PubMed]
43. Jin, Z.; Ding, S.; Sun, Q.; Gao, S.; Fu, Z.; Gong, M.; Lin, J.; Wang, D.; Wang, Y. High resolution spatiotemporal sampling as a tool for comprehensive assessment of zinc mobility and pollution in sediments of a eutrophic lake. *J. Hazard. Mater.* **2019**, *364*, 182–191. [CrossRef] [PubMed]
44. Pan, F.; Guo, Z.; Cai, Y.; Liu, H.; Wu, J.; Fu, Y.; Wang, B.; Gao, A. Kinetic Exchange of Remobilized Phosphorus Related to Phosphorus-Iron-Sulfur Biogeochemical Coupling in Coastal Sediment. *Water Resour. Res.* **2019**, *55*, 10494–10517. [CrossRef]
45. Liu, N.; Yang, Y.; Li, F.; Ge, F.; Kuang, Y. Importance of controlling pH-depended dissolved inorganic carbon to prevent algal bloom outbreaks. *Bioresour. Technol.* **2016**, *220*, 246–252. [CrossRef] [PubMed]
46. Liu, Y.; Guo, H.; Yang, P. Exploring the influence of lake water chemistry on chlorophyll a: A multivariate statistical model analysis. *Ecol. Model.* **2010**, *221*, 681–688. [CrossRef]
47. Yao, Y.; Li, D.; Chen, Y.; Liu, H.; Wang, G.; Han, R. High-resolution distribution of internal phosphorus release by the influence of harmful algal blooms (HABs) in Lake Taihu. *Environ. Res.* **2021**, *201*, 111525. [CrossRef]
48. Li, X.; Janssen, A.B.G.; de Klein, J.J.M.; Kroeze, C.; Strokal, M.; Ma, L.; Zheng, Y. Modeling nutrients in Lake Dianchi (China) and its watershed. *Agric. Water Manag.* **2019**, *212*, 48–59. [CrossRef]
49. Tyagi, S.; Sharma, B.; Singh, P. Water Quality Assessment in Terms of Water Quality Index. *Am. J. Water Resour.* **2013**, *1*, 34–38. [CrossRef]
50. Qin, J.; Wu, X.; Gao, W.; Yang, S.; Wang, Q.; Luo, W. Study on Temporal and Spatial Distribution Characteristics of Chl-a and Water Quality Factors in Qilu Lake and Their Correlations. *J. Anhui Agric. Sci.* **2012**, *40*, 345–348. (In Chinese with English Abstract). [CrossRef]
51. Zhou, J.; Han, X.; Brookes, J.D.; Qin, B. High probability of nitrogen and phosphorus co-limitation occurring in eutrophic lakes. *Environ. Pollut.* **2022**, *292*, 118276. [CrossRef] [PubMed]
52. Krueger, K.M.; Vavrus, C.E.; Lofton, M.E.; McClure, R.P.; Gantzer, P.; Carey, C.C.; Schreiber, M.E. Iron and manganese fluxes across the sediment-water interface in a drinking water reservoir. *Water Res.* **2020**, *182*, 116003. [CrossRef] [PubMed]
53. Hayes, N.M.; Patoine, A.; Haig, H.A.; Simpson, G.L.; Swarbrick, V.J.; Wiik, E.; Leavitt, P.R. Spatial and temporal variation in nitrogen fixation and its importance to phytoplankton in phosphorus-rich lakes. *Freshw. Biol.* **2019**, *64*, 269–283. [CrossRef]

water

MDPI

Article

Seasonal Water Quality Changes and the Eutrophication of Lake Yilong in Southwest China

Qingyu Sui, Lizeng Duan, Yang Zhang, Xiaonan Zhang, Qi Liu and Hucai Zhang *

Institute for Ecological Research and Pollution Control of Plateau Lakes, School of Ecology and Environmental Science, Yunnan University, Kunming 650500, China
* Correspondence: zhanghc@ynu.edu.cn

Abstract: To better understand the seasonal variation characteristics and trend of water quality in Lake Yilong, we monitored water quality parameters and measured nutrients, including the water temperature (WT), Chlorophyll-a (Chl-a), dissolved oxygen (DO) and pH from September 2016 to May 2020, total nitrogen (TN) and total phosphorus (TP) from October 2016 to August 2018. The results showed that the lake water was well mixed, resulting in no significant thermal stratification. The DO content was decreased in the northwest part of the lake during September and October, resulting in a hypoxic condition. It also varied at different locations of the lake and showed a high heterogeneity and seasonal variability. The Chl-a concentration in Lake Yilong demonstrated seasonal and spatial changes. It was maximum at the center and southwest area of the lake in January. However, in the northwest part of the lake, the maximum value appeared in September and October. The content of TN in the rainy season increased by 75% compared with that in dry season and TP content show a downward trend (from 0.11 mg/L to 0.05 mg/L). The comprehensive nutrition index evaluation shows that the water quality of Lake Yilong in 2016 was middle eutrophic (TLI = 60.56), and that in 2017 (TLI = 56.05) and 2018 (TLI = 56.38) was weak eutrophic, showing that the nutritional status has improved. TN remained at a high level (2.15 ± 0.48 mg/L), water quality needs further improvement. Based on our monitoring and analysis, it is recommended that human activities in the watershed of the lake should be constrained and managed carefully to maintain the water quality of the lake and adopt effective water quality protection and ecological restoration strategies and measures to promote continuous improvement of water quality, for a sustainable social development.

Keywords: water quality; Lake Yilong; spatial-temporal variations; anthropogenic activities

Citation: Sui, Q.; Duan, L.; Zhang, Y.; Zhang, X.; Liu, Q.; Zhang, H. Seasonal Water Quality Changes and the Eutrophication of Lake Yilong in Southwest China. *Water* **2022**, *14*, 3385. https://doi.org/10.3390/w14213385

Academic Editor: George Arhonditsis

Received: 8 September 2022
Accepted: 22 October 2022
Published: 25 October 2022

Publisher's Note: MDPI stays neutral with regard to jurisdictional claims in published maps and institutional affiliations.

1. Introduction

Lakes are an important inland water ecosystem with irreplaceable ecological functions, social benefits and economic value [1]. However, in recent decades, increased human activities have seriously affected the ecosystem of lakes, leading to a rapid deterioration in their water quality. Some lakes have such serious water pollution that it exceeds the carrying capacity of the lake itself. These changes reduce or even degrade the ecosystem of lakes, making them lose their adjustment ability. The spatial and temporal changes in the water quality parameters can directly reflect the water conditions of a lake, especially lake trophic status [2]. For example, Ndungu et al. [3] studied two lakes with different depths close to the equator, and the results showed that, being a shallow tropical lake, the mean water temperature in Lake Naivasha did not vary much between the sites. However, another lake in that study, Crescent Lake, showed lower temporal variability [3]. In lake Arrowhead, an oligotrophic alpine lake in southern California, strong thermal stratification during summer and early fall resulted in hypoxic hypolimnetic waters with dissolved oxygen [4]. In another example, it was found, by studying the seasonal variation of chl-a concentration in the Bay of Bengal, that in summer it is shown that the input of river nutrients, the reflux of nutrient water in the Arabian Sea and the coastal upwelling are the three main driving factors controlling the chl-a concentration in the surface layer [5]. István et al. [6] evaluated the changes in trophic conditions from 1985 to 2017 in the largest shallow freshwater lake in Central Europe, Lake Balaton, it is concluded that a drastic reduction in external phosphorus loads arriving in similar shallow lakes will lead to the eutrophication of lakes, although the time interval is very long [6].

Eutrophication is one of the most important and concerning water quality problems in freshwater ecosystems. The occurrence of water eutrophication is mainly caused by the increase of nitrogen, phosphorus and other nutrients in the water. The trophic level index (TLI) method is used to evaluate and select representative algae that can reflect the reservoir comprehensive index of quantity evaluation of chlorophyll a as the dominant factor parameters, TN, TP and other four indicators are used as water eutrophication assessment basic factor of price. TLI has been widely used to assess the nutritional status of lakes, such as Lake Wuliangsuhai [7] and Lake Chenghai [8].

In Yunnan-Guizhou Plateau, one of China's five lake distribution areas, there are many natural lakes with serious pollution. According to statistics, Lake Dian, Lake Qilu, Lake Xingyun, and Lake Yilong in Yunnan's plateau lakes are already in a eutrophication state, and the water quality of other lakes is not optimistic either [9,10]. Due to the aggravating pollution conditions, many scholars have investigated issues associated with lake pollution and proposed several treatment strategies. Zhang et al. [11] suggested that with the continuous eutrophication of Plateau Lake and excessive social-economic developments in the watershed, the lake's natural material sedimentation process has been interrupted, and its self-purification function has been lost, causing serious threats to water resources [11]. Through the qualitative and quantitative analysis of the organic matter in the sediments, it is found that the organic matter in the sediments of Yilong lake is mainly self-sustaining, and the macrophytes are absolutely dominant [12]. In this lake, the coverage rate of aquatic vegetation is high (about 60%), as in the northwest of the lake, submerged plants are flourishing (slightly swampy), forming a "grass lake area". In the east, the submerged plants are less distributed, and the algae density is high, forming an "algal lake area". The species of phytoplankton in Lake Yilong are mainly cyanobacteria that are easy to reproduce in the eutrophic water. Liu et al. [13] investigated the water quality and eutrophication of Lake Yilong and showed that the change in Chl-a concentration since 1980 was mainly driven by the long-term accumulation of nutrients in the lake, especially the effects of TP on the increase of algae biomass [14]. Further, an increase in temperature has been shown to further aggravate the effects on phytoplankton biomass [13].

Lake Yilong is an important water source for the neighboring Shiping and Jianshui counties. However, lagging environmental protection, economic development, population growth, urbanization and human activities are continuously affecting the lake and its ecosystem. A lack of awareness on environmental protection has led to eutrophication of lakes, decline in their water level, swamping, ecosystem degradation and more. However, most studies performed on Lake Yilong mainly focused on analyzing the reasons from the status quo and proposed countermeasures [15,16], lake area wetland protection [17], fish industry [18,19], wetland sustainability evaluation [20,21], investigated lake sediment proxy indicators, etc., and few analyses have done on the seasonal changes of water quality parameters to uncover the most fundamental causes of lake pollution [22]. As the lake area continues shrinking and the area of plants continues expanding, the water volume and quality of Lake Yilong will undoubtedly continue to deteriorate. Quantitative determination of the water quality of Lake Yilong is therefore an urgent task. In view of the above situation, this study comprehensively analyzed the water quality parameters of Lake Yilong, using indicators such as water temperature, chlorophyll a concentration, dissolved oxygen concentration, and pH value. We also combined the total nitrogen (TN) and total phosphorus (TP) concentrations at various sites and different months to obtain an overall understanding of the nutrition levels and water quality of water bodies at different areas of the lake, and the trophic level index (TLI) is applied to evaluate water quality quantitatively, providing a scientific basis for the evaluation of the nutritional status of Lake Yilong and water quality management.

2. Materials and Methods

2.1. Overview of the Study Area

Lake Yilong is located in the Shiping County of Yunnan Province, southwestern China, with geographical coordinates of $23°39'$–$23°42'$ N and $102°30'$–$102°38'$ E. Lake Yilong belongs to Nanpanjiang River system of pearl River Basin. The Landsat5 TM satellite data on 3 November 2009, showed that Lake Yilong had an area of 33.87 km^2 and a drainage area of 337.56 km^2, but as of April 2013, the lake area decreased by more than half, to only 12.23 km^2, reflecting the impact of climate conditions and human activities on the lake.

Climatically, Lake Yilong belongs to the northern subtropical dry monsoon and mid-tropical semi-humid climate zone. It is characterized by distinct dry and wet seasons, large daily temperature variations and small annual temperature differences. The rainy season lasts from May to October, while the dry season is from November to April of the following year. Lake Yilong has an average annual precipitation of 919.9 mm, maximum annual precipitation of 1160.4 mm and annual minimum precipitation of 613.2 mm. The lake water is mainly supplied by precipitation, surface runoff and groundwater. Among the surface runoff, only the surrounding moats supply perennial water, while the others are seasonal rivers. The initial watercourse was from the Xinjiehai River, which flows eastward through Jianshui and joins the Lu River. As the water level of Lake Yilong has dropped and the riverbed of Xinjiehai River has risen, Lake Yilong has become a shallow lake with no outflow [19]. In addition, there are 22 springs around the lake, most of which are located on the west bank, which are also important sources of lake water. The aquatic plants in Lake Yilong are prosperous, and the submerged plants are mainly *Potamogeton lucens Linn, Potamogeton wrightii Morong, and Hydrilla verticillata*, distributed throughout the lake. According to the water quality monitoring of Lake Yilong by the Honghe Prefecture Environmental Monitoring Station of the Environmental Protection Department of Yunnan Province, the water quality of the lake is poor, graded as category V. The water quality is heavily polluted and does not meet the requirements of Category III water environment criteria, with the lake being in a moderately eutrophic state. Its degree of swamping has been intensified, water depth is becoming shallower, and biodiversity is being seriously threatened [23].

2.2. Measurements and Sampling

Based on the shape and hydrological condition of the lake, 3 monitoring sites were set up in the southeast (YLH-1), center (YLH-2) and northwest (YLH-3) section of the lake, YLH-1 is the deepest and near Shiping County, YLH-3 is the shallowest (Figure 1). To investigate the water quality changes and causes in different areas of Lake Yilong, the lake was monitored and sampled from September 2016 to October 2017, after which the sampling time was randomly selected (January, March, Jun., and July, in 2018, while May, and November in 2020). The measurement was conducted once a month at the sampling point to understand the dynamic change of water quality. A complete year of data was used to identify seasonal variation characteristics, and other random sampling data were used to understand the lake's water quality dynamics. Through 20 times of monitoring and sampling at 3 points, we obtained approximately 1200 datasets (nutrients to environmental factors). To investigate the water quality parameter seasonal variation pattern at different depths and times, we averaged the data at 3 points using intervals of one meter.

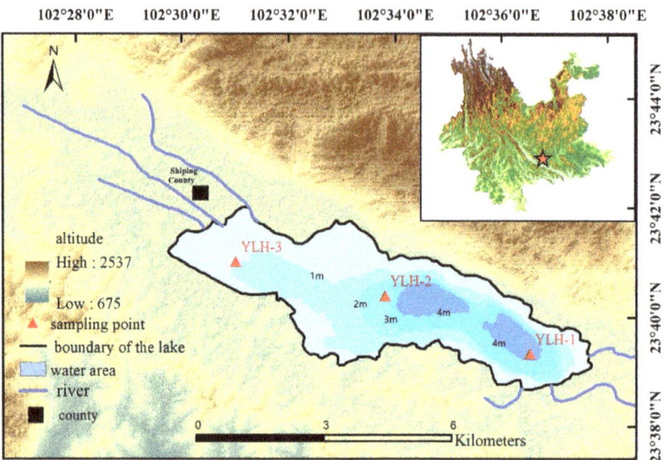

Figure 1. Water quality monitoring sites of Lake Yilong.

This study used a multi-parameter water quality analyzer (instrument model: YSI-6600V2, produced in the USA) for continuous on-site measurement. The instrument was calibrated before monitoring. During on-field investigations, the instrument was placed vertically in the lake to gradually measure and collect samples at a uniform speed. If necessary, the handle was connected to a cable to perform the reception of water quality parameters, including water temperature, pH, dissolved oxygen, conductivity, turbidity and chlorophyll a, at the indicated monitoring site. The instrument was cross-calibrated before each test to ensure data reliability and accuracy. The water was sampled using a low-temperature incubator and transported back to the laboratory on the same day. In addition, the monitoring of total nitrogen (TN) and total phosphorus (TP) in the water was performed following the Chinese standard method ("Water and Wastewater Monitoring and Analysis Method" Fourth Edition, China Environmental Science Press) [24]. The TN and TP were measured by a UV-spectrophotometer (UV-2600). TN was determined using an alkaline potassium persulfate digestion-UV spectrophotometric method (GB11894–89) at wavelengths of 220 nm and 275 nm, and the ammonium molybdate spectrophotometric method (GB11893–89) at wavelengths of 700 nm for TP [25,26].

2.3. Water Quality and Trophic State Assessment

The trophic level index (TLI), which reflecting lake trophic status [27] was used to evaluate the water quality of Lake Yilong. The aggregated TLI is calculated as the sum of individual TLIs, which are calculated according to lake and reservoir eutrophication evaluation methods and grading technical regulations, TN, TP, permanganate index (COD_{Mn}), DO, Chl-a, and Secchi disk transparency (SD) are the main indicators of nutrient levels in water [28,29]. Among them, the TLI is given as:

$$TLI = \sum_{j=1}^{m} w_j \cdot TLI(j) \tag{1}$$

where TLI ($\sum$) is the integrated trophic level index, $TLI(j)$ is the trophic level index of j, and W_j represents the weighting factor for the WQ parameter, including Chl-a, TN, TP, SD, and COD_{Mn} (j = 1, 2, 3, 4, 5).

$$w_j = \frac{r_{ij}^2}{\sum_{j=1}^{m} \gamma_{ij}^2} \tag{2}$$

where W_j is the correlative weighted score for the trophic level index of j and r_{ij} is a relative coefficient.

In Chinese lakes and reservoirs, the correlation coefficient for Chl-a to other parameters is presented in Table 1 [30], and the TLI ranges and corresponding trophic status are shown in Table 2.

Table 1. Correlation coefficient for Chl-a to other parameters in Chinese lakes and reservoirs.

Parameter	Chl-a	TN	TP	SD	COD$_{Mn}$
r_{ij}	1	0.82	0.84	−0.83	0.83
r_{ij}^2	1	0.6724	0.7056	0.6889	0.6889

Table 2. Trophic Level Index (TLI) value ranges and corresponding trophic status for Chinese lakes.

TLI	TLI < 30	30 ≤ TLI ≤ 50	50 < TLI < 60	60 < TLI ≤ 70	TLI >70
Eutrophic status	Oligotrophic	Mesotrophic	Weak–eutrophic	Middle-eutrophic	Hyper-eutrophic

The computational formula for each eutrophication index is:

$$TLI(Chl\text{-}a) = 10(2.5 + 1.086 \ln Chl\text{-}a) \tag{3}$$

$$TLI(TP) = 10(9.435 + 1.624 \ln TP) \tag{4}$$

$$TLI(TN) = 10(5.435 + 1.694 \ln TN) \tag{5}$$

$$TLI(COD_{Mn}) = 10(0.109 + 2.66 \ln COD_{Mn}) \tag{6}$$

$$TLI(SD) = 10(5.118 - 1.94 \ln SD) \tag{7}$$

2.4. Statistical Procedures

Microsoft Excel 2016 was used to assess the recorded data. Arcgis10.2. was used to generate the map of sampling locations. The Spearman method in origin Pro 2020 was used for correlation analysis (CA) to detect the relationship between Chl-a and various water quality parameters (pH, DO, TP, TN, etc.). In the water quality evaluation, the score value of the trophic level index of the lake is calculated using Microsoft Excel 2016.

3. Results and Discussion

3.1. Seasonal Variations in Water Temperature

The WT was not only crucial for understanding various physical and chemical processes and dynamic phenomena of lake water, but also an important factor affecting the metabolism of lake aquatic organisms. The physical characteristics, chemical reactions and biological activities of lake water are closely related to WT [31]. For example, water temperature is an important factor affecting the content of chlorophyll-a in water, and also a key factor for the growth of planktonic algae. The appropriate water temperature is conducive to the growth of planktonic algae, that is, in a certain temperature range, the rising water temperature can promote the photosynthesis of algae and the respiration of aquatic organisms [32–34]. Lake temperature changes with seasons. When wind and waves are not strong enough to disturb the entire water mass and energy cannot be fully exchanged, the vertical temperature difference increases [35]. The changes in WT are mainly caused by changes in seasonal solar radiation, which are synchronous and highly correlated with regional air temperature [36,37].

From September 2016 to November 2020, WT was inherently synchronized with seasonal change at each monitoring point (Figure 2b). The results showed that the WT of lake Yilong ranged from 13.1 °C to 29.8 °C during the whole monitoring period. Spatially, WT decreased gradually from the lake's surface to its bottom, especially during the summer (Figure 2a). The average WT at 3 sample sites at 0 m, 1 m, 2 m, 3 m and 4 m were 21.6 °C,

21.3 °C, 20.9 °C, 20.9 °C and 20.7 °C, in line with the general law that lake WT gradually decreases as the water depth increases. WT in different areas of Lake Yilong decreased with increased depth in the vertical direction, with a decreasing range of 1–2 °C. Seasonally, the changes in WT were obvious, with the highest WT recorded in June and July and the average temperature ~26 °C. The lowest WT was ~16 °C, recorded in November and January. The average WT in the center part of the lake (20.8 °C) was higher than in the southeast (20.4 °C) and northwest (20.6 °C) in January, March and May. The average WT in the northwest of the lake was higher than in other parts of the water body in September, October and November. As a shallow plateau lake, WT showed no significant thermal stratification in Lake Yilong. The change in WT was consistent with that of air temperature (Figure 3). Figure 3 shows the synchronous change of water and air temperature during the monitoring period without obvious hysteresis. The differences in altitude between the locality of the meteorological station and lake surface were the main reasons why the WT of Lake Yilong was higher than that recorded at the meteorological station. Another possible reason is that the WT was measured during the daytime and the air temperature was the average value for the whole day.

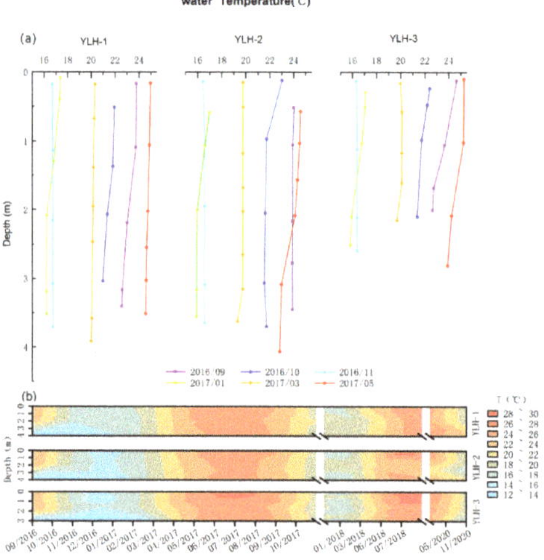

Figure 2. Seasonal and interannual variations in water temperature in lake Yilong. (**a**) Seasonal variations in water temperature; (**b**) Interannual variations in water temperature.

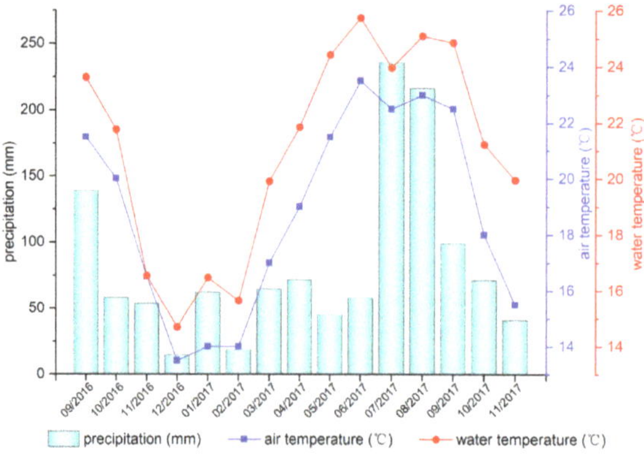

Figure 3. Change in precipitation, air temperature and water temperature in Lake Yilong from 2016–2017.

As Lake Yilong is a typical shallow water lake on plateau, WT in the lake shows strong seasonal variability. In August 2017, due to the heat energy exchange between surface water and air, the temperature increased, resulting in a significant increase in WT. In addition, this situation leads to a decrease in surface water density, which hinders heat exchange with bottom water, especially in summer. In winter, due to the disappearance of thermocline in shallow lakes, the exchange capacity of water bodies is enhanced, and surface water and bottom water show the same properties. In addition, the monitoring shows that there is no temperature stratification phenomenon in lake Yilong, which is significantly different from other deep-water bodies in Yunnan-Guizhou Plateau, such as Lake Lugu [38] and Lake Yangzong [39].

3.2. Seasonal Variations of Chlorophyll-a

Chlorophyll-a (Chl-a) is an important indicator of the presence of phytoplankton, with its content reflecting the amount of algae in the water. Thus, Chl-a is an important indicator used for investigating eutrophication, and it plays a key role in the evaluation of the nutritional status of the lake water. It was found that Chl-a in Lake Yilong changed with seasons [40–42].

The monitoring data shows the Chl-a concentration in Lake Yilong ranged from 1.3 to 137 µg/L (Figure 4b). Chl-a concentration in the vertical profile of the entire lake area increased with the increase of water depth (Figure 4a). From different areas, the Chl-a concentration of the vertical profile in the southwest reached its maximum value in January, its concentration at the southwest increased from 29.4 µg/L on the surface to 81.3 µg/L at the bottom (average value of 66.04 µg/L). This may have been due to the effects of precipitation and runoff in the rainy season. In March and May, the Chl-a content began to return to the average value in Lake Yilong. The lowest value of Chl-a on the surface of the center of the lake was recorded in November 2016 (1.3 µg/L). In January, the concentration reached a maximum of 68 µg/L, then began to decrease. The most obvious difference in Chl-a between the surface layer and the bottom of the lake was in the vertical section in October (13.2 µg/L at the bottom of the lake).

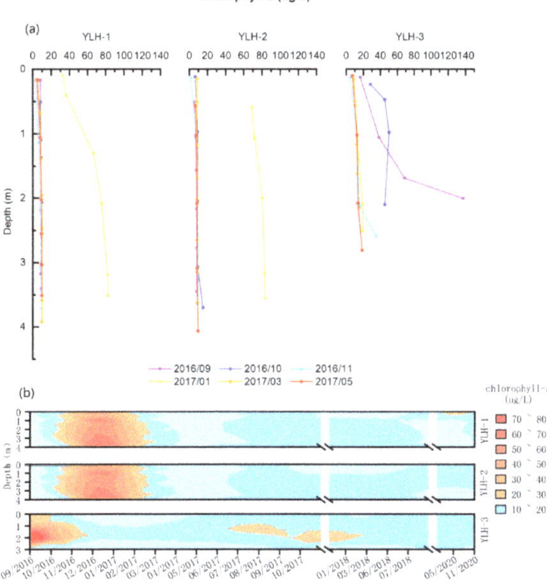

Figure 4. Seasonal and interannual variations in Chl-a in lake Yilong. (**a**) Seasonal variations in water temperature; (**b**) Interannual variations in water temperature.

It is worth paying attention to the monthly change in Chl-a at the northwest part of the lake. The vertical profile change was lowest in the surface layer. The Chl-a concentration increased with increase in water depth, and the concentration of Chl-a suddenly rose with every 1 m of water (Figure 4a). It is believed that the stratification of Chl-a occurred because, on the one hand, rivers were entering the northwest of the lake (Moat, ChengBei River, Cheng nan River), and the water quality was better than other rivers entering the lake. On the other hand, the introduction of large aquatic plants such as *Hunan lotus* improved water quality. Aquatic plants can maintain the water quality of lakes via various mechanisms, including providing zooplankton shelter, increasing the stability of sediments to reduce their resuspension, competing with phytoplankton for nutrients and releasing algae substances [12,13]. Therefore, it can be concluded that the purification effect of aquatic plants on the water is feasible, and biological engineering could be implemented in the future to protect Lake Yilong by purifying its water.

3.3. Seasonal Variations of Dissolved Oxygen

Dissolved oxygen (DO) refers to decomposed oxygen dissolved in water, mainly from aquatic plants' photosynthesis [43]. It is an important factor in maintaining the dynamic balance of the ecological environment of the water and is necessary for the survival of aquatic life. DO also participates in the transformation of some substances [44]. Compared with other parameters, the concentration of DO can reflect the pollution degree of lake water, especially the pollution degree of organic matter, which is an important indicator of water quality [45].

The monitoring results showed that the DO concentration in Lake Yilong was 0.39–14.55 mg/L (Figure 5b). In terms of long-term trends, the DO concentration of Lake Yilong was high in the dry season, with an average concentration of 8 mg/L, while it was lower in each lake area during the rainy season, with an average of concentration of 7 mg/L. From the vertical and horizontal changes of the DO curve (Figure 5a) in different seasons, we observed a higher concentration of DO on the surface of the lake, which decreased with depth in the vertical direction. Although the concentration decreased, the magnitude of the decrease varied in different lake areas and had obvious seasonality. Compared with other shallow lakes, the change in DO in the northwest part of the Lake Yilong showed a very special phenomenon, whereby the DO concentration at the surface was 5.08 mg/L in September 2016, 3.63 mg/L at a water depth of 1 m, and suddenly decreased to 0.39 mg/L at a water depth of 2 m, forming an anoxic environment. A similar situation also occurred in October and May but at different water depths. In October, the surface DO concentration was 5.66 mg/L. As the water depth increased, the DO concentration decreased from 0.65 mg/L at 1 m to 0.37 mg/L at the bottom of the lake. In May, the surface DO concentration was 7.12 mg/L, which was at a high concentration, but when it reached the bottom of the lake, it dropped sharply to 1.44 mg/L, indicating a hypoxic or anaerobic environment.

Lake Yilong is sometimes stratified and hypoxic at a water depth of about 1 m, with the possible causes being: first, the pollution of organic fertilizers in the farmland of the basin; second, the silt deposited at the bottom of the lake; third, the lack of water resources in the basin, and the poor quality of the mixing and refreshing water. As a shallow water lake, the seasonal water level fluctuations might also influence the DO contents. These could cause Algae to bloom, thereby affecting the concentration of DO. Whether the occurrence of a hypoxic environment in the northwestern part of Yilong Lake was accidental or occurs frequently remains to be determined.

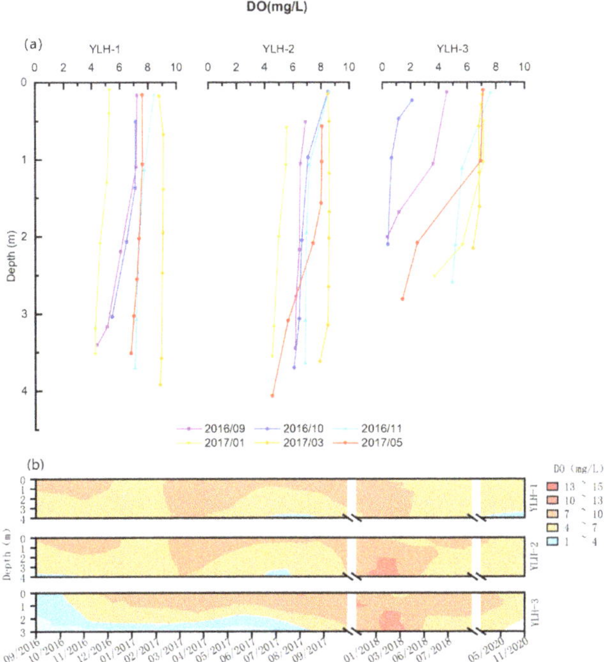

Figure 5. Seasonal and interannual variations in DO in lake Yilong. (**a**) Seasonal variations in water temperature; (**b**) Interannual variations in water temperature.

3.4. Seasonal Variations in pH

pH is an important factor determining the eutrophication and sediment phosphorus cycle of a lake. It is not only related to geological background, climatic conditions, terrestrial heat flow, etc., but also to the influx of nutrients and algae that are constantly changing during man-made eutrophication due to the interaction of factors such as reproduction and DO content. Lakes in the Yunnan Plateau are characterized by a high pH value, as the geological–geographical background of a Karst area and a subtropical climate. pH is mainly affected by the content of CO_2 in water, and CO_2 is affected by factors such as algae photosynthesis and aquatic organism respiration. Some studies reported that the pH value was closely related to algae growth. Algae photosynthesis affects the CO_2 buffer system, thereby affecting the pH value of the water, i.e., a water pH of ~8.5 represents the optimal pH for algae growth, while a pH >9.5 is worst for algae growth [46].

The monitoring results showed that the pH value of Lake Yilong varied from 7 to 9, and the vertical variation was relatively uniform (Figure 6b), indicating that the monthly pH at the monitoring sites was in favor of algae growth. Significantly, the pH value in the center of the lake was ~9 in September and October and ~8 in January, March and May. The lowest pH was recorded in November (pH: 7.9), while Chl-a content was at its maximum value (Figure 6a). The reasons for such observation could be related to decayed algae in November, as these organic algae were converted into CO_2, HNO_2 and HNO_3 under the action of biochemical oxidation, resulting in a decrease in pH. In addition, our monitoring data showed that the pH at the northwestern part of the lake was maintained at ~8 in September, October, March and May. The possible reason could be that the northwestern part of the lake had a higher density of planktonic algae, and suitable WT promoted the photosynthesis of algae and aquatic organisms. Respiration of CO_2 consumes CO_2, increasing the surface pH. At the end of November, aquatic plants and algae began to decay to produce organic substances. Under the action of biochemical oxidation, these organic substances are converted, resulting in lower pH.

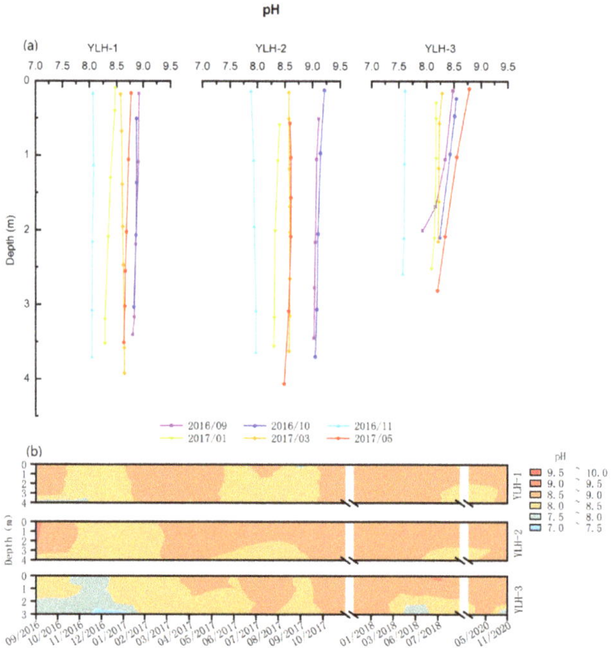

Figure 6. Seasonal and interannual variations in pH in lake Yilong. (**a**) Seasonal variations in water temperature; (**b**) Interannual variations in water temperature.

3.5. Assessment of Trophic State and Water Quality

The monitoring dataset (Figure 7) showed that the TN and TP concentrations ranged from 1.19 to 3.47 (2.15 ± 0.48) mg/L and 0.02 to 0.15 (0.06 ± 0.02) mg/L, respectively. The TN content in Lake Yilong demonstrated a certain change in different seasons. The TN content was lower in the dry season (March) of 2017, 2018, and higher in the rainy season (June), which was related to less Exogenous input in dry season. Notably, The TN content in the rainy season 2017 and 2018 increased compared to dry season (about 75%). The TP content decreased from 2016 to 2018. The average TP content 2016 to 2018 is 0.11 ± 0.02, 0.07 ± 0.02, 0.05 ± 0.01 mg/L. They were inferior to Class V (GB3838-2002) for a long time [47]. In addition, the comprehensive nutrition index evaluation shows that the water quality of Lake Yilong in 2016 was middle eutrophic (TLI = 60.56), and that in 2017 (TLI = 56.05) and 2018 (TLI = 56.38) was weak eutrophic.

3.6. Comprehensive Analysis

Based upon the correlation analysis of the water quality parameters (WT, Chl-a, DO, and pH), TN and TP (Figure 8a). The correlation between pH and DO is a high (r = 0.37, $p \leq 0.001$), showing a positive correlation. The correlation between TN and TP was relatively high (r = 0.22, $p \leq 0.05$). In general, Chl-a as a measure of algae stock and photosynthetic indicators have an effect on both pH and DO levels. Photosynthesis of algae consumes carbon dioxide in the water, causing the pH of the lake to increase and with it to increase DO. However, a different situation arose at Lake Yilong. The correlation analysis showed that there was no significant correlation between chlorophyll and pH and DO in Lake Yilong, which may be due to the high coverage of aquatic vegetation in Lake Yilong and the participation of aquatic plants in photosynthesis, which affected the pH value and DO of Lake Yilong.

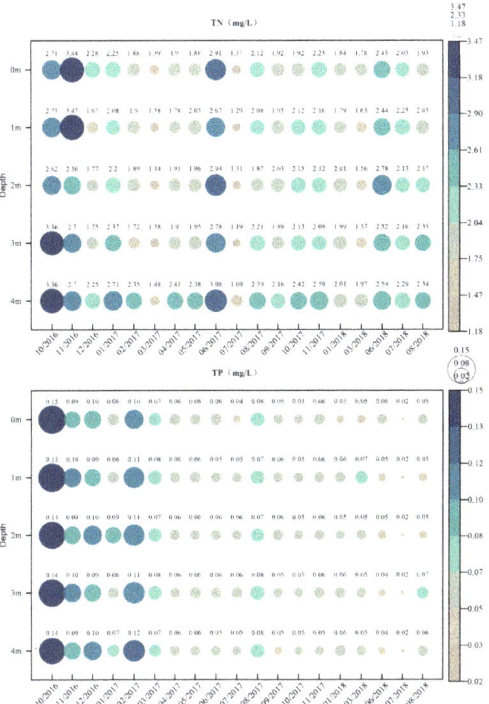

Figure 7. The change in TN and TP concentration in Lake Yilong from October 2016 to August 2018.

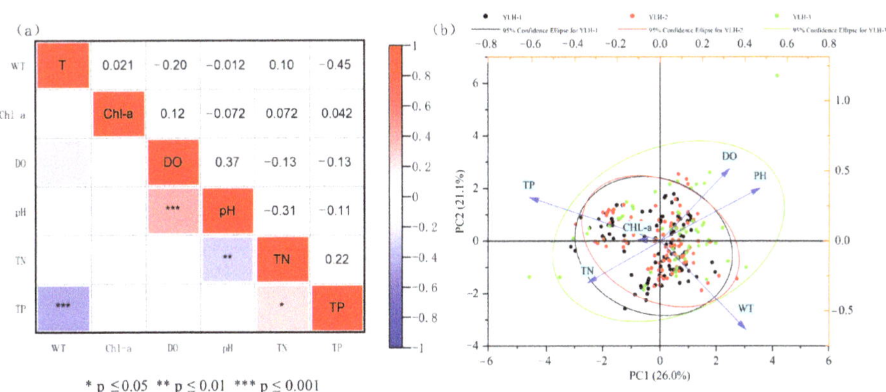

Figure 8. Correlations of each water parameter (**a**) and principal component (**b**) analysis of Lake Yilong.

The first two axes in a principal component analysis (PCA) of physicochemical variables from all sampling times and sites, explained 47.1 % of the total variation in the dataset (Figure 8b), including PC1 and PC2 accounted for 26.0% and 21.1% of all datasets, respectively (Figure 8b). The PC1 axis was related to total nutrient concentrations (TN, TP) and pH, while the PC2 axis was related to DO and WT. The results of principal component analysis show that there is a clear separation between the northwest lake area and the central and southwest lake areas, there was a general trend for northwest site to have higher DO and Chl-a concentrations than central site and southwest site. This is consistent with the previous results on DO and Chl-a.

3.7. Limitations and Implications

This study mainly analyzes the relevant water quality parameters of lake Yilong and the characteristics of TN and TP according to the detection data, discusses the hydrochemical properties of lake Yilong and evaluates the water quality. However, this paper also has the following shortcomings and defects: First, the number of monitoring points is limited, and the results may not reflect the whole situation of Lake Yilong well. In addition, due to force majeure, there is a lack of data in individual months in the whole testing year, which affects the continuity of data points. It is believed that with the improvement of monitoring systems and methods, this problem may be solved in the near future. Second, the change of water quality in Lake Yilong was not analyzed in combination with the characteristics of the basin and the impact of human activities (agricultural land use, cattle, urban area, industry, wastewater inflow). Third, in terms of water quality analysis, the impact of other factors that may affect water quality (such as large, submerged plants) has not been fully considered. Therefore, it is necessary to do further research on aquatic plants in order to get the situation in the whole lake in the future.

4. Conclusions

Monitoring of the water quality of a lake is the basis for understanding its development. Analysis of water quality parameter data based on different monitoring points and time of Lake Yilong, allowed the following conclusions to be drawn: The lake water was well mixed during the monitoring period, resulting in no significant thermal stratification.

The water of Lake Yilong began to experience a sudden drop in DO concentration in its northwest part in September and October at a water depth of about 1 m, forming an anoxic environment. The depth and magnitude of the reduction varied with different seasons. At present, it is believed to be caused by a variety of factors, of which the oxidation and decomposition of various organic pollutants in the sediments deposited at the bottom of the lake are considered the most important factors. Seasonal variation: The concentration of Chl-a in Lake Yilong demonstrated obvious changes in different seasons and lake monitoring sites. Analysis at the southwest and center of the lake showed that the concentration reached its maximum in January, while that of the northwest part occurred in September, with the maximum value recorded in October. This phenomenon may be related to the growth of aquatic plants in the lake. TN content in lake Yilong is higher in rainy season, which is related to temperature and precipitation process, and there is a risk of algal proliferation. Therefore, to protect the ecosystem of Lake Yilong, water quality monitoring should be conducted in rainy season. According to the assessment, Lake Yilong changed from middle eutrophic to weak eutrophic in 2016–2018. The eutrophication degree of the lake improved, but TN remained at a high level. Therefore, it is necessary to strengthen lake water quality monitoring and control the adverse effects of human activities on Lake Yilong to prevent water quality deterioration.

Author Contributions: H.Z.: Conceptualization, Supervision, Resources, Writing—review and editing, Foundations acquisition. Q.S.: Methodology, Software, Writing—Original draft preparation. L.D., X.Z., Y.Z. and Q.L.: Investigation, Data Curation. All authors have read and agreed to the published version of the manuscript.

Funding: This research was funded by the Yunnan Provincial Government Scientist workshop and the Special Project for Social Development of Yunnan Province (Grant No. 202103AC100001).

Data Availability Statement: The data that support the findings of this study are available from the corresponding author upon reasonable request.

Acknowledgments: We thank all the graduate students participated the field works and laboratory analyses.

Conflicts of Interest: The authors declare no conflict of interest.

References

1. Wang, S.M.; Dou, H.S. *Lakes of China*; Science Press: Beijing, China, 1998; pp. 1–2. (In Chinese)
2. Meshesha, T.W.; Wang, J.; Melaku, N.D. Modelling spatiotemporal patterns of water quality and its impacts on aquatic ecosystem in the cold climate region of Alberta, Canada. *J. Hydrol.* **2020**, *587*, 124952. [CrossRef]
3. Ndungu, J.; Augustin, D.; Hulscher, S.; Fulanda, B.; Kitara, N.; Mathooko, J.M. A multivariate analysis of water quality in lake Naivasha, Kenya. *Mar. Freshw. Res.* **2014**, *66*, 177–186. [CrossRef]
4. Saber, A.; James, D.E.; Hannon, I.A. Effects of lake water level fluctuation due to drought and extreme winter precipitation on mixing and water quality of an alpine lake, case study: Lake arrowhead, California. *Sci. Total Environ.* **2020**, *714*, 136762. [CrossRef] [PubMed]
5. Azam Chowdhury, K.M.; Jiang, W.S.; Liu, G.; Md, K.; Shaila, A. Dominant physical-biogeochemical drivers for the seasonal variations in the surface chlorophyll-a and subsurface chlorophyll-a maximum in the Bay of Bengal. *Reg. Stud. Mar. Sci.* **2021**, *48*, 102022. [CrossRef]
6. István, G.H.; Vinicius, D.B.; Péter, T.; József, K.; Székely, K. Spatiotemporal changes and drivers of trophic status over three decades in the largest shallow lake in Central Europe, Lake Balaton. *Ecol. Eng.* **2020**, *151*, 105861.
7. Lv, J.; Li, C.Y.; Zhao, S.N.; Sun, B.; Shi, X.H.; Tian, W.D. Evaluation of nutritional status in Wu Liangsuhai in frozen and non-frozen seasons. *J. Arid. Land Resour. Environ.* **2018**, *32*, 6. (In Chinese)
8. Hou, P.; Chang, F.; Duan, L.; Zhang, Y.; Zhang, H. Seasonal Variation and Spatial Heterogeneity of Water Quality Parameters in Lake Chenghai in Southwestern China. *Water* **2022**, *14*, 1640. [CrossRef]
9. Dong, Y.X.; Zhao, L.; Cheng, Y.H.; Yu, Y.H.; Zhao, R.; Yang, G.P. Succession of Nine Plateau Lakes and Regulation of Ecological Safety in Yunnan Province. *Ecol. Econ.* **2015**, *31*, 185–191. (In Chinese)
10. Bao, Y.F.; Cui, D.W. An analysis of comparative evaluation of the nutrition status of the nine plateau lakes in Yunnan province. *Environ. Sci. Surv.* **2012**, *65*, 357–364. (In Chinese)
11. Zhang, H.C. The Potential Endangers of the Tectonic Lake Water Leakage from Dian chi and Water Security. *Adv. Earth Sci.* **2016**, *31*, 849–857, (In Chinese with English Abstract)
12. Wu, H.; Chang, F.Q.; Zhang, H.C.; Duan, L.Z.; Zhang, X.N.; Peng, W.; Zhang, Y.; Liu, Q.; Liu, F.W. Changes of organic C and N stable isotope and their environmental implication during the past 100 years of Lake Yilong. *Chin. J. Ecol.* **2020**, *8*, 2478–2487. (In Chinese)
13. Liu, W.; Ning, P.; Sun, X.; Ma, Q. The Evolution Tendency of Water Quality and Aquatic System in Lake Yilong. *Environ. Sci. Technol.* **2018**, *41*, 281–287. (In Chinese)
14. Wang, Z.F.; Zhang, W.; Yang, Y.; Xu, Y.P.; Zhao, F.B.; Wang, L.Q. Characteristics of Phytoplankton Community and Its Relationship with Environmental Factors in Different Regions of Yilong Lake, Yunnan Province, China. *Environ. Sci.* **2019**, *40*, 2249–2257. (In Chinese)
15. Wu, T.; Wang, S.; Su, B.; Wu, H.; Wang, G. Understanding the water quality change of the Yilong lake based on comprehensive assessment methods. *Ecol. Indic.* **2021**, *126*, 107714. [CrossRef]
16. Yuan, Z.J.; Wu, D.; Niu, L.; Ma, X.; Li, Y.; Hillman, A.L.; Abbott, M.B.; Zhou, A. Contrasting ecosystem responses to climatic events and human activity revealed by a sedimentary record from Lake Yilong, southwestern China. *Sci. Total Environ.* **2021**, *783*, 146922. [CrossRef]
17. Chen, J.; Qin, J.; Zhou, Q.C.; Nie, J.F.; Zhou, H.X. Application of wetland ecological restoration technology in plateau lakes. *Environ. Sci. Technol.* **2016**, *39*, 158–168.
18. Li, X.J. Current Status and Countermeasures of Fishery Industry Development of Lake Yilong in Shiping County. *Mod. Agric. Technol.* **2014**, *03*, 328–333. (In Chinese)
19. Wang, Z.Z. Study on Algae Plants and Fish Productivity in Lake Yilong Yunnan Province. *J. Fish. China* **1997**, *21*, 93–96. (In Chinese)
20. Liu, P.; Chang, F.Q.; Wu, H.B. Health Evaluation of Wetland Ecosystem in Yi-long Lake Basin. *Wetl. Sci. Manag.* **2016**, *12*, 30–34. (In Chinese)
21. She, L.H.; Liu, P.C.; Zhang, F.Q.; Cai, F. Situation and protection countermeasures on lake Yilong wetland in Yunnan. *For. Constr.* **2016**, *04*, 15–18. (In Chinese)
22. Zhang, H. Application of Comprehensive Eutrophication State Index Method in Evaluation of Chaohu Lake Reservoir Eutrophication. *Anhui Agric. Sci. Bull.* **2018**, *24*, 4. (In Chinese)
23. Chen, S.S. *Climate Change and Human Activities Indicated by the Sediments of Yilong Lake*; Kun Ming Yunnan Normal University: Kunming, China, 2015. (In Chinese)
24. State Environmental Protection Administration (SEPA). *Water and Wastewater Monitoring and Analysis Method*, 4th ed.; China Environmental Science Press: Beijing, China, 2002; Volume 2, pp. 12–36. (In Chinese)

25. Li, Y.; Wang, L.; Yan, Z.; Chao, C.; Yu, H.; Yu, D.; Liu, C. Effectiveness of dredging on internal phosphorus loading in a typical aquacultural lake. *Sci. Total Environ.* **2020**, *744*, 140883. [CrossRef] [PubMed]
26. Cai, M.; Zhang, H.C.; Chang, F.Q.; Li, T.; Hu, J.J.; Duan, L.Z.; Zhang, Y. Nitrogen and Phosphorus Content Changes of the Lake Water Samples from the Typical Yunnan Plateau Lakes. *Resour. Environ. Yangtze Basin* **2019**, *12*, 3030–3037. (In Chinese)
27. Wang, M.; Liu, X.; Zhang, J. Evaluation method and classification standard on lake eutrophication. *Environ. Monitor. China* **2002**, *18*, 47–49. (In Chinese)
28. He, Y.F.; Li, H.C.; Zhu, Y.J.; Yang, D.G. Status and spatial-temporal variations of eutrophication in lake Changhu, Hubei province. *J. Lake Sci.* **2015**, *27*, 853–864. (In Chinese)
29. Luo, L.; Zhang, H.; Luo, C.; McBride, C.; Muraoka, K.; Zhou, H.; Hou, C.; Liu, F.; Li, H. Tributary Loadings and Their Impacts on Water Quality of Lake Xingyun, a Plateau Lake in Southwest China. *Water* **2022**, *14*, 1281. [CrossRef]
30. China National Environmental Monitoring Centre. *Lake (Reservoir) Eutrophication Evaluation Method and Classification Technical Provisions, 2001*; China National Environmental Monitoring Centre: Beijing, China, 2002. (In Chinese)
31. Zhao, L.L.; Zhu, G.W.; Chen, Y.F.; Li, W.; Zhu, M.Y.; Yao, X.; Cai, L.L. Characteristics of vertical stratification of water temperature in Lake Tai and its influencing factors. *Adv. Water Sci.* **2011**, *22*, 844–850.
32. Liu, Y.; Yang, F.; Gao, Y.X.; Zhu, Y.M.; Kou, M.; Zhao, Y.; Qian, W.H. Temporal and Spatial Variation Characteristics of Chlorophyll a and Analysis of Related Environmental Factors in the Estuary Area of Lake Ge. *J. Ecol. Rural Environ.* **2021**, *37*, 733–739.
33. Song, Y.L.; Zhang, J.S. Spatiotemporal distribution of chlorophyll a and its influencing factors in SHIYAN reservoirs. *Environ. Sci.* **2017**, *38*, 3302–3311.
34. Tian, P.; Li, Y.L.; LI, Y.J.; Li, H.; Wang, L.J.; Song, L.X.; Ji, D.B.; Zhao, X.X. Influence of Three Gorges Reservoir Operation on the Vertical Distribution of Chlorophyll a and Environmental Factors in Tributary Water. *Environ. Sci.* **2022**, *43*, 295–350. (In Chinese)
35. Shi, L.P.; Sun, Q.Y. Brief Talk on Effect of Aquatic Plant Purifying Water body of Yilong Lake. *Yunnan Environ. Sci.* **2005**, *24*, 40–42. (In Chinese)
36. Li, D.; Chang, F.; Wen, X.; Duan, L.; Zhang, H. Seasonal Variations in Water Quality and Algal Blooming in Hypereutrophic Lake Qilu of Southwestern China. *Water* **2022**, *14*, 2611. [CrossRef]
37. Hideki, T.; Hironori, F.; Mirai, W.; Seiji, H. Effects of temperature and oxygen on 137Cs desorption from bottom sediment of a dam lake. *Appl. Geochem.* **2022**, *140*, 105303.
38. Chang, F.; Hou, P.; Wen, X.; Duan, L.; Zhang, Y.; Zhang, H. Seasonal Stratification Characteristics of Vertical Profiles and Water Quality of Lake Lugu in Southwest China. *Water* **2022**, *14*, 2554. [CrossRef]
39. Xu, W.; Duan, L.; Wen, X.; Li, H.; Li, D.; Zhang, Y.; Zhang, H. Effects of Seasonal Variation on Water Quality Parameters and Eutrophication in Lake Yangzong. *Water* **2022**, *14*, 2732. [CrossRef]
40. Zhao, L.; Wang, M.; Liang, Z. Identification of Regime Shifts and Their Potential Drivers in the Shallow Eutrophic Lake Yilong, Southwest China. *Sustainability* **2020**, *12*, 3704. [CrossRef]
41. Wei, X.; Tang, G.M. Changes of Nutrients and Aquatic Ecosystem of Yilong Lake in Recent Twenty Decades. *Yunnan Environ. Sci.* **2014**, *33*, 9–14. (In Chinese)
42. Zhang, H.X. *Evolution of Water Ecological Environment Quality of Typical Lakes and Reservoirs and Its Response to Climate Change*; Chinese Academy of Environmental Sciences: Beijing, China, 2019. (In Chinese)
43. Zhang, L.; Liu, Y.J.; Ge, F.J.; Xue, P.; Li, X.; Zhang, X.Y.; Zhang, S.X.; Zhou, Q.H.; Wu, Z.B.; Liu, B.Y. Spatial characteristics of nitrogen forms in a large degenerating lake: Its relationship with dissolved organic matter and microbial community. *J. Clean. Prod.* **2022**, *371*, 133617. [CrossRef]
44. O'Reilly, C.M.; Alin, S.R.; Plisnier, P.D. Climate Change Decreases Aquatic Ecosystem Productivity of Lake Tanganyika, Africa. *Nature* **2003**, *424*, 766–768. [CrossRef]
45. Niu, Z.Y.; Shen, X.X.; Chai, M.W.; Xu, L.H.; Li, R.L.; Qiu, G.Y. Characteristics of Water Quality Changes in the Futian Mangrove National Natural Reserve. *Acta Sci. Rum Nat. Univ. Pekin.* **2018**, *54*, 137–145.
46. Mi, W.M.; Shi, J.Q.; Yang, Y.J.; Yang, S.Q.; He, S.H.; Wu, Z.X. Changes of algae communities in Meixi River, a tributary of the Three Gorges Reservoir Area, and its relationship with environmental factors. *Environ. Sci.* **2020**, *41*, 1636–1647.
47. GB3838–2002; Environmental Quality Standard for Surface Water. State Environmental Protection Administration (SEPA): Beijing, China, 2002; pp. 1–8. (In Chinese)

Article

Release of Endogenous Nutrients Drives the Transformation of Nitrogen and Phosphorous in the Shallow Plateau of Lake Jian in Southwestern China

Yang Zhang, Fengqin Chang *, Xiaonan Zhang, Donglin Li, Qi Liu, Fengwen Liu and Hucai Zhang *

Institute for Ecological Research and Pollution Control of Plateau Lakes, School of Ecology and Environmental Science, Yunnan University, Kunming 650504, China
* Correspondence: changfq@ynu.edu.cn (F.C.); zhanghc@ynu.edu.cn (H.Z.)

Abstract: Eutrophication remediation is an ongoing priority for protecting aquatic ecosystems, especially in plateau lakes with fragile ecologies and special tectonic environments. However, current strategies to control the phosphorus (P) and nitrogen (N) levels in eutrophication sites have been mainly guided by laboratory experiments or literature reviews without in-field analyses of the geochemical processes associated with the hydrological and eutrophic characteristics of lakes. This study analyzed the water quality parameters of 50 sites at Lake Jian in May 2019, a moderate eutrophication shallow plateau lake, based on dissolved/sedimentary nitrogen, phosphorous and organic matter, grain size, C/N ratios and stable isotope ratios of δ^{13}C or δ^{15}N in sediments. The results showed that the average total nitrogen (TN) and total phosphorus (TP) concentrations in the lake water were 0.57 mg/L and 0.071 mg/L, respectively. The TN and TP contents of surface sediment ranged from 2.15 to 9.55 g/kg and 0.76 to 1.74 g/kg, respectively. Stable isotope and grain source analysis indicated that N in sediments mainly existed in organic matter form and P mainly occurred as inorganic mineral adsorption. Endogenous pollution contributed to >20% of TN. Furthermore, our findings showed that phosphorus was the nutrient that limited eutrophication at Lake Jian, unlike other eutrophic shallow lakes. In contrast, the nutrient levels in the sediment and input streams belonged entirely to the N-limitation state. The distinctness in release intensity of N and P could modify the N/P limitation in the lake, which affects algae growth and nutrient control. These results demonstrated that reducing exogenous nutrients might not effectively mitigate lake eutrophication due to their endogenous recycling; thus, detailed nutrient monitoring is needed to preserve aquatic ecosystems.

Keywords: plateau lake; eutrophication; nutrient limitation transformation

Citation: Zhang, Y.; Chang, F.; Zhang, X.; Li, D.; Liu, Q.; Liu, F.; Zhang, H. Release of Endogenous Nutrients Drives the Transformation of Nitrogen and Phosphorous in the Shallow Plateau of Lake Jian in Southwestern China. *Water* **2022**, *14*, 2624. https://doi.org/10.3390/w14172624

Academic Editor: Bommanna Krishnappan

Received: 14 March 2022
Accepted: 26 July 2022
Published: 26 August 2022

Publisher's Note: MDPI stays neutral with regard to jurisdictional claims in published maps and institutional affiliations.

1. Introduction

Eutrophication, i.e., the excessive richness of nutrients in a lake or other water body, is one of the principal ecological disasters during the evolution of lakes [1]. Overloaded N, P and other biogenic elements from exogenous pollution in a water body directly cause an abnormal increase in the primary productivity of aquatic ecosystems, leading to eutrophication [2]. Nutrient control of the eutrophicated lake is one of the remediation strategies based on the nitrogen and phosphorus limitation of algae growth [3]. Previous studies demonstrated that in severely affected eutrophicated lakes, the nutrient limitation theory might not work [3]. However, some scholars believe that it is necessary to control all nutrient elements [2]. In recent years, considering the costs and excessive amount of nitrogen or phosphorus causing algae growth, the single nutrient limitation strategy has been proposed as an attempt to remediate the eutrophication of lakes [4].

Lake sediments are important potential reservoirs of N, P and biogenic elements that accumulate in lake water through physical diffusion, convection, resuspension, and other geobiochemical procedures after discharge [5]. Their recycling can also affect lake water quality, resulting in endogenous pollution. Meanwhile, long-term accumulation and degradation of organic matter (OM) produced by terrestrial materials, phytoplankton, and aquatic plant residues are also the main sources for the release of N and P [6,7]. Before effectively solving endogenous pollution, eutrophication might become more serious, even with exogenous pollution cut off [8].

Despite large amounts of investments, including >USD 10 billion annually and globally, to control N and P influx in lakes, the eutrophication problem is still spreading and growing, leading to recurrent cyanobacteria blooming even after remediation [3]. This phenomenon indicates that the current nutrient influx controls are insufficient. The main reason could be that we lack a full understanding of the internal cycling nutrients and geological or biogeochemical conditions of lakes. The previous eutrophication control paradigm was based mainly on laboratory or field experimental results, which tended to simulate the effects of external input without considering biogeochemical processes and geological background in the catchment of lakes [4,6,7]. Thus, more effort is needed to improve our understanding of the nutrient dynamics in lakes and the recharge procedures from catchment to achieve the desired eutrophication mitigation solutions, especially in plateau lakes with fragile ecological environments.

During the past decade, there has been a growing consensus that the concentration of nutrients in the upper sedimentary or water samples provides a basic general understanding of catchment development and the distribution of nutrients in a specific area because their concentrations do not account for the local geological types and other bio-geochemical controls [9]. Variations in grain size and composition of the sediment samples must be considered when documenting spatial variations in elemental concentrations. Samples rich in chemically reactive fine-grained (<63 μm) sediments are likely to contain higher concentrations than a sample dominated by sand, even if both originated from the natural rock unit or contaminated soils [10]. Moreover, isotopes are also useful tools for tracking the sources of nutrients and estimating their inner cycle processes [11]. To address the recycling issues and sedimentary variations in lakes, many researchers propose the use of multi-geochemical indexes, including C, N and P isotopes and grain size, as well as nutrient fractions, to determine the concentration of nutrients [12].

Lake Jian is an ecologically and environmentally protected region in northwest Yunnan that plays an important role in maintaining the regional ecological and environmental function and biodiversity [13,14]. Recently, developments in the catchment have drastically reduced the volume of Lake Jian, causing the lake to gradually wither [15]. Moreover, the nutrient element content has rapidly increased through agricultural production and other anthropogenic impacts, which have caused eutrophication and reduced the environmental function of lakes [16,17]. Most previous studies have focused on paleolimnology and paleoclimatology [18], ecological responses [13,16] and persistent environmental pollutants [19,20]. No comprehensive investigations have been conducted to reconstruct the reasons for eutrophication and estimate exogenous and endogenous pollution. Therefore, current nutrient levels, especially the levels of N and P in lake catchment systems, the source of pollutants and the internal reaction of biogenic elements, must be accurately and comprehensively explored.

Herein, 50 water and sediment samples from Lake Jian and its catchment rivers were analyzed to investigate the potential source and inner cycle of nutrient dynamics, specifically isotopes and grain size, in relation to lake trophic status and relative importance of N and P limitation.

2. Materials and Methods

2.1. Overview of Lake Jian

Lake Jian is a plateau fault lake located in the southern Hengduan Mountains in Dali prefecture, northwest Yunnan, Southwest China (Figure 1). It covers an area of 6.23 km^2, with an average water depth of 2.3 m. As one of the important upstream sources of Southeast Asian fluvial systems, more than five rivers originate from the basin but have only one outlet, the Heihui River, which eventually joins the Yangbi River to become the principal branch of the Lantsang (Mekong) river and crosses six countries. Lake Jian has a maximum volume of 45.32×10^6 m^3, reflecting a decrease of more than 60% in recent years [14]. The mean air temperatures in winter and summer are 15 °C and 28 °C, respectively, and the mean annual rainfall is 786 mm.

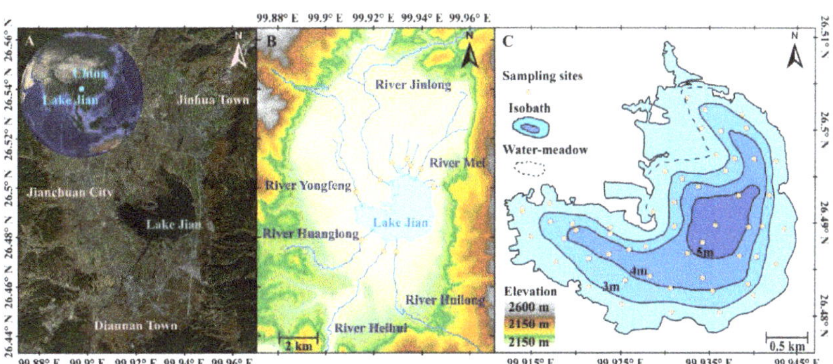

Figure 1. Location and geographic information (**A**), watershed system and digital elevation model of the catchment (**B**), lake isobaths (**C**) and sampling site ((**B**) rivers and streams; (**C**) lake water and sediment) of Lake Jian.

2.2. Sampling and Detecting

In May 2019, 50 surface sediment (each 0.5 cm thick), surface water and bottom water were sampled from Lake Jian, and 10 water samples from main streams and rivers into its catchment were collected using a gravity corer and condensate trap (Figure 1). All the samples were kept in brown polyethylene bottles and frozen in a refrigerator.

The nutrient element indices in water, including total nitrogen (TN) and total phosphorus (TP), were determined by the alkaline potassium persulfate digestion-UV spectrophotometric method (GB 11894-89) and ammonium molybdate spectrophotometric method (GB 11893-89) formulated by the State Environmental Protection Administration of China (SEPAC). The results were taken from the average values determined at three parallel times, and the measurement errors were less than 1%. Water quality parameters, including depth, temperature, chlorophyll a (Chl-a), pH, and dissolved oxygen (DO), were measured using 6600-V2 YSI during sampling. TN and TP in the surface sediment were measured by sulfuric acid digestion-Kjeldahl determination (GB 7173-87) and the Mo-Sb colorimetric method (GB 7852-87), following the standard of SEPAC with resulting measurement errors <5%. Samples for grain-size analysis were pre-treated using H_2O_2 and HCl to remove organic matter and carbonates. Grain size distribution between 0.02 μm and 2000 μm was measured using a Malvern Mastersizer 2000 laser grain-size analyzer, before the samples were deionized and dispersed by $Na(PO_3)_6$. Three types of bulk (<4 mm, 4–64 mm and >64 mm) size fractions were analyzed, and the analysis was focused on the fine-grained fraction (closely related to element enrichment in the surface soils). OM in the surface sediment was measured by the loss on ignition method. C/N, δ^{15}N, and δ^{13}C in the surface sediment were detected with a Thermo MAT-253, with analytical errors of less than 0.2%.

2.3. Statistical Analysis

When considering the geochemical background and sources of nutrients, their relationship with grain size composition of samples were used for calculation [9]. Generally, higher concentrations of major elements in the earth's crust were observed in chemically reactive fine-grained (<64 μm) sediments, even in the samples obtained from natural and anthropogenic conditions [10]. Thus, the nutrients strongly correlated with the fine-grained fraction were chosen as exogenous sources to identify elements that could account for the differences in grain size and composition [21]. The LTS robust regression model was defined by a logarithmic regression [22], and geochemical conditions in the catchment area and possible sources were estimated by the regression analyses in Lake Jian (details in Section 3.5). The N-, and P-limitations of lake eutrophication for algae growth complied with the formulation of the "Redfield ratio", e.g., N-limitation, N/P < 10; P-limitation, N/P > 20 [23].

For δ^{15}N and δ^{13}C analyses, δ notation was used to represent isotopic ratio differences between samples and standard materials. The formulas are expressed as follows:

$$\delta^{13}N = \frac{(Rsp_N - Rst_N)}{Rst_N} \times 1000\text{‰} \tag{1}$$

$$\delta^{15}C_{org} = \frac{(Rsp_{C_{org}} - Rst_{C_{org}})}{Rst_{CC_{org}}} \times 1000\text{‰} \tag{2}$$

Here, δ^{15}N and δ^{13}C represent the differences (‰) from the Vienna PDB standard and atmospheric N_2, and Rst and Rsp represent the stable isotope ratio of δ^{13}C or δ^{15}N in standard samples. Based on the mass conservation hybrid model and linear mixed model, the contribution of N and OM from different sources could be estimated as follows [24]:

$$\delta^{13}N = \sum_{x=1}^{n} W_x \times \delta^{13}N_x \tag{3}$$

$$\delta^{13}C_{org} = \sum_{x=1}^{n} W_x \times \delta^{13}C_{org} \tag{4}$$

Here, δ^{15}Nx and δ^{13}Corgx represent the corresponding isotope ratio detected in different end members of samples [25,26]. In this study, the main sources of the sedimentary OM were considered to be plankton (with typical stable compositions of C/N: 5~8, δ^{13}C: −32~−23‰ and δ^{15}N: 5~8‰), macrophytes (10~30, −27~−20‰, −15~20‰), soil (8~15, −32~−9‰, 2~5‰), terrestrial C3 plants (15~40, −32~−22‰, −6~5‰), terrestrial C4 plants (15~40, −16~−9‰, −6~5‰), and sewage (6.6~13, −26.7~−22.9‰, 7~25‰). The principal sources of sedimentary N were as follows: agricultural fertilizer (with δ^{15}N from −4~4‰), domestic sewage (10~20‰), soil erosion (3~8‰), terrestrial OM (with avg. of 2‰), and endogenous OM (6.5‰) [25]. The W_x was the contribution probability of different pollutant sources, computed by an iterative calculation model, which is as follows:

$$W_x = \frac{\left[\left(\frac{100}{i}\right) + (n_S - 1)\right]!}{\left(\frac{100}{i}\right)!(n_S - 1)!} \tag{5}$$

Here, i represents the increment coefficient of the calculation model, and nS represents the number of N sources. Isotope sources were calculated using IsoSource [24]. Statistical analysis, one-way analysis of variance (ANOVA) and Pearson correlation (PC) were implemented using PAST v4.0 [27].

3. Results and Discussion

3.1. Nutrient Level and Water Quality of Lake Jian

The spatial distributions of mean TN (range: 0.05–0.99 mg/L, average: 0.57 mg/L) and TP (range: 0.003–0.173 mg/L, average: 0.071 mg/L) contents in surface water were as follows: western areas > central part > eastern areas (Figure 2). Moreover, the mean contents of TN and TP in the sites around the lakeside were higher than in the lake basin. Similarly, the contents of TN and TP in the bottom water of Lake Jian also showed broad variations, which ranged between 0.09 and 0.76 mg/L (0.29 mg/L) and 0.003 and 0.140 mg/L (0.087 mg/L). Generally, the highest TN and TP values were observed at the western end of the lake near the entrance of the Yongfeng River, but the minimum values were detected in the central and eastern lake areas. Additionally, higher mean concentrations of TN and TP were observed at the surface water compared with bottom water (Table 1). Except for some sites in the western part of Lake Jian, the whole lake was in a state of middle eutrophication, belonging to Level III water quality (SEPAC standard). Significant differences in nutrient levels from the entering rivers and streams were observed. TN (2.81 mg/L) and TP (0.428 mg/L) levels in the Yongfeng River, which crosses a population center carrying domestic sewage, were significantly higher than in the Huilong River (TN: 1.93 mg/L, TP: 0.065 mg/L), Mei River (0.52 mg/L, 0.062 mg/L) and Jinlong River (1.58 mg/L, 0.016 mg/L) and streams entering the basin (0.14–0.51 mg/L, 0.015–0.049 mg/L), and traversing rural villages and farms in the catchment basin. High nutrient element levels were observed in the influx river, which was consistent with that accumulated in the lake area near an estuary, explaining the spatial discrepancy of nutrient elements in water to some extent. Nutrient elements of only one outlet, the Haiwei River, were 1.55 mg/L and 0.052 mg/L, respectively.

Figure 2. Environmental indexes of Lake Jian, including TN/TP in surface water (**A,F**), bottom water (**B,G**) and surface sediment (**C,H**), spatial distribution of C/N ratios (**D**), OM contents (**I**), δ^{13}C (**E**), δ^{15}N (**J**), fine-grain size (**K**) and sand fraction (**L**) composition.

Table 1. Nutrient element levels, water quality indices and geochemical information.

Environmental Parameters		Range	Environmental Parameters		Range
TN	Surface water	0.05~0.99 mg/L	TP	Surface water	0.003~0.173 mg/L
	Bottom water	0.09~0.76 mg/L		Bottom water	0.003~0.140 mg/L
	Surface sediment	2.15~9.55 g/kg		Surface sediment	0.76~1.74 g/kg
	Output river	1.55 mg/L		Output river	0.052 mg/L
	Rivers and streams	0.14~2.81 mg/L		Rivers and streams	0.015~0.428 mg/L
	Pre-rivers and streams [a]	0.04~1.85 mg/L		Pre-rivers and streams [a]	0.020~0.360 mg/L
	Pre-surface water [b]	0.13~0.56 mg/L		Pre-surface water [b]	0.020~0.035 mg/L
Chl-a	Surface water	16.90~63.68 µg/L	pH	Surface water	8.34~9.23
	Bottom water	12.50~85.73 µg/L		Bottom water	8.21~9.17
DO	Surface water	6.53~10.96 mg/L	EC	Surface water	250~284 µS/cm
	Bottom water	1.57~8.87 mg/L		Bottom water	252~275 µS/cm
OM	Surface sediment	4.25~21.30%	C/N	Surface sediment	10.76~16.53
δ^{15}N	Surface sediment	2.45~5.00‰	δ^{13}C	Surface sediment	−30.08~−24.08‰
Clay	Surface sediment	26.80~72.28%	Silt	Surface sediment	27.49~67.38%
Fine-grain size	Surface sediment	90.32~100%	Sand	Surface sediment	0~9.68%

Note: [a] [28]; [b] [16].

3.2. Spatial Distribution of Environmental Parameters in Surface Sediments

The spatial variability in TP, TN and OM concentrations in the surface sediment of Lake Jian is shown in Figure 2 (additional details in Table 1). OM concentrations in Lake Jian ranged from 4.25% to 21.30% (averaging 11.57%), with higher values observed on the western and eastern shores (>14.05%). There were lower contents of OM in the middle of the lake basin.

Generally, the distribution of TN in surface sediment is similar to that of OM, with a significant inward decreasing trend from the lakeshore (4.68–9.55 g/kg) to the deepest basin (2.15–4.03 g/kg). The mean concentration of TP decreased from the western lake margin (1.31 g/kg) to the middle (1.17 g/kg) and eastern parts of the lake (1.16 g/kg). In Lake Jian, the ratio of C/N was characterized by obvious spatial variation, with a range from 10.76 to 16.53, and an average value of 12.59. Higher C/N ratios were found in the sediment samples near Yongfeng River estuary (16.53) and the Mei River (15.03). Meanwhile, the C/N ratio was lower in the lake basin at its deepest portions (10.76~11.08), compared with other areas (11.17~13.05). The spatial distribution of δ^{13}C and δ^{15}N of surface sediment from Lake Jian is presented in Figure 2; their values ranged from −30.08 to −24.08‰ (averaging −28.28‰) and from 2.45 to 5.00 ‰ (3.84 ‰), respectively. Δ^{13}C exhibited a great spatial gradient, with its highest values recorded at the central lake bay area and lowest values recorded in the estuary of the western lake, while there was an overall decreasing trend from the eastern (−27.60‰) to the western part of the lake (−28.68‰). In contrast, δ^{15}N values showed regular spatial variation by isobath, with the most depleted δ^{15}N values occurring at the lakeshore (3.91–5.00‰), compared to the central basin (2.45–3.69‰).

The spatial variability in grain size fraction components is shown in Figure 2 (additional details in Table 1). To sort the particle size fractions, the percentage of clay components in the surface sediment samples of Lake Jian fell between 26.80 and 72.28%, with an average of 51.32%. There were obvious spatial differences between the central region and those close to the lake shore. The spatial change characteristics showed an increasing trend from the lake shore (49.46%) to the central region (56.73%). The silt component ranged between 27.50 and 67.38% (averaging 46.98%). The characteristics of spatial variation show that there was a decreasing trend from the lake shore to the lake center, i.e., the eastern lake area (54.78%) > the western lake area (48.32%) $\geq$ the central lake basin (41.75%). The percentage of sand components was between 0 and 9.68% (1.70%). Obvious high values were observed at the west lake bay near the entrance of the Jinlong River. Some high-value areas were also observed from the east bay to the central lake area. The mean contents of sand in Lake Jian were as follows: the western lake area (2.22%) > the central lake area (1.52%) > the eastern lake area (1.20%).

3.3. Source of Sedimentary OM and N

In this study, five OM sources (plankton, macrophytes, sewage, soil OM, terrestrial C3 and terrestrial C4) and N sources (agricultural fertilizer, domestic sewage, exogenous release, soil erosion, terrestrial input) were chosen to evaluate sources of pollutants (Figure 3).

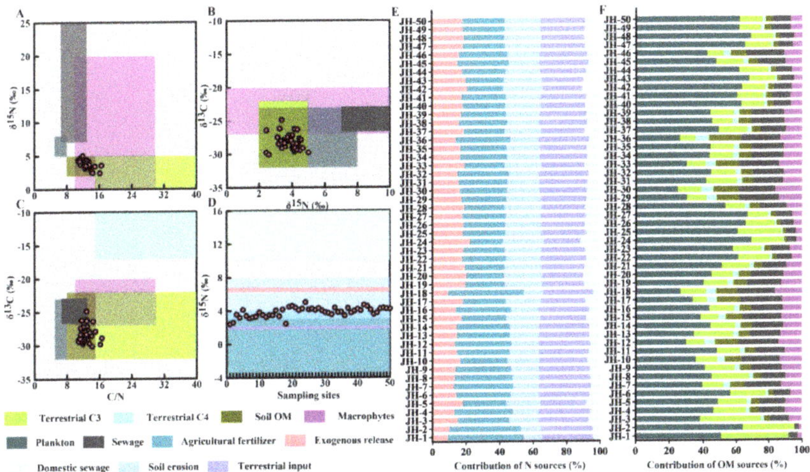

Figure 3. Scatterplots of δ^{15}N/CN ratios (**A**), δ^{15}N/δ^{13}C (**B**) and δ^{13}C/CN ratios (**C**), site-special range of δ^{15}N (**D**) (isotopes boundary data from [25,26]), as well as the contribution of N (**E**) and organic matter sources (**F**).

Scatterplots of the δ^{15}N and C/N (Figure 3A) in Lake Jian showed that sedimentary OM was mainly divided into two sources, macrophytes and soil OM. In addition, the results of δ^{13}C vs. δ^{15}N analyses (Figure 3B) showed that sedimentary OM sediments were primarily derived from terrestrial C3 plants and soil OM. The site-specific distribution of δ^{13}C and C/N in Lake Jian was within the ranges of soil OM (Figure 3C). Based on the background range of source-specific environments (Figure 3D), δ^{15}N in Lake Jian was mainly introduced from agricultural fertilizer and soil erosion.

In the increment coefficient of a calculation model for sources analysis, the mean and median values from the typically stable composition were used as the calculating δ^{13}C values of plankton ($-27.5‰$), macrophytes ($-23.5‰$), sewage ($-24.8‰$), soil OM ($-27.5‰$), terrestrial C3 ($-27.0‰$) and terrestrial C4 ($-12.5‰$). Similarly, the C/N ratios of plankton, macrophytes, sewage, soil OM, terrestrial C3 and terrestrial C4 were 6.5, 20.0, 9.8, 11.5, 27.5 and 27.5, respectively [25]. Meanwhile, the δ^{15}N values of plankton ($6.5‰$), macrophytes ($2.5‰$), sewage ($16.0‰$), soil OM ($3.5‰$), terrestrial C3 ($-0.5‰$) and terrestrial C4 ($-0.5‰$) were set as the source information into the IsoSource database [25,26]. For the multiple linear regression analysis in this study, an increase in the models was set as 1%, and the output data calculated were all reliable under the tolerance index of 3 ‰. The results shown in Figure 3 indicate that the contribution of plankton in sedimentary OM was 25.4–68.8% (average: 48.5%), 1.5–16.7% (9.7%) for macrophytes, 2.3–25.5% (15.7%) for sewage, 0.6–18.8% (6.7%) for soil OM, 8.5–40.7% (16.6%) for terrestrial C3 plant and 0.1–7.7% (2.8%) for terrestrial C4 plant. For the N sources calculation model, δ^{15}N mean values of agricultural fertilizer, domestic sewage, exogenous release, soil erosion, and terrestrial input were $0.0‰$, $15.0‰$, $6.5‰$, $5.5‰$ and $2.0‰$, respectively. The calculated mean contribution of agricultural fertilizer, domestic sewage, exogenous release, soil erosion, terrestrial input accounted for 28.4% (range: 19.3–44.8%), 7.5% (3.9–11.6%), 16.8% (9.6–22.7%), 19.3% (11.5–24.6%), and 28.0% (21.9–31.1%), respectively.

3.4. Exogenous Pollution due to N and P

The PC results showed that the correlation of TN (coefficient of correlation R = 0.70, $p < 0.001$) and TP (R = 0.99, $p < 0.001$) between surface water and bottom water samples in Lake Jian were significant (Figure 4A), indicating that the exchange and mixing of the upper and lower water body were sufficient. Considering a strong disturbance, the stable condition of solid nutrient elements in the surface sediment was destroyed, significantly exacerbating the migration and release of nutrient elements, especially in the shallow Lake Jian [29]. Higher dissolved nutrient element contents in the bottom water were found in Lake Jian, which was 1.2–1.4 times that in surface water (Figure 4C,D).

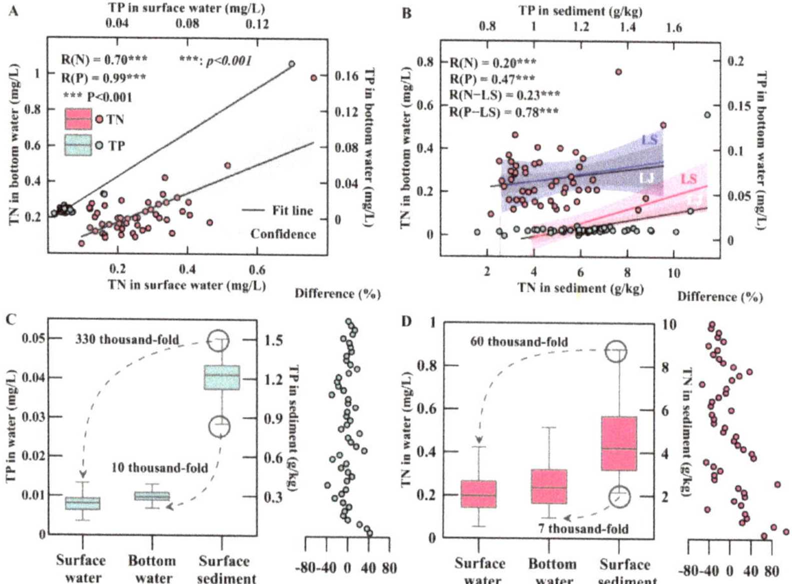

Figure 4. Relationship of TN and TP in surface water/bottom water (**A**) and bottom water/surface sediment (**B**), the contents levels of TP (**C**) and TN (**D**) in Lake Jian, with difference (%) in nutrient concentration between median value and site-specific data in surface sediment (**C,D**).

Except for the influence of deposition and gravity, vertical divergence in dissolved nutrient elements could be affected by the migration process near the sediment boundary by a series of bio-geochemical and geophysical processes [30,31]. The nutrient elements in the surface sediment were weakly correlated with TN (R = 0.20) and TP (R = 0.47) in the bottom water. All these results indicated that an increase in the concentrations of N and P in bottom sediments led to only a slight increase in the concentrations of these elements in waters and a slow release and migration from sediments into the lake water [32]. Recently, endogenous pollution of nutrient elements through release and migration has become one of the most important problems for pollution remediation and protection [33,34]. Previous studies have shown that exogenous pollution in lake systems was serious and difficult to control. In major eutrophic lakes worldwide, the endogenous load of nutrient elements exceeded 20–40% [35]. Noticeably, the TN (3.83–5.87 g/kg) and TP (1.16–1.31 g/kg) content in the surface sediment of Lake Jian were extremely high compared with other eutrophic lakes in China [36,37], indicating that endogenous pollution was likely to occur. The distribution difference between dissolved nutrient elements in water and sedimentary solid nutrient elements revealed that the release/migration process occurred at the sediment interface caused by endogenous pollution.

However, we found that the release intensity of N and P was remarkably different, and the fluctuation degree of TN (4.85 ± 1.02 g/kg, with a maximum distance of 21.03%) content was much higher than that of TP (1.24 ± 0.07, 6.07%) in surface sediment (Figure 4C,D). Under the same disturbed conditions and with similar input backgrounds, the results suggest that the deposition process of TP was more stable in Lake Jian. Furthermore, compared with nutrient element fluxes in the whole lake, the relative level of P content (10.10–326.76 × 10^3 times the amount in the water body) in surface sediment was exceptionally higher than that of N (7.72~59.17 × 10^3 times). Moreover, TP contents in the surface sediment and bottom water were more positively correlated (R = 0.44) compared with TN (0.24), especially in lakeside areas (P: 0.78, N: 0.23), indicating that the deposition process was relatively stable and there was no obvious endogenous release process (Figure 4B). On the other hand, the ANOVA result of N showed that the deposition state of solid N and dissolved N in lake water were two distinct processes ($p < 0.001$), and the mean relative level of N in surface sediment was nearly six times lower than that of P, indicating that the release of N could be more pronounced.

3.5. Endogenous Load and Different Forms of Nutrient Elements Pollutants

The pattern of endogenous pollution in a lake ecosystem is modulated by the distribution of distinct forms of nutrients and by environmental background limitations. Generally, under natural conditions, the different forms of N can be divided into organic nitrogen and inorganic nitrogen. Inorganic nitrogen mainly includes nitrate nitrogen, ammonia nitrogen, and nitrite nitrogen, introduced by agricultural activities, domestic sewage, industrial smelting, and other anthropogenic processes [38]. Additionally, through the absorption and transformation processes of microorganisms and aquatic animals and plants, organic nitrogen is abundant in the ecosystem, and could be in the form of biological debris and biological chain [39,40]. Both the degradation of organic nitrogen and the process of digestion/denitrification that involves inorganic nitrogen could lead to the migration and release of N in sediments [41]. However, under natural conditions, solid inorganic P in sediments produced by human activities were found to be mainly composed of ferric hydroxides, iron-manganese oxidizing material, and the AL–OM complex, which could reduce and decompose only in an anaerobic environment [42,43]. In comparison, the endogenous pollution of P was more limited than that of N, and irrespective of the effects of pH, temperature, visibility, dissolved oxygen, and other hydro-chemical environmental factors, or physical factors, such as biological disturbance intensity and wind/wave disturbance, it had more severe impacts on N release [43,44].

Nutrient elements and grain size could be influenced by additional factors, including provenance, degree of weathering, diagenesis, and biogenic production, requiring site-specific empirical data [45]. The main streams and their tributaries drain different catchments composed of variable bedrock geology (Figure 5C). Most river courses are predominantly of the Carboniferous system (gravelly coarse-grained and quartz sandstone), with minor basic metamorphics [21]. The upper reach of the Jinglong River was influenced by many tributaries draining Pleistocene, Jurassic, and Triassic sandstone and shales. The upper courses of the Huanglong and Huilong Rivers predominantly drain from Neogene and Quaternary formations. Overall, similar provenances in near-source areas did not influence the composition of minerals and phyllosilicates, which are typically the main carriers of trace and inorganic nutrient elements [21]. Areas of different provenance in our dataset did, nonetheless, display two opposing trends of N/FG and P/FG ratios, although we observed minor excursions from the general trend for FG, excluding the influence of samples from lake inlets and high-pollution outliers (Figure 5A,B). This result suggests that N concentrations in the surface sediment showed a relatively low sensitivity to provenance changes. Moreover, the occurrence form, combined with fine-grained minerals, was not the main speciation of sedimentary inorganic nitrogen contaminants. In contrast, high positive correlations of TP and FG in the surface sediment are observed in Lake Jian, indicating that the occurrence form combined with fine-grained minerals was the main speciation of P, i.e., ferric hydroxides, iron–manganese oxidizing material inorganic fine-grained complexes [43].

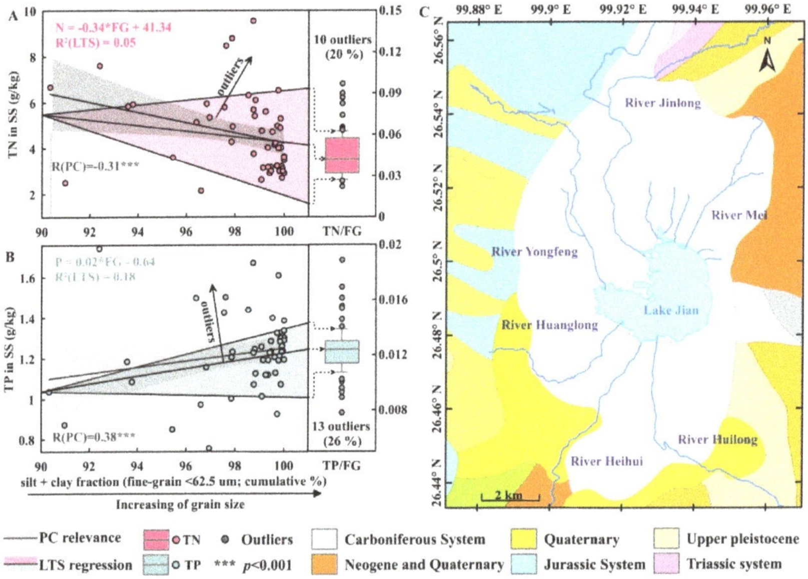

Figure 5. Bivariate plots of N (**A**) and P (**B**) vs. fine grain (FG), as well as Tukey's boxplots showing the geometrical form of absorbed mineral particle size based on the LTS robust regression, and geological stratigraphic map of the catchment of Lake Jian (**C**).

The results of biogenic element distribution showed that relative OM and TN levels in the surface sediment of Lake Jian were significantly higher than those of other lakes in Yunnan Province [25]. In addition to nutrient element levels, other factors, such as vegetation coverage and a well-developed river system, also increased TOC contents. Specifically, OM and TN contents in Lake Jian (R = 0.95) exhibited a significant linear correlation (Figure 6D), suggesting that more organic nitrogen was stored in Lake Jian. The situation was different with the PC results of TP and OM (0.36), which revealed that inorganic P was the main occurrence form of P in Lake Jian, which was then verified by traceability assessment. The OM in >75% of the sediment samples derived mainly from endogenous processes (plankton: 25.4–68.8%, averaging 52.53%; macrophytes: 3.7–16.6%, 8.99%), and only samples from entering river estuaries were related to domestic sewage (2.3–25.5% 17.47%), agricultural irrigation (soils OM: 0.6–14.4%, 7.37%), and crop planting (terrestrial C3: 8.5–40.7%, 18.35%; terrestrial C4: 0.1–6.0%, 3.1%). The source estimation of N showed that more than 16.8% of the sediment samples were derived from the sourcing process, which was remarkable for a lake with a low eutrophication level. Furthermore, we found that the level and distribution of N in surface sediment were less related to its material source (R$-$N/δ^{15}N = 0.23) and occurrence form (R$-$N/δ^{13}C = 0.28). ANOVA demonstrated that the sedimentary state of N was an independent process ($p < 0.001$). All this evidence indicates that the processes of N migration and release were determined by the current N level of the lake.

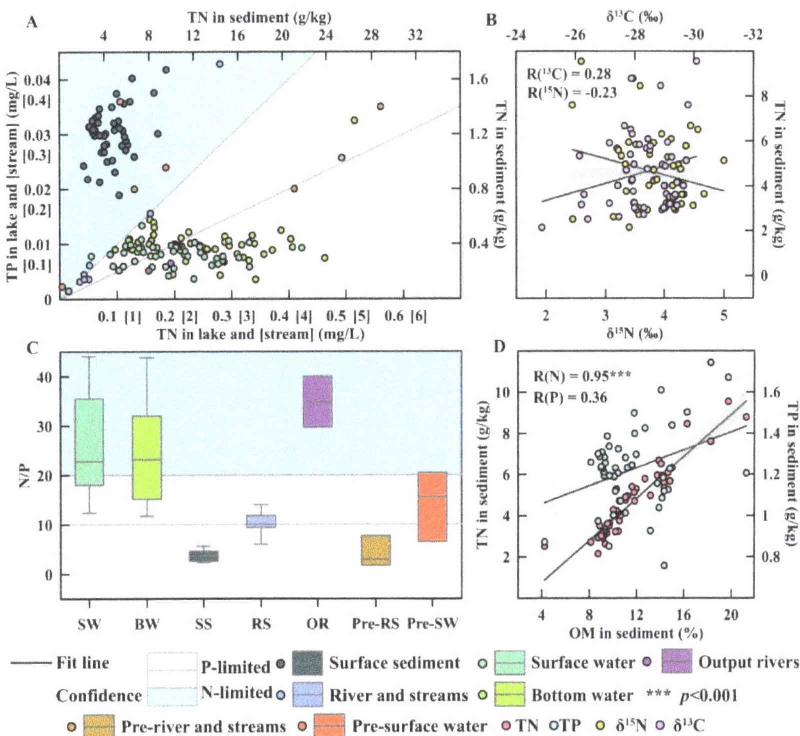

Figure 6. Ratio of TN and TP (**A**), correlation between TN and $\delta^{15}N/\delta^{13}C$ (**B**), TN/TP range (**C**), and correlation between nutrient elements in surface sediment and bottom water (**D**) in Lake Jian.

3.6. Transformation of N/P-Limitation Driven by Endogenous Pollution

Intriguingly, we found that the algal growth pattern of the Lake Jian water body was limited mainly by P (N/P ratios ranged from 5.7 to 64.2 times, averaging 26.5 times), while the nutrient elements levels in the sediment belonged entirely to the N-limitation state (2.1–7.3 times, 3.8 times). Generally, the strong exogenous pollution of N in the sediment, caused by its multiple unstable migrated forms and pathways, led to the enrichment of P [46,47]. These processes result in more dissolved N released into water, changing nutrient limitation. The hydrological results and lake monitoring records also support this conclusion (Figure 6A). There was also a great difference between the inflow of rivers and streams (5.8–13.2 times, 10.1 times) and the output river (29.5–45.3 times), meaning that whatever the river input via irrigation agriculture or city sewage, P was the dominating nutrient element and eutrophication was limited mainly by N. However, after the water entered Lake Jian, the exogenous pollution and dynamic processes changed the relative levels and structure of N/P, altering nutrient element limitation. Compared with previous monitoring data, the status of nutrient elements in Lake Jian has gradually changed from N-limitation (2000–2010) to P-limitation (2010–2018) in recent years [15,28,48]. This phenomenon was affected by the differences in cyclic processes and occurrences between N and P [49]. Due to the bio-accumulation of algae and aquatic plants, a large amount of external N in the lake is bounded by OM, enriching the sediments with biological death, completing the process of re-concentration of N in the lacustrine system [50]. Inevitably, combined with OM degradation and improved lake productivity, these processes re-intensify endogenous pollution and the enrichment condition of N, ultimately causing the transformation of nutrient element limitation.

We hypothesized that except for the influence of exogenous pollution, the primary cause of nutrient elements limiting their transformation in Lake Jian was their geophysical structure and environmental conditions in the catchment area. As a shallow lake, oxygen in the water-sediment interface in Lake Jian could be sufficiently supplied by air–water exchange and dynamic disturbance, alleviating anaerobic conditions by OM degradation [3]. Normally, the unstable states of P from anthropogenic activities that enter sediments by dynamic processes are principally excited as Al-Fe binding states, Ca-OM binding states, and residual P complexes [3,30]. Among them, hydrolysis and reduction of the Al-Fe-P complex, including ferric hydroxide and phosphoric iron, are the pathways for inorganic P release [46,49]. Under aerobic conditions, these materials are stable and immobile, inhibiting the migration and release of P.

A series of previous studies indicated that the external loss process of N (denitrification process) was strictly limited by oxygen content, and the loss process of P (dynamic deposition) was weakened, resulting in the formation of an N limitation state in shallow lakes, due to stronger water disturbance [3,37,49]. However, we observed dramatic OM levels and poor DO states at some sites, which were closely related to swamp formation processes at Lake Jian. In recent years, due to the impact of land reclamation and climate change, the lake area has decreased by more than 40% [14], resulting in most lakeside areas being exposed directly to the impact of the surrounding runoff. The original lakeside ecosystem has gradually evolved into a swampy area with dense vegetation. Despite high vegetation coverage having significantly alleviated the nutrient elements imported from external sources, it also played an important role in improving the transparency and oxygen content of the water body, limiting the occurrence of endogenous pollution processes, especially P release, which depend upon an anaerobic environment. However, due to the lack of exogenous pollution controls in early rising times, the huge algae masses produced by eutrophication and aquatic plants adsorbed innumerable nutrient elements in the lake sediments. Through the biological enrichment process, pollutants are enriched and transferred, especially the production of organic nitrogen, which could complete the migration and release process through various pathways, resulting in dissolved nitrogen frequently existing in the lake. Under an aerobic environment, the obstruction of gasification and efflux processes, such as denitrification, were also aggravated, finally achieving a P-limited state.

4. Conclusions

Based on the influence of environmental conditions, we observed several differences in the observed release/migration intensity of N and P in the sediments of Lake Jian, which resulted in the transformation of nutrients as limiting elements. We call this phenomenon the "pump diaphragm effect," which is the self-enrichment process of N, leading to the swamping of shallow plateau lakes. More remarkably, the water exchange cycle period was relatively long because of the closed environment and structural characteristics, which rendered the ecosystem of plateau lakes fragile and water self-restoration processes slow. Once the environmental background begins to fluctuate, the ecological and environmental health functions of plateau lakes could be irreversibly damaged. Our research results provide a sound theoretical basis and supporting data for improving the treatment of shallow, swamped plateau lakes. While decreasing/eliminating endogenous pollution, attention should also be focused on nutrient element circulation processes in the lake itself. More importantly, strategies that can control the nitrogen and phosphorus levels should be more carefully formulated due to complex and variable lake conditions; thus, the best strategy may be to implement the "one lake with one governance" strategy to fully understand the special characteristics of lakes and establish protocols for long-term eutrophication detection, assessment and management.

Author Contributions: Conceptualization, Y.Z. and H.Z; Formal analysis, D.L.; Funding acquisition, F.C. and H.Z.; Methodology, Y.Z.; Project administration, X.Z.; Resources, H.Z.; Software, Q.L. and F.L.; Supervision, F.C. and H.Z.; Writing—original draft, Y.Z.; Writing—review and editing, H.Z. All authors have read and agreed to the published version of the manuscript.

Funding: This research was funded by the Yunnan Provincial Government Scientist workshop and the Special Project for Social Development of Yunnan Province (Grant No. 202103AC100001, Funder: Hucai Zhang).

Institutional Review Board Statement: Not applicable.

Informed Consent Statement: Not applicable.

Data Availability Statement: Not applicable.

Acknowledgments: Special thanks are given to the team of the Key Laboratory of Plateau Lake Ecology and Global Change and the Institute for Ecological Research and Pollution Control of Plateau Lakes, who provided support with the sample collection and analysis.

Conflicts of Interest: The authors declare no conflict of interest.

References

1. Reitzel, K.; Ahlgren, J.; Gogoll, A.; Rydin, E. Effects of aluminum treatment on phosphorus, carbon, and nitrogen distribution in lake sediment: A 31P NMR study. *Water Res.* **2006**, *40*, 647–654. [CrossRef] [PubMed]
2. Hai, X.; Paerl, H.W.; Qin, B.; Zhu, G.; Gao, G. Nitrogen and Phosphorus Inputs Control Phytoplankton Growth in Eutrophic Lake Taihu, China. *Limnol. Oceanogr.* **2010**, *55*, 420–432. [CrossRef]
3. Qin, B.; Zhou, J.; Elser, J.J.; Gardner, W.S.; Deng, J.; Brookes, J.D. Water Depth Underpins the Relative Roles and Fates of Nitrogen and Phosphorus in Lakes. *Environ. Sci. Technol.* **2020**, *54*, 3191–3198. [CrossRef] [PubMed]
4. Ma, J.; Qin, B.; Wu, P.; Zhou, J.; Niu, C.; Deng, J.; Niu, H. Controlling cyanobacterial blooms by managing nutrient ratio and limitation in a large hyper-eutrophic lake: Lake Taihu, China. *J. Environ. Sci.* **2015**, *27*, 80–86. [CrossRef] [PubMed]
5. Qin, B.Q. The nutrient forms, cycling and exchange flux in the sediment and overlying water system in lakes from the middle and lower reaches of Yangtze River. *Sci. China Ser. D Earth Sci.* **2006**, *49*, 1–13. [CrossRef]
6. Zhan, Q.; Stratmann, C.N.; Geest, H.G.V.D.; Veraart, A.J.; Brenzinger, K.; Lürling, M.; Domis, L.N.D.S. Effectiveness of phosphorus control under extreme heatwaves: Implications for sediment nutrient releases and greenhouse gas emissions. *Biogeochemistry* **2021**, *156*, 421–436. (In English) [CrossRef]
7. Chen, X.; Wang, Y.; Sun, T.; Huang, Y.; Chen, Y.; Zhang, M.; Ye, C. Effects of Sediment Dredging on Nutrient Release and Eutrophication in the Gate-Controlled Estuary of Northern Taihu Lake. *J. Chem.* **2021**, *2021*, 1–13. (In English) [CrossRef]
8. Lee, J.K.; Oh, J.M. A Study on the Characteristics of Organic Matter and Nutrients Released from Sediments into Agricultural Reservoirs. *Water* **2018**, *10*, 980. (In English) [CrossRef]
9. Grygar, T.M.; Nováková, T.; Bábek, O.; Elznicová, J.; Vadinová, N. Robust assessment of moderate heavy metal contamination levels in floodplain sediments: A case study on the Jizera River, Czech Republic. *Sci. Total Environ.* **2013**, *452*, 233–245. [CrossRef]
10. Grygar, T.M.; Elznicová, J.; Tůmová, Š.; Faměra, M.; Balogh, M.; Kiss, T. Floodplain architecture of an actively meandering river (the Ploučnice River, the Czech Republic) as revealed by the distribution ofpollution and electrical resistivity tomography. *Geomorphology* **2016**, *254*, 41–56. [CrossRef]
11. Duan, L.; Zhang, H.; Chang, F.; Li, D.; Liu, Q.; Zhang, X.; Liu, F.; Zhang, Y. Isotopic constraints on sources of organic matter in surface sediments from two north–south oriented lakes of the Yunnan Plateau, Southwest China. *J. Soils Sediments* **2022**, *22*, 1597–1608. [CrossRef]
12. Yuan, H.Z.; Shen, J.; Liu, E.F.; Wang, J.J.; Meng, X.H. Assessment of nutrients and heavy metals enrichment in surface sediments from Taihu Lake, a eutrophic shallow lake in China. *Environ. Geochem. Health* **2011**, *33*, 67–81. [CrossRef]
13. Chen, G.Z.; Jin, J.J.; Zhang, F.F.; Qiu, Y.P.; Li, Z.H. Fish diversity change and fauna evolution in Jianhu Lake, Yunnan, China. *Chin. J. Ecol.* **2018**, *37*, 3691–3700. (In Chinese) [CrossRef]
14. Zhang, Q.; Peng, E.R.; Zhao, Q.Q.; Shi, J. Research Progress of Jianhu Wetland in Northwest Yunnan. *Environ. Sci. Surv.* **2019**, *38*, 30–35. (In Chinese) [CrossRef]
15. Guo, Y.; Wang, Y.; Zheng, Y.; Liu, Y.; Wen, G.; Zhan, P. Temporal and spatial evolution of landscapes in Jianhu Lake Basin of northwestern Yunnan Province. *J. Zhejiang A F Univ.* **2018**, *35*, 695–704. (In Chinese) [CrossRef]
16. Liu, S.; Chen, G.J.; Wang, J.Y.; Deng, Y. Diatom community response to long-term enviromental changes and ecological evaluation in Jianhu Lake, northwest Yunnan. *Quat. Sci.* **2018**, *38*, 939–952. (In Chinese)
17. Guo, Y.; Zheng, Y.; Wang, Y.; Liu, Y. Evolution of Jianhu Lake and its eco-environmental effects in the northwestern Yunnan Province. *Environ. Eng.* **2017**, *4*, 105–112. (In Chinese) [CrossRef]
18. Yang, R.; Wu, D.; Li, Z.; Yuan, Z.; Niu, L.; Zhang, H.; Chen, J.; Zhou, A. Holocene-Anthropocene transition in northwestern Yunnan revealed by records of soil erosion and trace metal pollution from the sediments of Lake Jian, southwestern China. *J. Paleolimnol.* **2021**, *68*, 91–102. (In English) [CrossRef]

19. Li, B.; Wang, H.; Yu, Q.G.; Wei, F.; Zhang, Q. Ecological Assessment of Heavy Metals in Sediments from Jianhu Lake in Yunnan Province, China. *Pol. J. Environ. Stud.* **2020**, *29*, 4139–4150. (In English) [CrossRef]
20. Ouyang, M.; Yu, Q.; Zhao, X.; Li, J.; Li, B.; Wang, J.; Zhang, Y. The distribution characteristics and the risk assessment of organochlorine pesticides in the sediments of Jianhu Lake. *Acta Sci. Circumstantiae* **2019**, *39*, 4075–4087. (In Chinese) [CrossRef]
21. Bábek, O.; Grygar, T.M.; Faměra, M.; Hron, K.; Nováková, T.; Sedláček, J. Geochemical background in polluted river sediments: How to separate the effects of sediment provenance and grain size with statistical rigour? *Catena* **2015**, *135*, 240–253. [CrossRef]
22. Reimann, C.; Garrett, R.G. Geochemical background—Concept and reality. *Sci. Total Environ.* **2005**, *350*, 12–27. [CrossRef] [PubMed]
23. Redfield, A.C. The biological control of chemical factors in the environment. *Am. Sci.* **1958**, *46*, 205–221. [CrossRef]
24. Gregg, P.J.W. Source partitioning using stable isotopes: Coping with too many sources. *Oecologia* **2003**, *136*, 261–269. [CrossRef]
25. Wu, H.; Zhang, H.; Chang, F.; Duan, L.; Zhang, X.; Peng, W.; Liu, Q.; Zhang, Y.; Liu, F. Isotopic constraints on sources of organic matter and environmental change in Lake Yangzong, Southwest China. *J. Asian Earth Sci.* **2021**, *217*, 104845–104856. [CrossRef]
26. He, Z.; Cai, J.; Ni, Z.; Huang, Y.; Zhao, J.; Wang, S. Seasonal characteristics of nitrogen sources from different ways and its contribution to water nitrogen in Lake Erhai. *Acta Entiae Circumstantiae* **2018**, *38*, 1939–1948. (In Chinese) [CrossRef]
27. Hammer, Ø.; Harper, D.; Ryan, P.D. Past: Paleontological statistics software package for education and data analysis. *Palaeontol. Electron.* **2001**, *4*, 1–9. [CrossRef]
28. Guo, Y.; Wang, Y.; Liu, Y.G. The Dynamic Changes and Simulating Prediction of Land Use in Jianhu Lake Basin from 1990 to 2015. *J. Southwest For. Univ. Nat. Sci.* **2016**, *6*, 87–93. (In Chinese) [CrossRef]
29. Sun, X.; Zhu, G.; Luo, L.C. Experimental study on phosphorus release from sediments of shallow lake in wave flume. *Sci. China Ser.* **2006**, *49*, 92–101. [CrossRef]
30. Rudneva, I.I.; Zalevskaya, I.N.; Shaida, V.G.; Memetlaeva, G.N.; Scherba, A.V. Biogenic Migration of Nitrogen and Phosphorus in Crimean Hypersaline Lakes: A Seasonal Aspect. *Geochem. Int.* **2020**, *58*, 1123–1134. [CrossRef]
31. Long, E.R.; Macdonald, D.D.; Smith, S.L.; Calder, F.D. Incidence of adverse biological effects within ranges of chemical concentrations in marine and estuarine sediments. *Environ. Manag.* **1995**, *19*, 81–97. [CrossRef]
32. Xu, C.R. Research on Nitrogen and Phosphorus Release from Sediments in Small Inland Freshwater Lakes. *Adv. Mater. Res.* **2013**, *864*, 248–255. [CrossRef]
33. Wang, Z.; Bao, L.I.; Liang, R. Comparative study on endogenous release of nitrogen and phosphorus in Nansi Lake, China. *Acta Sci. Circumstantiae* **2013**, *33*, 487–493. (In Chinese). Available online: https://www.actasc.cn/hjkxxb/ch/reader/view_abstract.aspx?file_no=20120602001&flag=1 (accessed on 14 March 2022).
34. Jiang, Y.S.; Xiao-Chen, L.I.; Xing, Y.H.; Jiang, R.X. Impacts of Disturbance on Release of Total Nitrogen and Total Phosphorus from Surficial Sediments of Dongping Lake. *Environ. Sci. Technol.* **2010**, *33*, 41–44. [CrossRef]
35. Bennett, E.M.; Carpenter, S.R.; Caraco, N.F. Human Impact on Erodable Phosphorus and Eutrophication: A Global Perspective. *Bioscience* **2001**, *51*, 227–234. [CrossRef]
36. GaChter, R.; Wehrli, B. Ten Years of Artificial Mixing and Oxygenation: No Effect on the Internal Phosphorus Loading of Two Eutrophic Lakes. *Environ. Sci. Technol.* **1998**, *32*, 3659–3665. [CrossRef]
37. Wu, F.; Zhan, J.Y.; Deng, X.Z.; Lin, Y.Z. Influencing factors of lake eutr0phicati0n in China A case study in 22 lakes in China. *Ecol. Environ. Sci.* **2012**, *21*, 94–100. (In Chinese) [CrossRef]
38. Paerl, H.W. Assessing and managing nutrient-enhanced eutrophication in estuarine and coastal waters: Interactive effects of human and climatic perturbations. *Ecol. Eng.* **2006**, *26*, 40–54. [CrossRef]
39. Heike, K.; Patrick, G.H. Sequestration of organic nitrogen in the sapropel from Mangrove Lake, Bermuda. *Org. Geochem.* **2001**, *32*, 733–744. [CrossRef]
40. Hadas, O.; Altabet, M.A.; Agnihotri, R. Seasonally varying nitrogen isotope biogeochemistry of particulate organic matter in Lake Kinneret, Israel. *Limnol. Oceanogr.* **2009**, *54*, 75–85. [CrossRef]
41. Zhu, H.; Yan, B.; Pan, X.; Yang, Y.; Wang, L. Geochemical characteristics of heavy metals in riparian sediment pore water of Songhua River, Northeast China. *Chin. Geogr. Sci.* **2011**, *21*, 195–203. [CrossRef]
42. Kleeberg, A.; Kozerski, H.P. Phosphorus release in Lake Großer Müggelsee and its implications for lake restoration. *Springer Neth.* **2021**, *119*, 9–26. [CrossRef]
43. Sun, X.J.; Qin, B.Q.; Zhu, G.W.; Zhang, Z.P.; Gao, Y.X. Release of Colloidal N and P from Sediment of Lake Caused by Continuing Hydrodynamic Disturbance. *Environ. Sci.* **2007**, *28*, 1223–1232. (In Chinese) [CrossRef]
44. Wenchao, L.I.; Yin, C.; Chen, K.; Qinglong, W.U.; Pan, J. Discussion on Phosphorous Release from Lake Sediment. *J. Lakeence* **1999**, *11*, 296–303. (In Chinese) [CrossRef]
45. Bloemsma, M.R.; Zabel, M.; Stuut, J.; Tjallingii, R.; Collins, J.A.; Weltje, G.J. Modelling the joint variability of grain size and chemical composition in sediments. *Sediment. Geol.* **2012**, *280*, 135–148. [CrossRef]
46. Correll, D.L. The Role of Phosphorus in the Eutrophication of Receiving Waters: A Review. *J. Environ. Qual.* **1998**, *27*, 261–266. [CrossRef]
47. Bogdanović, D. The role of phosphorus in eutrophication. *Zb. Matice Srp. Prir. Nauk.* **2006**, *111*, 75–86. [CrossRef]
48. Liu, S.; Chen, G.J.; Wang, J.Y. Long-term response model and ecological assessment of diatom community in Jianhu Lake. *Quat. Sci.* **2018**, *38*, 939–952. (In Chinese)

49. Pettersson, K. Mechanisms for internal loading of phosphorus in lakes. *Hydrobiologia* **1998**, *373*, 21–25. [CrossRef]
50. Jones, D.L.; Shannon, D.; Murphy, D.V.; Farrar, J. Role of dissolved organic nitrogen (DON) in soil N cycling in grassland soils. *Soil Biol. Biochem.* **2004**, *36*, 749–756. [CrossRef]

Article

Nutrient Thresholds Required to Control Eutrophication: Does It Work for Natural Alkaline Lakes?

Jing Qi [1], Le Deng [1], Yongjun Song [1], Weixiao Qi [2,*] and Chengzhi Hu [1,3]

1   Key Laboratory of Drinking Water Science and Technology, Research Center for Eco-Environmental Sciences, Chinese Academy of Sciences, Beijing 100085, China
2   State Key Joint Laboratory of Environmental Simulation and Pollution Control, School of Environment, Tsinghua University, Beijing 100084, China
3   University of Chinese Academy of Sciences, Beijing 100049, China
*   Correspondence: wxqi@mail.tsinghua.edu.cn

Abstract: The responses of phytoplankton to nutrients vary for different natural bodies of water, which can finally affect the occurrence of phytoplankton bloom. However, the effect of high alkalinity characteristic on the nutrient thresholds of natural alkaline lake is rarely considered. Bioassay experiments were conducted to investigate the nutrient thresholds and the responses of phytoplankton growth to nutrients for the closed plateau Chenghai Lake, Southwest China, which has a high pH background of up to 9.66. The growth of the phytoplankton community was restricted by phosphorus without obvious correlation with the input of nitrogen sources. This can be explained by the nitrogen fixation function of cyanobacteria, which can meet their growth needs for nitrogen. In addition, nitrate nitrogen (NO_3-N) could be utilized more efficiently than ammonia nitrogen (NH_4-N) for the phytoplankton in Chenghai Lake. Interestingly, the eutrophication thresholds of soluble reactive phosphorus (SRP), NH_4-N, and NO_3-N should be targeted at below 0.05 mg/L, 0.30 mg/L, and 0.50 mg/L, respectively, which are higher than the usual standards for eutrophication. This can be explained by the inhibition effect of the high pH background on phytoplankton growth due to the damage to phytoplankton cells. Therefore, the prevention of phytoplankton blooms should be considered from not only the aspect of reducing nutrient input, especially phosphorus input, but also maintaining the high alkalinity characteristic in natural alkaline lake, which was formed due to the geological background of saline-alkali soil.

Keywords: nutrient threshold; alkaline lake; pH; phytoplankton blooms

Citation: Qi, J.; Deng, L.; Song, Y.; Qi, W.; Hu, C. Nutrient Thresholds Required to Control Eutrophication: Does It Work for Natural Alkaline Lakes? *Water* **2022**, *14*, 2674. https://doi.org/10.3390/w14172674

Academic Editors: Hucai Zhang and Jingan Chen

Received: 21 July 2022
Accepted: 26 August 2022
Published: 29 August 2022

Publisher's Note: MDPI stays neutral with regard to jurisdictional claims in published maps and institutional affiliations.

1. Introduction

Harmful phytoplankton blooms in natural water have aroused great concern due to their negative effects on water quality and aquatic ecosystems globally [1,2]. This poses a serious threat to the safety of drinking water, food webs, and the overall sustainability of freshwater ecosystems [3,4]. It has been reported that megafauna may be endangered by cyanotoxins released by harmful phytoplankton [5]. In addition, the expansion of phytoplankton blooms could be triggered by climate change and eutrophication [6]. It is widely believed that the reduction of nutrient input is fundamental for the control of harmful phytoplankton blooms. However, the responses of phytoplankton to nutrients vary in different natural bodies of water [3,7]. Research indicates that the thresholds for regime shifts between turbid-water and clear-water conditions in shallow lakes vary depending on basins and climates [8]. Therefore, it is important to determine the nutrient thresholds in water bodies, especially for natural water with special water quality backgrounds.

Nutrient thresholds are regarded as the critical levels of nutrients that control population shifts, such as the sudden and long-term dominance of phytoplankton blooms. The determination of nutrient thresholds is a quantifiable and meaningful approach [9]. N and P are the main material bases for phytoplankton growth, and their relationship with phytoplankton biomass is one of the important aspects in studying eutrophication [10]. It is well known that addition of N and P can not only stimulate phytoplankton growth, but also promote phosphorus release from sediment or nitrogen fixation from atmosphere [11,12]. The critical values of total nitrogen (TN) and total phosphorus (TP) for eutrophication of lakes were reported to be 0.8 mg/L and 0.05 mg/L, respectively [13]. It has been proved that the understanding and utilization of ecological thresholds is the key to the successful management of water environments [14]. The control of nutrients below threshold levels is more practical, achievable, and cost-effective than reducing them to historical levels [15].

Due to the great differences in the environment and geographical location at different altitude, plateau lakes have unique hydrological chemistry characteristics, such as high alkalinity [16]. Chenghai Lake is a typical representative of plateau lakes in southwest China with high alkalinity characteristic, which was naturally formed due to the geological background of saline-alkali soil [17,18]. Increasing salinization due to climate change might negatively affect inland water sources [19]. According to the water quality conditions of Chenghai Lake in 2018–2019, the average pH value is up to 9.42. In addition, the average values of TN and TP are 1.12 and 0.06 mg/L, respectively, which have exceeded the generally recognized concentrations for eutrophication occurrence [10,13,20–22]. However, the current cyanobacteria biomass in Chenghai Lake is still at a slight bloom level without large-scale phytoplankton blooms [18]. Therefore, the nutrient thresholds of natural alkaline lake might be higher than other lakes, which needs to be clarified. It has been reported that pH range has some inhibition or promotion effects on phytoplankton in various environmental backgrounds [23,24]. However, the effect of high alkalinity characteristic on the nutrient thresholds of natural alkaline lake is rarely considered. Therefore, it is important to determine the nutrient thresholds and explore the effect of high alkalinity, which are fundamental for the prevention of phytoplankton blooms in natural alkaline lakes.

Based on the aforementioned considerations, this study aims to: (1) determine the nutrient thresholds of NH_4-N, NO_3-N, and SRP in Chenghai Lake; (2) investigate the binding and synergistic interactions between NH_4-N, NO_3-N, and SRP; (3) explore the effect of high alkalinity characteristic on nutrient thresholds of natural alkaline lake.

2. Materials and Methods

2.1. Study Area and Field Method

Chenghai Lake is located in the Yunnan Plateau, southwestern China (26°27′–26°38′ N; 100°38′–100°41′ E) with an altitude of 1503 m (Figure 1). The lake covers an area of about 72.9 square kilometers and has an average water depth of 23.7 m, and an annual average water temperature of 17.8 °C. Chenghai is a typical closed-type deep-water lake, surrounded by mountains on the east, west and north, and the terrain is flat on the south [25].

Monthly sampling was conducted from June 2018 to May 2019 for all 15 sampling sites (Figure 1). The in-situ data of conductivity (EC), pH, and temperature (WT) were measured on site with a hand-held multi-parameter meter (MYRON L 6P, California, USA). TN, NH4-N, NO3-N, TP, and total dissolved phosphorus (TDP) were analyzed according to standard methods [25]. The phytoplankton samples were settled for 48 h after being fixed with Lugol iodine solution (2%) [26]. Cell density and community composition were determined under microscope with a Sedgwick-Rafter counting chamber [27].

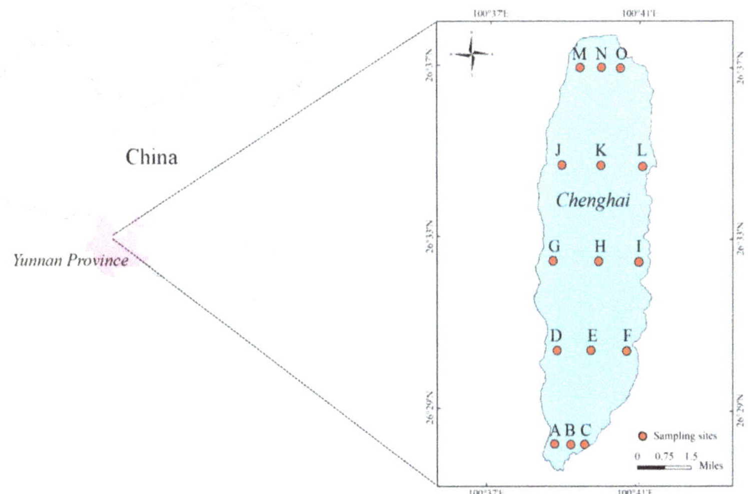

Figure 1. Location of the study area and sampling sites.

2.2. Nutrient Limitation Bioassay Experiments

Water samples from site H were used for nutrient threshold and addition bioassay experiments. Zooplankton were removed by screening with a 200-μm grid to minimize the effects of grazing [28]. The water samples were cultivated in a lighted incubator within 2 °C of the in situ temperature. The light intensity was maintained at 100 μmol photon m^{-2} s^{-1} with a 14:10-h light-dark cycle. The initial physical, chemical and biological properties of the water sample used for nutrient limitation bioassay experiments are given in Table S1. For all the experiments, SRP, NO$_3$-N, and NH$_4$-N were added as K$_2$HPO$_4$·3H$_2$O, NaNO$_3$, and NH$_4$Cl, respectively. The experiment was carried out over 15 days to ensure that the phytoplankton had sufficient time to adapt and grow.

The nutrient threshold bioassay experiment was conducted to explore the nutrient thresholds of SRP, NO$_3$-N, and NH$_4$-N. Various concentrations of SRP (0.017, 0.02, 0.03, 0.04, 0.05, 0.06, 0.08, 0.10, 0.50, 1.00 mg/L P) and fixed NO$_3$-N (10 mg/L N) were used in the SRP threshold experiment. Various concentrations of NO$_3$-N (0.04, 0.10, 0.20, 0.30, 0.40, 0.50, 0.80, 1.00, 1.50, 2.00 mg/L N) and fixed P (5 mg/L N) were used in the NO$_3$-N threshold experiment. Various concentrations of NH$_4$-N (0.04, 0.10, 0.20, 0.30, 0.40, 0.50, 0.80, 1.00, 1.50, 2.00 mg/L N) and fixed P (5 mg/L N) were used in the NH$_4$-N threshold experiment.

A separate nutrient addition bioassay experiment was conducted with treated water samples containing individual or combined SRP, NO$_3$-N, and NH$_4$-N, which could assess the individual or combined effects of these nutrients on the growth of phytoplankton. A total of 30 treatments for three scenarios were designed in this experiment (Table 1).

The growth rate (μ) for all treatments was calculated according to the exponential growth equation [9]: $\mu = \ln(X_2/X_1)/(T_2-T_1)$, where X_1 is the phytoplankton density at the initial incubation time point (T_1), and X_2 is the phytoplankton density at the last time point (T_2). The Monod kinetic equation was used to calculate the maximum growth rate ($\mu_{\max}$) and half-saturation constant (Ku) [29]. The nutrient threshold can be estimated according to the change points on the response curves.

The difference in growth response among various treatments was analyzed by one-way ANOVA [30]. The Tukey's least significant difference procedure was used to compare multiple treatments after the event [31]. Statistical analysis was conducted using the SPSS 13.0 statistical software package (SPSS Inc., Chicago, IL, USA), and the significance level of the test used was $p < 0.05$ [32].

Table 1. Basic Schemes for Nutrient Addition Bioassay Experiment.

	NO.	**Nutrient Addition**	
Control	0	-	
A	1-1	0.5 mg N/L NH_4Cl	Without P
	1-2	1.0 mg N/L NH_4Cl	
	1-3	2.0 mg N/L NH_4Cl	
	1-4	4.0 mg N/L NH_4Cl	
	2-1	0.5 mg N/L NH_4Cl	5.0 mg P/L, K_2HPO_4
	2-2	1.0 mg N/L NH_4Cl	
	2-3	2.0 mg N/L NH_4Cl	
	2-4	4.0 mg N/L NH_4Cl	
B	1-1	0.5 mg N/L $NaNO_3$	Without P
	1-2	1.0 mg N/L $NaNO_3$	
	1-3	2.0 mg N/L $NaNO_3$	
	1-4	4.0 mg N/L $NaNO_3$	
	2-1	0.5 mg N/L $NaNO_3$	5.0 mg P/L, K_2HPO_4
	2-2	1.0 mg N/L $NaNO_3$	
	2-3	2.0 mg N/L $NaNO_3$	
	2-4	4.0 mg N/L $NaNO_3$	
C	1-1	0.02 mg P/L K_2HPO_4	Without N
	1-2	0.5 mg P/L K_2HPO_4	
	1-3	1.0 mg P/L K_2HPO_4	
	1-4	2.0 mg P/L K_2HPO_4	
	2-1	0.02 mg P/L K_2HPO_4	10 mg N/L, NH_4Cl
	2-2	0.5 mg P/L K_2HPO_4	
	2-3	1.0 mg P/L K_2HPO_4	
	2-4	2.0 mg P/L K_2HPO_4	
	3-1	0.02 mg P/L K_2HPO_4	10 mg N/L, $NaNO_3$
	3-2	0.5 mg P/L K_2HPO_4	
	3-3	1.0 mg P/L K_2HPO_4	
	3-4	2.0 mg P/L K_2HPO_4	
	4	2.0 mg P/L K_2HPO_4	5.0 mg N/L NH_4Cl + 5.0 mg N/L $NaNO_3$

2.3. Effect of pH Range on the Growth of Phytoplankton

According to the water quality conditions of Chenghai Lake in 2018–2019, the average pH value is up to 9.42 with the highest value of 9.66. The effect of pH range on nutrient thresholds was investigated with 3 different gradients of pH value (9.17, 8.50, and 7.50). The initial pH value of the water samples from Chenghai Lake used in this study was 9.17, and all the other gradients of pH value were adjusted before cultivation. The treatments were cultivated under a light intensity of 100 μmol photon m^{-2} s^{-1} with a 14:10-h light-dark cycle. The pH of each treatment was controlled during the whole experimental period. The phytoplankton density was recorded regularly every day.

3. Results and Discussion

3.1. Seasonal Variation of Phytoplankton Density and Water Quality

The water quality parameters for each sampling point of Chenghai Lake from June 2018 to May 2019 are shown in Figure 2. As we can see from the monthly changes, most water quality parameters showed seasonal variations. As shown in Figure 2b, the phytoplankton density was relatively lower in autumn and spring. The highest phytoplankton density of 6.46×10^7 cells/L was found in the northern part of Chenghai Lake in May. The nutritional status for Chenghai Lake was mainly maintained at a slight bloom level according to the phytoplankton density [33].

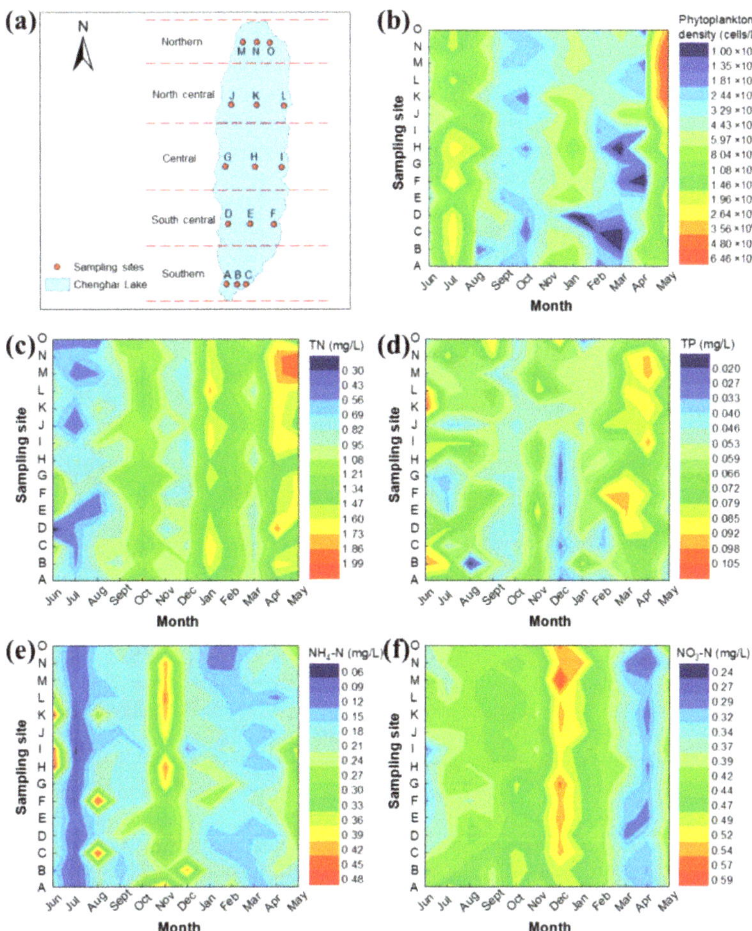

Figure 2. Monthly variation of phytoplankton density and water quality indexes in Chenghai Lake for all the 15 sampling sites. (**a**) Sampling sites delineation in Chenghai Lake. The spatial and temporal distribution characteristics of (**b**) phytoplankton density, (**c**) TN, (**d**) TP, (**e**) NH_4-N, and (**f**) NO_3-N in Chenghai Lake.

Nitrogen and phosphorus are the most important nutrients for phytoplankton growth. The TN value showed a continuous upward trend during the monthly sampling, varying from 0.30 to 1.99 mg/L. The highest TN content of 1.99 mg/L also appeared in the north of Chenghai Lake in May. What's more, the TP value was in the range from 0.02 to 0.11 mg/L. Most of the measured TN and TP values were higher than the commonly reported critical values for eutrophication [10,13,20–22]. In addition, NH_4-N and NO_3-N are also provided to illustrate the variation of different nitrogen sources (Figure 2e,f). The highest NH_4-N and NO_3-N values were determined to be as high as 0.43 mg/L and 0.59 mg/L, respectively. Therefore, the reduction of external nutrients is essential to ensure acceptable water quality, which can finally reduce the internal nitrogen and phosphorus load [15,34].

3.2. Nutrient Thresholds Required to Control Eutrophication in Chenghai Lake

Nitrogen and phosphorus are important nutrients that limit the growth of phytoplankton, and it is particularly important to formulate nutrient thresholds to control the occurrence of blooms [35]. The bioassay experiment is an effective tool to explore the growth response of phytoplankton under different nutrient concentrations [36]. To examine the nutrient thresholds of SRP, NO_3-N, and NH_4-N, the growth rate response of phytoplankton to different nutrient concentrations was explored. Figure 3 shows the growth curves fitted by nonlinear regression for SRP, NO_3-N, and NH_4-N, respectively. Figure 3a shows that the growth rate of phytoplankton increased with SRP addition from 0 to 0.05 mg/L P, while it remained constant with the further increase in SRP addition from 0.05 to 1.0 mg/L P. The change point on the response curve was found to be 0.05 mg/L P, which indicated that the growth of phytoplankton would be no longer restricted by P when SRP enrichment exceeded 0.05 mg/L P. Therefore, the threshold of SRP could be determined to be 0.05 mg/L P. Similarly, eutrophication thresholds of NH_4-N and NO_3-N were found to be 0.3 mg/L N and 0.5 mg/L N, respectively (Figure 3b,c).

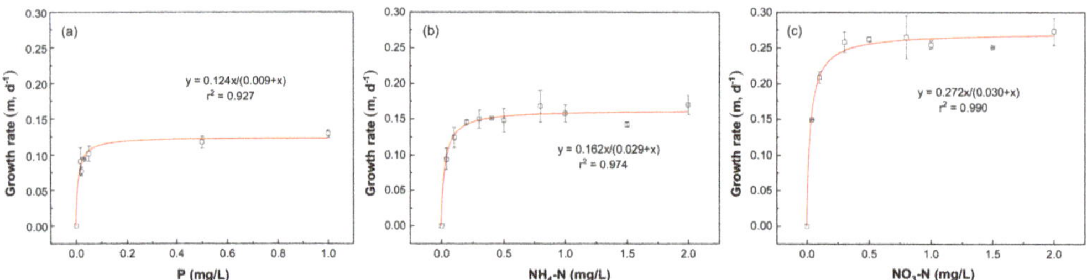

Figure 3. Growth kinetics of phytoplankton in response to (**a**) SRP, (**b**) NH_4-N, and (**c**) NO_3-N concentrations.

The maximum growth rates (μ_{max}) for SRP, NH_4-N, and NO_3-N were found to be 0.124, 0.162, and 0.272 d^{-1}, respectively, according to the Monod equation [13]. The half-saturation constant (Ku) for SRP, NH_4-N, and NO_3-N were determined to be 0.009, 0.029, and 0.030 mg/L, respectively. The higher μ_{max}/Ku ratio of NO_3-N compared with NH_4-N could demonstrate the higher utilization efficiency of NO_3-N in Chenghai Lake.

The growth responses of phytoplankton to increasing concentrations of SRP, NH_4-N, and NO_3-N are shown in Figure 4. The addition of NH_4-N and NO_3-N alone showed little effect on the growth of phytoplankton compared with the control group. As we can see from Figure 4a, the growth rate reached a peak when the concentration of NH_4-N reached 1.0 mg/L N together with P addition. This indicated that phosphorus could be the main limiting factor for the growth of phytoplankton. As for NO_3-N (Figure 4b), the growth rate of phytoplankton remained almost unchanged with the increase in concentration from 0.5 mg/L N to 4.0 mg/L N which was higher than the control group. However, the growth rates of phytoplankton could be promoted by increased SRP concentration with or without nitrogen sources (Figure 4c). These results indicated that the growth of the phytoplankton community in Chenghai Lake was restricted by phosphorus without obvious correlation with the input of nitrogen sources. Similar results have been reported in Meiliang Bay of Taihu Lake, and phosphorus was also found to restrict the growth of phytoplankton [20]. This can be explained by the nitrogen fixation function of cyanobacteria, which can meet their growth needs for nitrogen [37].

Figure 5 shows the growth rate of phytoplankton in response to individual or combined NH_4-N, NO_3-N, and SRP additions according to the nutrient addition bioassay experiment. N addition alone showed little effect on the growth rate of phytoplankton compared with the control. A significant stimulatory effect can be found with the combined addition of NO_3-N and SRP. This can directly prove the stimulatory effect of NO_3-N on phytoplankton growth, which is consistent with the results of nutrient threshold bioassay experiments. In addition, the stimulatory effect of SRP addition compared with the control can be further enhanced with the combined addition of NH_4-N or NO_3-N. However, the growth rate of phytoplankton with combined addition of "2 mg/L SRP + 5.0 mg/L NH_4-N + 5.0 mg/L NO_3-N" was not higher than the group with "2 mg/L SRP + 10.0 mg/L NH_4-N" or "2 mg/L SRP + 10.0 mg/L NO_3-N". This indicated that the form of nitrogen source could not show similar stimulatory effect on the growth of phytoplankton under the sufficient concentration of phosphorus.

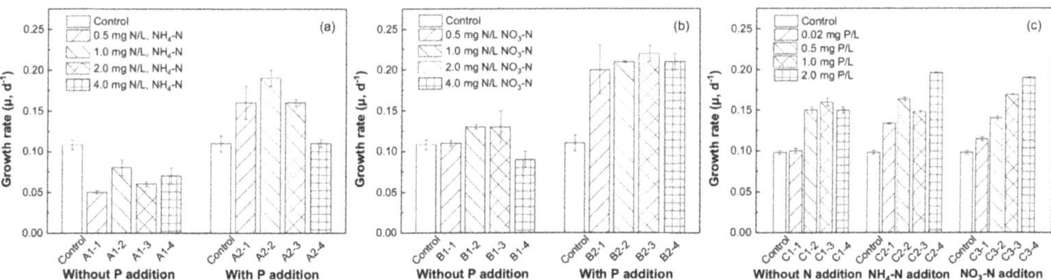

Figure 4. Phytoplankton growth responses to various concentrations of (**a**) NH$_4$-N, (**b**) NO$_3$-N, and (**c**) SRP additions.

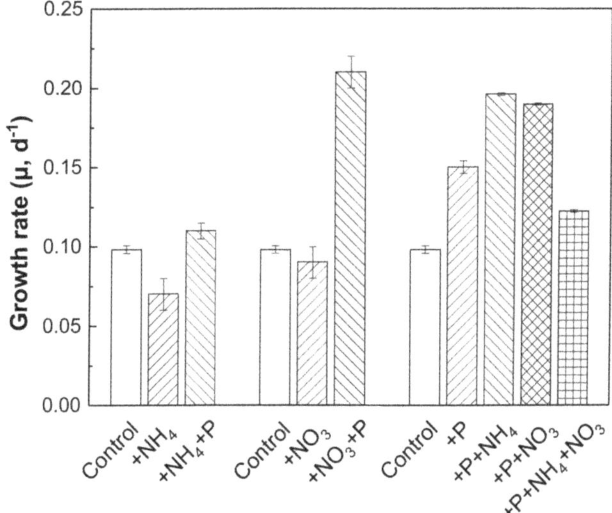

Figure 5. Comprehensive effects of NH$_4$-N, NO$_3$-N, and SRP on phytoplankton growth.

3.3. Effect of High Alkalinity Background on Nutrient Thresholds

According to the above results of nutrient limitation bioassay experiments, the nutrient thresholds of Chenghai Lake are higher than other reported lakes. Only a slight level of bloom was found in Chenghai Lake with a high-nutrient environment. These results indicate that the growth of phytoplankton might be influenced by other factors in addition to nutrients. The pH range in natural water bodies has been reported to show inhibition or promotion effects on phytoplankton [23,24]. Combined with the high alkalinity characteristic in Chenghai Lake [18], the effects of pH range on the growth of phytoplankton in Chenghai Lake were also explored in this study.

As we can see from Figure 6, the phytoplankton density increased with decrease in the pH value from 9.17 to 7.50 during the cultivation. These results indicated that a relatively lower water body pH could promote the growth of phytoplankton, while the high pH background of Chenghai Lake (pH = 9.17) could limit the growth of phytoplankton. It has been reported that a high pH condition in freshwater could not promote the growth and reproduction of cyanobacteria [38]. Most phytoplankton species cannot grow properly under high alkalinity conditions, especially when pH value exceeds 9 [39,40]. Thus, the relatively higher nutrient thresholds of Chenghai Lake can be explained by the inhibition effects of the high pH background on phytoplankton growth. High pH background might alter the transport processes of membrane and the metabolic functions of cells and change the relative composition of amino acids in cellular, which can finally affect the growth of phytoplankton [40]. As a means to prevent and control phytoplankton blooms, many researchers have undertaken pH adjustment as a method in actual lake management [41–43]. Therefore, it is necessary to maintain the high alkalinity characteristic of Chenghai Lake, which is helpful for the prevention of phytoplankton bloom.

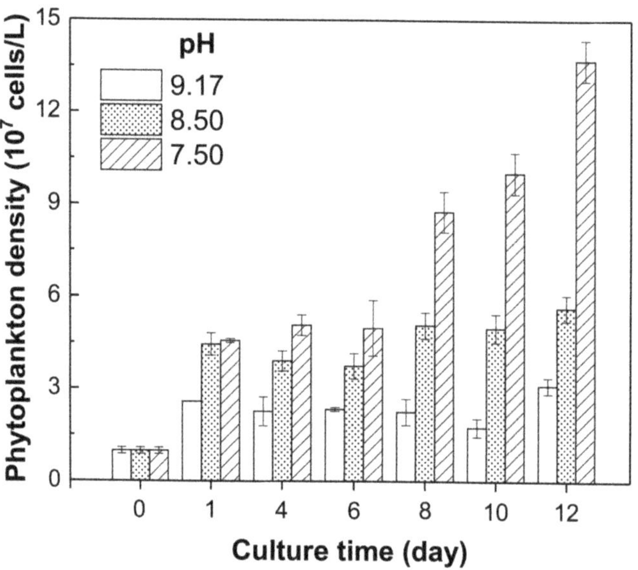

Figure 6. Effect of different pH levels on phytoplankton growth in Chenghai Lake.

4. Conclusions

The results in this study indicated that Chenghai Lake was always maintained at a slight bloom level according to the phytoplankton density. Phosphorus was found to be the main limiting factor for the growth of phytoplankton. In addition, the utilization efficiency of NO_3-N was higher than that of NH_4-N in Chenghai Lake. The eutrophication thresholds of SRP, NH_4-N, and NO_3-N were determined to be 0.05 mg/L, 0.30 mg/L, and 0.50 mg/L, respectively. The higher nutrient thresholds can be explained by the high pH range in Chenghai Lake, which would inhibit phytoplankton growth. In addition to the reduction of nutrient input, the maintenance of high alkalinity characteristic is also necessary for the prevention of phytoplankton blooms in natural alkaline lake.

Supplementary Materials: The following supporting information can be downloaded at: https://www.mdpi.com/article/10.3390/w14172674/s1, Table S1: Physical and chemical indexes of water samples collected at site H.

Author Contributions: Conceptualization, J.Q. and W.Q.; methodology, J.Q. and L.D.; software, L.D.; validation, J.Q. and L.D.; formal analysis, J.Q.; investigation, J.Q., L.D., and Y.S.; resources, C.H.; data curation, J.Q.; writing—original draft preparation, J.Q.; writing—review and editing, W.Q. and C.H.; supervision, J.Q. and W.Q; project administration, J.Q. and W.Q.; funding acquisition, J.Q. All authors have read and agreed to the published version of the manuscript.

Funding: This work was financially supported by the Funds for National Key R&D Program of China (Grant No. 2018YFE0204101), and the National Natural Science Foundation of China (Grant No. 52170014, and 51808531).

Conflicts of Interest: The authors declare that they have no known competing financial interests or personal relationships that could have appeared to influence the work reported in this paper.

References

1. Paerl, H.W.; Xu, H.; McCarthy, M.J.; Zhu, G.; Qin, B.; Li, Y.; Gardner, W.S. Controlling harmful cyanobacterial blooms in a hyper-eutrophic lake (Lake Taihu, China): The need for a dual nutrient (N & P) management strategy. *Water Res.* **2011**, *45*, 1973–1983.
2. Zhang, L.; Wang, Z.S.; Wang, N.; Gu, L.; Sun, Y.F.; Huang, Y.; Chen, Y.F.; Yang, Z. Mixotrophic Ochromonas Addition Improves the Harmful Microcystis-Dominated Phytoplankton Community in In Situ Microcosms. *Environ. Sci. Technol.* **2020**, *54*, 4609–4620. [CrossRef]
3. Wang, H.; Zhang, Z.Z.; Liang, D.F.; Du, H.B.; Pang, Y.; Hu, K.M.; Wang, J.J. Separation of wind's influence on harmful cyanobacterial blooms. *Water Res.* **2016**, *98*, 280–292. [CrossRef]
4. Olson, N.E.; Cooke, M.E.; Shi, J.H.; Birbeck, J.A.; Westrick, J.A.; Ault, A.P. Harmful Algal Bloom Toxins in Aerosol Generated from Inland Lake Water. *Environ. Sci. Technol.* **2020**, *54*, 4769–4780. [CrossRef]
5. Wang, H.; Xu, C.; Liu, Y.; Jeppesen, E.; Svenning, J.-C.; Wu, J.; Zhang, W.; Zhou, T.; Wang, P.; Nangombe, S.; et al. From unusual suspect to serial killer: Cyanotoxins boosted by climate change may jeopardize megafauna. *Innovation* **2021**, *2*, 100092. [CrossRef]
6. Huisman, J.; Codd, G.A.; Paerl, H.W.; Ibelings, B.W.; Verspagen, J.M.H.; Visser, P.M. Cyanobacterial blooms. *Nat. Rev. Microbiol.* **2018**, *16*, 471–483. [CrossRef]
7. Su, X.M.; Steinman, A.D.; Tang, X.M.; Xue, Q.J.; Zhao, Y.Y.; Xie, L.Q. Response of bacterial communities to cyanobacterial harmful algal blooms in Lake Taihu, China. *Harmful Algae* **2017**, *68*, 168–177. [CrossRef]
8. Wang, H.J.; Wang, H.Z.; Liang, X.M.; Wu, S.K. Total phosphorus thresholds for regime shifts are nearly equal in subtropical and temperate shallow lakes with moderate depths and areas. *Freshw. Biol.* **2014**, *59*, 1659–1671. [CrossRef]
9. Kwak, D.-H.; Kim, M.-S. Flotation of algae for water reuse and biomass production: Role of zeta potential and surfactant to separate algal particles. *Water Sci. Technol.* **2015**, *72*, 762–769. [CrossRef]
10. Cao, X.F.; Wang, J.; Liao, J.Q.; Sun, J.H.; Huang, Y. The threshold responses of phytoplankton community to nutrient gradient in a shallow eutrophic Chinese lake. *Ecol. Indic.* **2016**, *61*, 258–267. [CrossRef]
11. Ma, S.N.; Wang, H.J.; Wang, H.Z.; Li, Y.; Liu, M.; Liang, X.M.; Yu, Q.; Jeppesen, E.; Søndergaard, M. High ammonium loading can increase alkaline phosphatase activity and promote sediment phosphorus release: A two-month mesocosm experiment. *Water Res.* **2018**, *145*, 388–397. [CrossRef]
12. Ma, S.N.; Wang, H.J.; Wang, H.Z.; Zhang, M.; Li, Y.; Bian, S.-J.; Liang, X.-M.; Søndergaard, M.; Jeppesen, E. Effects of nitrate on phosphorus release from lake sediments. *Water Res.* **2021**, *194*, 116894. [CrossRef]
13. Xu, H.; Paerl, H.W.; Qin, B.; Zhu, G.; Hall, N.S.; Wu, Y. Determining Critical Nutrient Thresholds Needed to Control Harmful Cyanobacterial Blooms in Eutrophic Lake Taihu, China. *Environ. Sci. Technol. Environ. Sci. Technol.* **2015**, *49*, 1051–1059. [CrossRef]
14. Huo, S.L.; Ma, C.Z.; Xi, B.D.; Zhang, Y.L.; Wu, F.C.; Liu, H.L. Development of methods for establishing nutrient criteria in lakes and reservoirs: A review. *J. Environ. Sci.* **2018**, *67*, 54–66. [CrossRef]
15. Newell, S.E.; Davis, T.W.; Johengen, T.H.; Gossiaux, D.; Burtner, A.; Palladino, D.; McCarthy, M.J. Reduced forms of nitrogen are a driver of non-nitrogen-fixing harmful cyanobacterial blooms and toxicity in Lake Erie. *Harmful Algae* **2019**, *81*, 86–93. [CrossRef]
16. Yuan, W.Z.; Su, X.S.; Cui, G.; Wang, H. Microbial community structure in hypolentic zones of a brine lake in a desert plateau, China. *Environ. Earth Sci.* **2016**, *75*, 1–14. [CrossRef]
17. Wang, Y.J.; Peng, J.F.; Cao, X.F.; Xu, Y.; Yu, H.W.; Duan, G.Q.; Qu, J.H. Isotopic and chemical evidence for nitrate sources and transformation processes in a plateau lake basin in Southwest China. *Sci. Total Environ.* **2020**, *711*, 134856.
18. Yan, D.N.; Xu, H.; Yang, M.; Lan, J.H.; Hou, W.G.; Wang, F.S.; Zhang, J.X.; Zhou, K.E.; An, Z.S.; Goldsmith, Y. Responses of cyanobacteria to climate and human activities at Lake Chenghai over the past 100 years. *Ecol. Indic.* **2019**, *104*, 755–763. [CrossRef]
19. Jeppesen, E.; Beklioglu, M.; Ozkan, K.; Akyurek, Z. Salinization Increase due to Climate Change Will Have Substantial Negative Effects on Inland Waters: A Call for Multifaceted Research at the Local and Global Scale. *Innovation* **2020**, *1*, 100030. [CrossRef]
20. Wang, M.; Wu, X.F.; Li, D.P.; Li, X.; Huang, Y. Annual variation of different phosphorus forms and response of algae growth in Meiliang bay of Taihu Lake. *Huanjing Kexue* **2015**, *36*, 80–86.
21. Li, B.; Yang, G.S.; Wan, R.R. Multidecadal water quality deterioration in the largest freshwater lake in China (Poyang Lake): Implications on eutrophication management. *Environ. Pollut.* **2020**, *260*, 114033. [CrossRef]
22. Zhang, Y.; Song, C.L.; Ji, L.; Liu, Y.Q.; Xiao, J.; Cao, X.Y.; Zhou, Y.Y. Cause and effect of N/P ratio decline with eutrophication aggravation in shallow lakes. *Sci. Total Environ.* **2018**, *627*, 1294–1302.
23. Ni, M.; Yuan, J.L.; Liu, M.; Gu, Z.M. Assessment of water quality and phytoplankton community of Limpenaeus vannamei pond in intertidal zone of Hangzhou Bay, China. *Aquac. Rep.* **2018**, *11*, 53–58. [CrossRef]
24. Su, X.M.; Steinman, A.D.; Xue, Q.J.; Zhao, Y.Y.; Tang, X.M.; Xie, L.Q. Temporal patterns of phyto- and bacterioplankton and their relationships with environmental factors in Lake Taihu, China. *Chemosphere* **2017**, *184*, 299–308. [CrossRef]
25. Zhang, Y.L.; Huo, S.L.; Zan, F.Y.; Xi, B.D.; Zhang, J.T.; Wu, F.C. Historical records of multiple heavy metals from dated sediment cores in Lake Chenghai, China. *Environ. Earth Sci.* **2015**, *74*, 3897–3906. [CrossRef]

26. Banerjee, M.; Bar, N.; Basu, R.K.; Das, S.K. Comparative study of adsorptive removal of Cr (VI) ion from aqueous solution in fixed bed column by peanut shell and almond shell using empirical models and ANN. *Environ. Sci. Pollut. Res. Int.* **2017**, *24*, 10604–10620. [CrossRef]

27. Laskar, H.S.; Gupta, S. Phytoplankton community and limnology of Chatla floodplain wetland of Barak valley, Assam, North-East India. *Knowl. Manag. Aquat. Ecosyst.* **2013**, *411*, 06. [CrossRef]

28. Hwang, S.J.; Kim, H.S.; Shin, J.K.; Oh, J.M.; Kong, D.S. Grazing effects of a freshwater bivalve (Corbicula leanaPrime) and large zooplankton on phytoplankton communities in two Korean lakes. *Hydrobiologia* **2004**, *515*, 161–179. [CrossRef]

29. Monod, J. Technique, Theory and Applications of Continuous Culture. *Ann. Inst. Pasteur* **1950**, *79*, 390–410.

30. Swamy, M.; Norlina, W.; Azman, W.; Suhaili, D.; Sirajudeen, K.N.S.; Mustapha, Z.; Govindasamy, C. Restoration of glutamine synthetase activity, nitric oxide levels and amelioration of oxidative stress by propolis in kainic acid mediated excitotoxicity. *Afr. J. Tradit. Complementary Altern. Med.* **2014**, *11*, 458–463. [CrossRef]

31. Valla, L.; Janson, H.; Wentzel-Larsen, T.; Slinning, K. Analysing four Norweigian population-based samples using the six-month version of the Ages and Stages Questionnaire showed few relevant clinical differences. *Acta Paediatr.* **2016**, *105*, 924–929. [CrossRef]

32. Schmadel, N.M.; Harvey, J.W.; Alexander, R.B.; Schwarz, G.E.; Moore, R.B.; Eng, K.; Gomez-Velez, J.D.; Boyer, E.W.; Scott, D. Thresholds of lake and reservoir connectivity in river networks control nitrogen removal. *Nat. Commun.* **2018**, *9*, 1–10. [CrossRef]

33. Jiang, Z.; Du, P.; Liu, J.; Chen, Y.; Zhu, Y.; Shou, L.; Zeng, J.; Chen, J. Phytoplankton biomass and size structure in Xiangshan Bay, China: Current state and historical comparison under accelerated eutrophication and warming. *Mar. Pollut. Bull.* **2019**, *142*, 119–128. [CrossRef]

34. Wu, J.L.; Gagan, M.K.; Jiang, X.Z.; Xia, W.L.; Wang, S.M. Sedimentary geochemical evidence for recent eutrophication of Lake Chenghai, Yunnan, China. *J. Paleolimnol.* **2004**, *32*, 85–94.

35. Abell, J.M.; Oezkundakci, D.; Hamilton, D.P. Nitrogen and phosphorus limitation of phytoplankton growth in New Zealand lakes: Implications for eutrophication control. *Ecosystems* **2010**, *13*, 966–977. [CrossRef]

36. Janssen, A.B.G.; de Jager, V.C.L.; Janse, J.H.; Kong, X.; Liu, S.; Ye, Q.; Mooij, W.M. Spatial identification of critical nutrient loads of large shallow lakes: Implications for Lake Taihu (China). *Water Res.* **2017**, *119*, 276–287. [CrossRef]

37. Peterson, B.J.; Barlow, J.P.; Savage, A.E. The Physiological State with Respect to Phosphorus of Cayuga Lake Phytoplankton. *Limnol. Oceanogr.* **1974**, *19*, 396–408. [CrossRef]

38. Zhang, Y.F.; Gao, Y.H.; Kirchman, D.L.; Cottrell, M.T.; Chen, R.; Wang, K.; Ouyang, Z.X.; Xu, Y.Y.; Chen, B.S.; Yin, K.D.; et al. Biological regulation of pH during intensive growth of phytoplankton in two eutrophic estuarine waters. *Mar. Ecol. Prog. Ser.* **2019**, *609*, 87–99. [CrossRef]

39. Hama, T.; Inoue, T.; Suzuki, R.; Kashiwazaki, H.; Wada, S.; Sasano, D.; Kosugi, N.; Ishii, M. Response of a phytoplankton community to nutrient addition under different CO_2 and pH conditions. *J. Oceanogr.* **2016**, *72*, 207–223. [CrossRef]

40. Middelboe, A.L.; Hansen, P.J. Direct effects of pH and inorganic carbon on macroalgal photosynthesis and growth. *Mar. Biol. Res.* **2007**, *3*, 134–144. [CrossRef]

41. Chakraborty, P.; Acharyya, T.; Babu, P.V.R.; Bandyopadhyay, D. Impact of salinity and pH on phytoplankton communities in a tropical freshwater system: An investigation with pigment analysis by HPLC. *J. Environ. Monit.* **2011**, *13*, 614–620. [CrossRef] [PubMed]

42. Chen, C.Y.; Durbin, E.G. Effects of pH on the growth and carbon uptake of marine phytoplankton. *Mar. Ecol. Prog.* **1994**, *109*, 83–94. [CrossRef]

43. Shi, D.; Xu, Y.; Morel, F.M.M. Effects of the pH/pCO(2) control method on medium chemistry and phytoplankton growth. *Biogeosciences* **2009**, *6*, 1199–1207. [CrossRef]

Article

Synergistic Effects and Ecological Responses of Combined In Situ Passivation and Macrophytes toward the Water Quality of a Macrophytes-Dominated Eutrophic Lake

Wei Yu [1,2], Haiquan Yang [1,*], Yongqiong Yang [2,*], Jingan Chen [1], Peng Liao [1], Jingfu Wang [1], Jiaxi Wu [1,3], Yun He [1,4] and Dan Xu [1]

1. State Key Laboratory of Environmental Geochemistry, Institute of Geochemistry, Chinese Academy of Sciences, Guiyang 550081, China; yuwei@mail.gyig.ac.cn (W.Y.); chenjingan@vip.skleg.cn (J.C.); liaopeng@mail.gyig.ac.cn (P.L.); wangjingfu@vip.skleg.cn (J.W.); wujiaxi@mail.gyig.ac.cn (J.W.); heyun@mail.gyig.ac.cn (Y.H.); xudan@mail.gyig.ac.cn (D.X.)
2. School of Geography and Environmental Sciences, Guizhou Normal University, Guiyang 550025, China
3. College of Resource and Environmental Engineering, Guizhou University, Guiyang 550025, China
4. College of Eco-Environmental Engineering, Guizhou Minzu University, Guiyang 550025, China
* Correspondence: yanghaiquan@vip.skleg.cn (H.Y.); yyongqiong@163.com (Y.Y.)

Citation: Yu, W.; Yang, H.; Yang, Y.; Chen, J.; Liao, P.; Wang, J.; Wu, J.; He, Y.; Xu, D. Synergistic Effects and Ecological Responses of Combined In Situ Passivation and Macrophytes toward the Water Quality of a Macrophytes-Dominated Eutrophic Lake. *Water* **2022**, *14*, 1847. https:// doi.org/10.3390/w14121847

Academic Editors: Dibyendu Sarkar, Rupali Datta, Prafulla Kumar Sahoo and Mohammad Mahmudur Rahman

Received: 29 April 2022
Accepted: 6 June 2022
Published: 8 June 2022

Publisher's Note: MDPI stays neutral with regard to jurisdictional claims in published maps and institutional affiliations.

Abstract: Combined use of in situ passivation and macrophytes is a valuable technology that exerts remarkable effects on aquatic systems. However, the effectiveness and ecological functions of this combined technology for macrophytes-dominated eutrophic (MDE) lakes with organophosphorus-controlled internal phosphorus (P) loading were poorly understood. In this study, aquatic simulation experiments were performed to study the combination of La-modified materials (LMM; La-modified bentonite (LMB), and La/Al co-modified attapulgite (LAA)) with macrophytes (*Myriophyllum verticillatum* L. (MVL), *Hydrilla verticillata (Linn. f.) royle* (HVR), and *Ceratophyllum demersum* L. (CDL)) for the control of P mobility in the water column, and to investigate the passivator effects on the physiological characteristics of macrophytes. The mineralization of organophosphates ($BD-Po$, $HCl-Po$, and $Res-Po$) is an important factor for maintaining high internal P loadings and overlying water P concentrations in the experiments. Compared with individual treatment groups, the reduction of internal P release flux and porewater SRP concentrations was more obvious in the combined treatments. Moreover, the redox-sensitive P forms transformation is more pronounced in the surface sediments. In the LAA+M group, internal P release flux was reduced by 55% and 55% compared with individual passivators and macrophytes retreatment groups, respectively. In contrast, the LMB+M group decreased by 16% and 46%, respectively. Simultaneously, LMM had less effect on macrophytes traits compared with individual macrophytes group and enhanced the absorption of phosphate by macrophytes. The phosphate content of macrophytes in the LAA+M and LMB+M groups increased by 24% and 11%, respectively, in comparison with the individual macrophytes group. Results concluded that the combination of passivator and macrophytes enhanced the effect of ecological restoration and exerts a synergistic effect on internal P pollution with macrophytes.

Keywords: La-modified material; macrophyte; sediments; phosphorus; eutrophication

1. Introduction

Eutrophication due to excessive enrichment of nutrients such as nitrogen and phosphorus (P) is one of the most important global water quality issues [1,2]. From the perspective of controllability and effectiveness, reducing external and internal P loading has generally been accepted as a key method to mitigate lake eutrophication [3,4]. However, studies have substantiated that the release of P in sediments significantly affected the concentration of P in the lake water as well as the migration and transformation of P at sediment–water interface (SWI) [5,6]. Even when the input of external P is reduced, internal P loading from sediments may sustain eutrophication in lakes [6,7]. Nevertheless, internal sediment loading remains an important impediment to the restoration and management of aquatic ecosystems within eutrophic lakes.

Effective eutrophication management in macrophytes-dominated lakes requires reduction of external and internal P loading, with reduced internal loading mainly relying on sediments dredging, in situ passivation-based remediation, and aquatic macrophytes community restoration [3,8]. The transition between a clear state in shallow lakes dominated by macrophytes and a turbid state dominated by phytoplankton is very abrupt and difficult to reverse [1,9]. Nutrient reductions alone minimally affect the turbidity of the water column, although appropriate disturbance and ecosystem use may shift the water column to a stable and clear state [3,10]. Further, lakes with high-density macrophytes coverage tend to have a higher clarity than lakes with the same nutrient status but little or no macrophytes coverage [3,9,10]. These observations indicate that constructing a healthy and stable aquatic biological system is key to eutrophication management in macrophyte-dominated lakes.

Aquatic macrophytes community restoration approaches are operationally time-consuming and only slowly achieve pronounced remediation outcomes, while its performance can even be inhibited by continuous high internal loading and seasonal climate change [3,8]. Therefore, improving the physical and chemical conditions of sediments is a necessary measure for water ecological restoration. However, sediments dredging often has problems such as high costs and remediation outcomes, which limit the long-term remediation [11,12]. Therefore, more and more researchers have begun to pay attention to the development of in situ passivation technology. Currently, the passivator for the efficient control of internal P pollution is mainly based on lanthanum-modified materials (LMM), including La-modified bentonite (LMB) [13], La/Al-modified attapulgite (LAA) [14], and La-modified zeolite [15].

The migration and transformation of lake P at SWI are often closely related to the change in different P forms, and the release risk of internal P mainly depends on the forms of P in sediments [16,17]. According to the organophosphorus remineralization and Fe–P coupled pathway of P cycling, sedimentary P dynamics are mainly controlled by depositional flux of organophosphorus and iron-bound P [18,19]. However, a large amount of organophosphorus was often stored in the sediments of macrophytes-dominated lakes due to the strong biological action, and the P cycle may be controlled by microbial decomposition or remineralization [17,20]. Hypoxic diffusion due to enhanced microbial decomposition or remineralization is common in many shallow lakes with high organophosphorus content, and the remediation effects of in situ passivation and macrophytes techniques in these environments are readily affected by sedimentation processes [19,21]. However, ecological remediation effects of in situ passivation and macrophytes techniques on MDE lakes with organophosphorus-dominated internal P loading remain largely unclear. Moreover, changes in microscopic morphology, elemental composition, and surface physical properties of surface sediments in response to the combined use of these techniques also remain unclear.

The objective of this study was to promote the effective application of combined in situ passivator and macrophytes in the MDE lakes with organophosphorus-controlled internal P loading. To this end, a 60-day simulation experiment was conducted to evaluate these aims starting in August 2021 in the heavily polluted areas of Lake Caohai, which is currently experiencing a sudden change in water turbidity, large-scale extinction of macrophytes, and severe eutrophication. Specifically, two highly efficient LMM with three pollution-tolerant, native macrophytes seedlings were used in the experiment. Moreover, the microstructures of sediments before and after sediments remediation were investigated along with the mobility and ecological effects of P in the water column using a combination of modified Hupfer sequential extraction schemes. These measurements were combined with in situ, dynamic, high-resolution composite diffusive gradient in thin films (DGT) technology, scanning electron microscope-X-ray energy dispersive (SEM-EDS) analysis, and X-ray diffraction (XRD) analysis. These results provide a baseline for restoring aquatic ecosystems in MDE lakes, in addition to evaluating the ecological importance and scientific applicability of LMM.

2. Materials and Methods

2.1. Mesocosm Establishment

The mesocosm experiment was conducted from August to October 2021 at national positioning observation and research station of Caohai wetland ecosystem near Lake Caohai in Weining county, China. Six groups of aquariums were set up (three repetitions): (1) control group without any materials added; (2) macrophytes planted group (M); (3) La/Al co-modified attapulgite added group (LAA); (4) La-modified bentonite group (LMB) added; (5) La/Al co-modified attapulgite added + macrophytes planted group (LAA+M); (6) La-modified bentonite added + macrophytes planted group (LMB+M). Mesocosm experiments were conducted in plexiglass boxes with lengths, width, and height of 40 cm, 40 cm, and 60 cm, respectively. Triplicate parallel samples were used in each group, with nine seedlings planted in each macrophyte's additional group. Dominant macrophytes in seedlings that are native to the lake were used, including three strains *Myriophyllum verticillatum* L. (MVL), *Hydrilla verticillata (Linn. f.) royle* (HVR), and *Ceratophyllum demersum* L. (CDL).

The bottoms of the incubators were lined with mixed homogeneous sediments collected from a heavily polluted area in eastern Lake Caohai to a thickness of 12 cm. Lake water (collected at the Jiangjiawan wharf) was added to the upper layer by the siphon method to achieve a water depth of about 45 cm. All experiment was conducted in the laboratory of the Lake Caohai ecological station. After letting stand for 72 h, two LMM were added (the mass ratio of LMM to biologically available P in the sediments was 100:1), and macrophytes were transplanted. After the start of the experiment, nutrient concentrations in the overlying water of the incubator were regularly measured with the molybdenum blue method (minimum detection limit of 0.01 mg·L^{-1}) [22] and a multiparameter water quality analyzer (YSI EXO-2). Parameters including temperature (T), pH, dissolved oxygen (DO), and oxidation-reduction potential (ORP) were measured on-site in the overlying water. The incubations were conducted for 60-day.

2.2. DGT Deployments and Analysis

Zr-Oxide and AgI DGT probes were vertically inserted as previously described [23] at 30-day and at the end of the incubations. The probes were taken out after letting stand for 24 h. The probes were washed with ultrapure water, sealed, and refrigerated for storage. Additional details regarding the structures and functioning of DGT probes are described elsewhere [24]. The concentrations of unstable elements (C_{DGT}) were calculated as following (Equation (1)):

$$C_{DGT} = \frac{M\Delta g}{DAt} \tag{1}$$

where C_{DGT} is the concentration of the target compound (mg·L^{-1}); M is the accumulated amount of Zr-Oxide DGT thin film over the sampling time (ug); Δg is the thickness of the diffusive layer (cm); D is the molecular diffusive coefficient of the phosphate in the diffusion layer (cm^2·s^{-1}); A is the film area of each slice (cm^2); t is the sample standing time (s) [23].

Both Zr-Oxide (one-dimensional) and AgI DGT (two-dimensional) probes were purchased from Nanjing Zhigan Environmental Technology Co., Ltd. (Nanjing, China). To accomplish two-dimensional high-resolution imaging and vertical profile measurement of DGT-labile S in sediments, the developed AgI-fixed film (600 dpi) images were first converted into grayscale using the CanoScan 5600 F (Canon, Tokyo, Japan) and ImageJ software programs, followed by comparison against a calibration curve. The grey values were then converted into the cumulative mass per unit area (M) of S in the AgI-binding gel using the following Equation (2):

$$M = -7.23 \times ln(\frac{220 - G}{171}) \tag{2}$$

where G is the grayscale intensity of the AgI binding gels. DGT-labile S concentrations could then be calculated from Equation (1).

In order to measure unstable P and Fe concentrations in sediments, the Zr-Oxide fixed film was cut with a razor blade from the exposure window and then sectioned into long strips (with 2 mm width and 20 mm length) with a ceramic microtome. The film was placed in a centrifuge tube with 0.8 mL of 1 M NaOH or 1 M HNO$_3$ extraction solution and left to stand at room temperature for 16 h for extraction. P and Fe concentrations in the extracts of each section were determined by microplate spectrophotometry with a minimum detection limit of 0.01 mg·L^{-1} and with the concentrations of DGT-labile P and DGT-labile Fe calculated using Equation (1). In order to measure the apparent diffusion flux of P at the WSI in each experimental treatment, Fick's first law of diffusion was used to estimate the specific calculation method [24].

2.3. Sediments Characteristics

After the experiment, sediments (0–8 cm) were collected from each incubator at intervals of 1 cm, followed by cryopreservation. Sediments porewater was obtained by centrifugation at 4000 r/min (MKE-VCK-22R) in the laboratory, and the TP, DIP, and DTP concentrations were measured by the molybdenum blue method [25]. Partial surface sediment samples were used to determine sediments porosity after sediments were freeze-dried (Techconp FD-3-85-MP), and the micro-area surface morphology and mineral compositions were analyzed by scanning electron microscopy (SEM, JEOL, resolution 0.6 nm) and X-ray energy spectroscopy at a resolution of 127.9 eV (EDS, JSM-IT800). The remaining sediment samples were ground and sieved (120 mesh). Followed by sealing for storage. Qualitative phase analysis of sediments was conducted by pre-treating samples (200 mesh), followed by analysis at the Guizhou Provincial Geology and Mineral Centre laboratory. X-ray diffraction instrument (XRD, Rigaku Ultima IV; tube pressure of 40 kV, tube flow of 40 mA; scanning speed of 2°/min) was used for qualitative phase measurements.

The content of sediments TP was determined by the molybdenum blue method after extraction with 3.5 M HCl [25]. The P fraction was obtained using a modified Hupfer sequential extraction method [26], yielding an extraction recovery of >90%. The extracted P forms include weakly adsorbed P (NH$_4$Cl–Pi), iron–manganese-bound inorganic P (BD–Pi), organophosphorus (BD–Po), iron–aluminum-bound inorganic P (NaOH–Pi), biodetritus organophosphorus (NaOH–Po), calcium-bound inorganic P (HCl–Pi), organophosphorus (HCl–Po), residual P (Res–Po) [7,26]. The release or accumulation rates (r) of different P forms during incubation were then estimated using the following formula (3) [27]:

$$r = \frac{P_{60d} - P_{0d}}{P_{0d}} \tag{3}$$

where P_{60d} (mg·kg^{-1}) is the P concentration determined after 60 days of incubation; P_{0d} (mg·kg^{-1}) is the P concentration measured at time zero (before incubation); and r is the proportion of P_{60d} in P_{0d}. Negative values represent reduced P levels over the incubation, indicating release of the fraction, while positive values represent increased P levels over the incubation, indicating accumulation of the fraction.

The release sedimentation rate (R) of P forms in sediments cores was calculated using the method proposed by [28]. The fast-release sedimentation rate (R_1) was calculated from the percentage change in P content of the first and second layers at the top of the columnar sediment, while the slow-release sedimentation rate (R_2) was calculated from the top of the first layer and the bottom of the last layer of the columnar sediment. These measurements were used to calculate the percent change in P concentration as following (Equations (4) and (5)):

$$R_1 = \frac{(S_1 - S_2)}{\sum_{i=1}^{n}|S_1 - S_8|} \times 100\%, \tag{4}$$

$$R_2 = \frac{(S_1 - S_8)}{\sum_{i=1}^{n} |S_1 - S_8|} \times 100\%, \tag{5}$$

where S_1 is the P concentration of the top layer of the columnar sediments; S_2 is the P concentration of the second layer at the top of the columnar sediments; S_8 is the P concentration of the bottom-most layer of the columnar sediments. n represents the number of extracting P forms from sediments. When R_1 and R_2 are positive, that P component is primarily released. When R_1 and R_2 are negative, that P component is primarily retained.

2.4. Macrophytes Traits

Macrophyte growth characteristics (i.e., biomass, length, phosphate content) were measured at the start and end of the experiment. In addition, an additional 10 shoots were selected to measure the water content at the start, which was used to calculate the initial dry weight. All macrophytes were collected, washed three times, and then placed in a 45 °C oven to dry after the experiment. Macrophyte dry weights and lengths were measured with a precision electronic balance and a ruler, respectively. TP content in macrophytes was determined by the molybdenum blue method when macrophytes were extracted with 3.5 M HCl [22]. The relative growth rates (RGR) of different macrophytes in the incubator were calculated with the following formula (Equation (6)) [3]:

$$RGR = \ln\left(\frac{W_{60d}}{W_{0d}}\right)/\text{days}, \tag{6}$$

where W_{60d} (g) and W_{0d} (g) are the total biomass (DW) after 60 day and 0 days of incubation in each incubator, respectively.

3. Results

3.1. Water Quality

LMM and macrophytes, whether used in combination or individually, significantly reduced water column P concentrations (Figures 1 and 2), with the LAA+M group exhibiting the best performance. Aquatic ecosystems in all incubators generally reached steady states by the 15 days of incubation. In addition, SRP concentrations increased after reaching equilibrium stability in all LMM treatments. TP and DTP concentrations were reduced after 60 days of restoration in the M, LAA, LMB, LAA+M, and LMB+M groups by 16% and 30%, 25% and 37%, 33% and 53%, 45%, and 67%, and 38% and 53%, compared with the control group, respectively. In addition, SRP concentrations in the overlying water were below the limit of detection (<0.01 mg·L^{-1}). A comparison of the P removal rates from the water column among treatments indicated that the LMM group was most efficient in controlling water column P concentrations and that its efficiency was further improved when used in conjunction with macrophytes.

The start TP, DTP, and SRP concentrations in surface porewaters before the experiment were 3.80 mg·L^{-1}, 3.09 mg·L^{-1}, and 2.59 mg·L^{-1}, respectively, with SRP accounting for about 68% of TP concentrations. After 30 days of restoration, porewater concentrations of TP, DTP, and SRP were significantly lower than the initial porewater concentrations, while P concentrations were still relatively high in the surface porewater (0–5 cm) of the control group and peak concentrations (TP, DTP, and SRP) were observed at 4 cm depth (Figure 2). Vertical changes of porewater P concentrations in different treatment groups were basically the same as those of the control group, and the peak concentrations occurred at deeper depths in the LAA+M (5 cm) and LMB+M (6 cm) groups. All of the treatments, except for the LAA+M and LMB+M groups, did not show significant changes in porewater P concentrations at the end of the experiment compared with the initial porewater P concentrations. Further, SRP was the dominant porewater P form among profiles, accounting for about 47% of TP concentrations and about 71% of DTP concentrations. The LAA+M group achieved the best restoration outcome, with porewater concentrations of TP, DTP, and SRP decreasing by 53%, 53%, and 62%, compared with the control group, respectively.

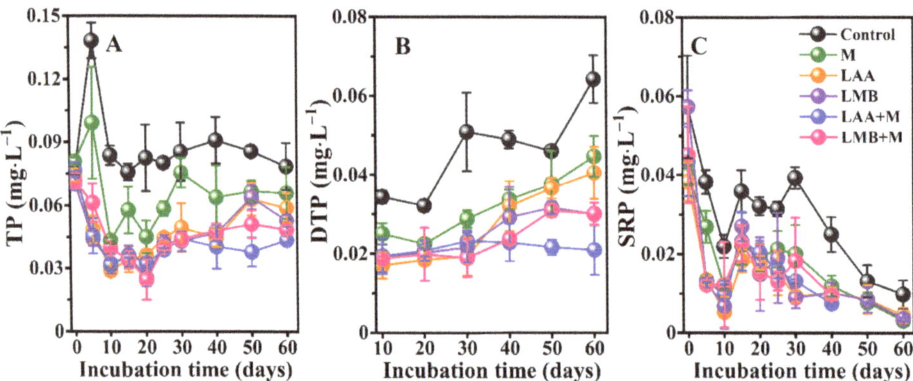

Figure 1. The changes of TP, DTP, and SRP concentrations in the overlying water of the six treatments during the experiment. Vertical bars indicate standard deviation. (**A–C**) represent TP, DTP, and SRP in the overlying water, respectively.

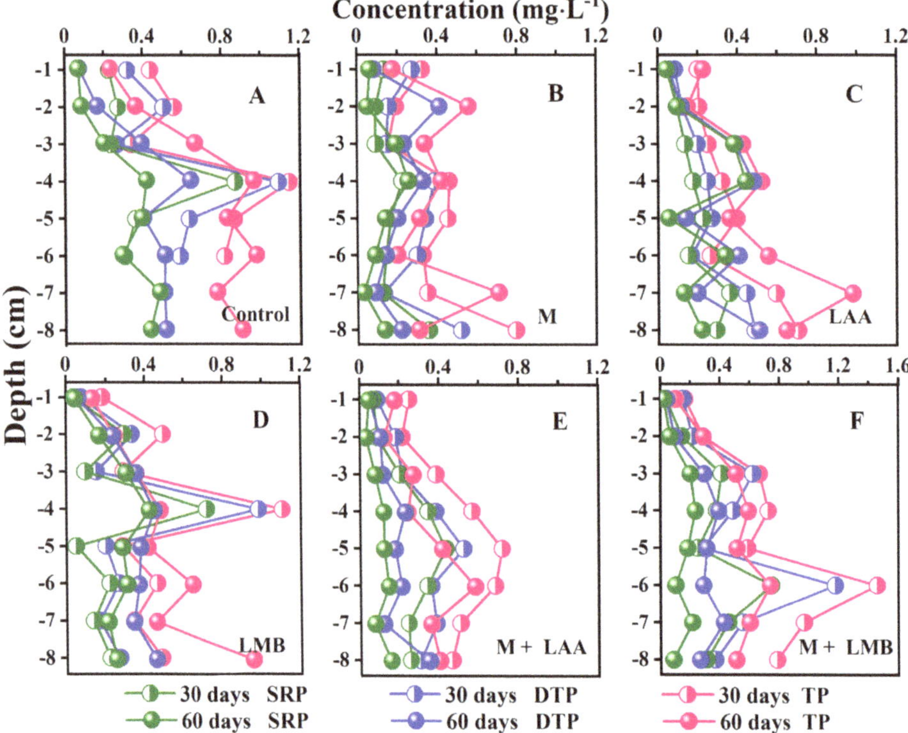

Figure 2. The changes of TP, DTP, and SRP concentrations in the porewater of the six treatments during the experiment. All data are triplicates. (**A–F**) indicate the concentration changes of TP, DTP, and SRP in sediment pore water of the Control, M, LAA, LMB, LAA+M, and LMB+M experimental groups on the 30-day and 60-day, respectively.

3.2. Physicochemical Characterization of Sediments

SEM-EDS elemental mapping revealed that C, O, Si, Al, and Ca are the primary chemical elements in Lake Caohai sediments, and $CaCO_3$ and SiO_2 are the main minerals in the sediments (Figure 3A,E). SEM imaging revealed that minerals in the surface sediments were in crystal form, ranging from 1 to 10 μm in size (Figure 3C,F). Surface sediments treated with LMM passivators exhibited significantly higher content of La^{3+} compared with the control sediments (Figure 3B,C,F,G) owing to the high dosage used in the preparation of LMM (Figure 3D,H).

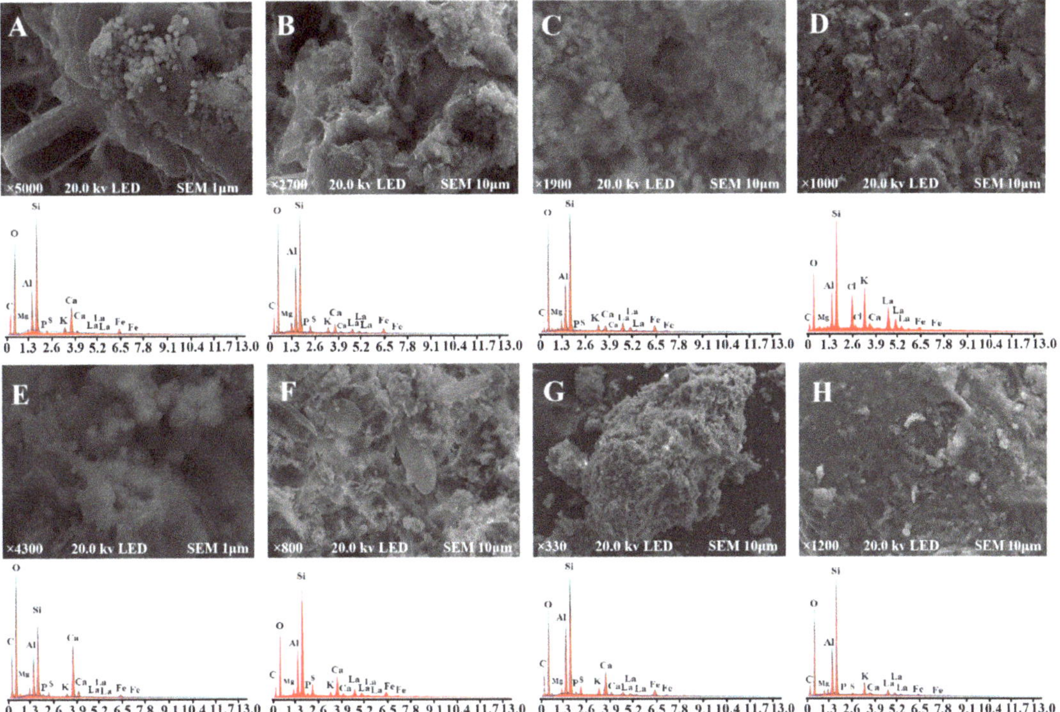

Figure 3. SEM images and spectra of chemical composition of Control (**A**), LAA (**B**), LAA+M (**C**), Original LAA (**D**), M (**E**), LMB (**F**), LMB+M (**G**) and Original LMB (**H**). Original LAA and Original LMB are substances that are not mixed with sediments respectively.

XRD peak positions exhibited large differences between LMM treatment profiles and those from other treatment groups (Figure 4). Bragg equation and Scherrer formula calculations revealed a lack of changes in the characteristic peak position and d-values of LMM-treated surface sediments compared with those from the control group, indicating that moderate LMM exposure would not change the crystalline structure of original lake sediments but would reduce sediments grain sizes (Table S1). Quantitative analysis also revealed that the LMM group did not lead to changes in the basic mineral composition of the original lake sediments (taranakite) but led to a sharp decrease in $Ca(PO_3)_2$ mass fractions (Table 1).

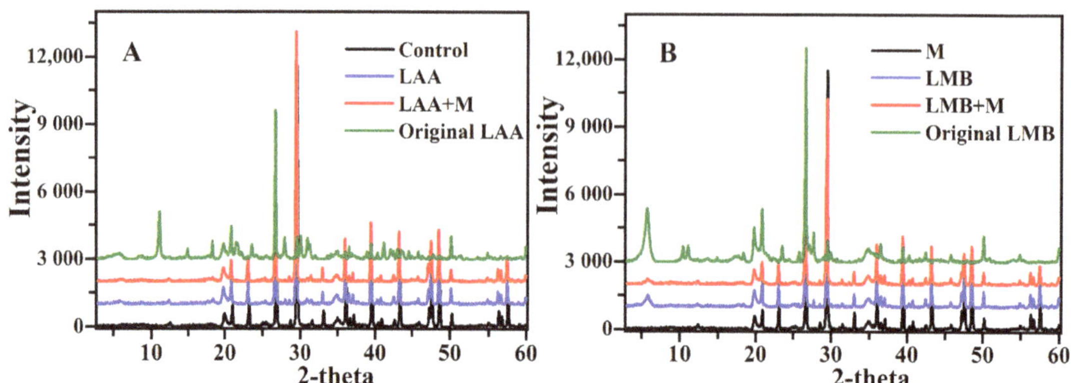

Figure 4. (**A**) statistical XRD patterns of control, LAA, LAA+M, and Original LAA. (**B**) statistical XRD patterns of M, LMB, LMB+M, and Original LMB. Original LAA and Original LMB are substances that are not mixed with sediments respectively.

Table 1. Chemical composition (mass fraction) of surface sediments after restoration experiments.

Material	Control	M	LAA	LMB	LAA+M	LMB+M
$CaCO_3$	55.6%	57.8%	45.5%	46.1%	58.6%	46.3%
SiO_2	18.1%	18.0%	22.0%	25.6%	19.2%	24.0%
$K_{0.77}Al_{1.93}(Al_{0.5}Si_{3.5})O_{10}(OH)_2$	15.7%	14.1%	19.5%	19.4%	13.9%	16.7%
$Al_2(Si_2O_5)(OH)_4$	6.0%	4.7%	7.8%	4.9%	3.6%	3.4%
FeS_2	3.1%	3.3%	2.4%	2.6%	2.9%	2.7%
$Ca(PO_3)_2$	1.4%	2.2%	1.0%	1.6%	0.5%	0.9%

3.3. Sediments P

Sediments TP concentrations ranged from 572.7 to 927.2 $mg \cdot kg^{-1}$, with a mean of $733.7 \pm 86.0\ mg \cdot kg^{-1}$, which is indicative of moderate pollution levels (Figure S1A), and that is generally consistent with those measured in a recent study [17]. Sediments P primarily existed in the following inorganic (Pi) and organic (Po) forms, NaOH–Pi, HCl–Pi, BD–Po, NaOH–Po, and Res–Po, with mean concentrations of $174.4 \pm 31.5\ mg \cdot kg^{-1}$, $99.8 \pm 16.0\ mg \cdot kg^{-1}$, $91.0 \pm 34.6\ mg \cdot kg^{-1}$, $115.4 \pm 45.3\ mg \cdot kg^{-1}$, and $121.0 \pm 14.5\ mg \cdot kg^{-1}$, respectively (Figure 5 and Figure S2). The eight extractable forms of sediments P exhibited varying degrees of transport and transformation during the incubation period. After 60 days of restoration, the concentrations of BD–Pi, NaOH–Pi, NaOH–Po, HCl–Pi, HCl–Po, and Res–Po in the water column increased by 7%, 52%, 40%, 3%, 11%, and 23%, compared with concentrations before the restoration, respectively, while NH_4Cl–Pi and BD–Po concentrations decreased by 63% and 43%, respectively.

The NaOH–Pi concentrations of sediments cores were significantly lower in each treatment than in the control group, while the HCl–Po concentrations were significantly higher. The BD–Po, BD–Pi, and NaOH–Pi concentrations in all LMM-treated sediments gradually increased with increasing depth, while the NaOH–Po, HCl–Pi, and HCl–Po concentrations decreased with depth. In addition, the vertical profile of BD–Po concentrations in sediments subjected to combined treatments significantly differed from those in other sediments, with the means of both R_1 and R_2 being negative for the combined treatment sediments. Moreover, the vertical sediment profiles of each P form indicated that the low P concentrations of bottom sediments were primarily due to decreased concentrations of BD–Po, NaOH–Po, and HCl–Pi that exhibited average R_2 values after 60 days of restoration of 10%, 13%, and 11%, respectively.

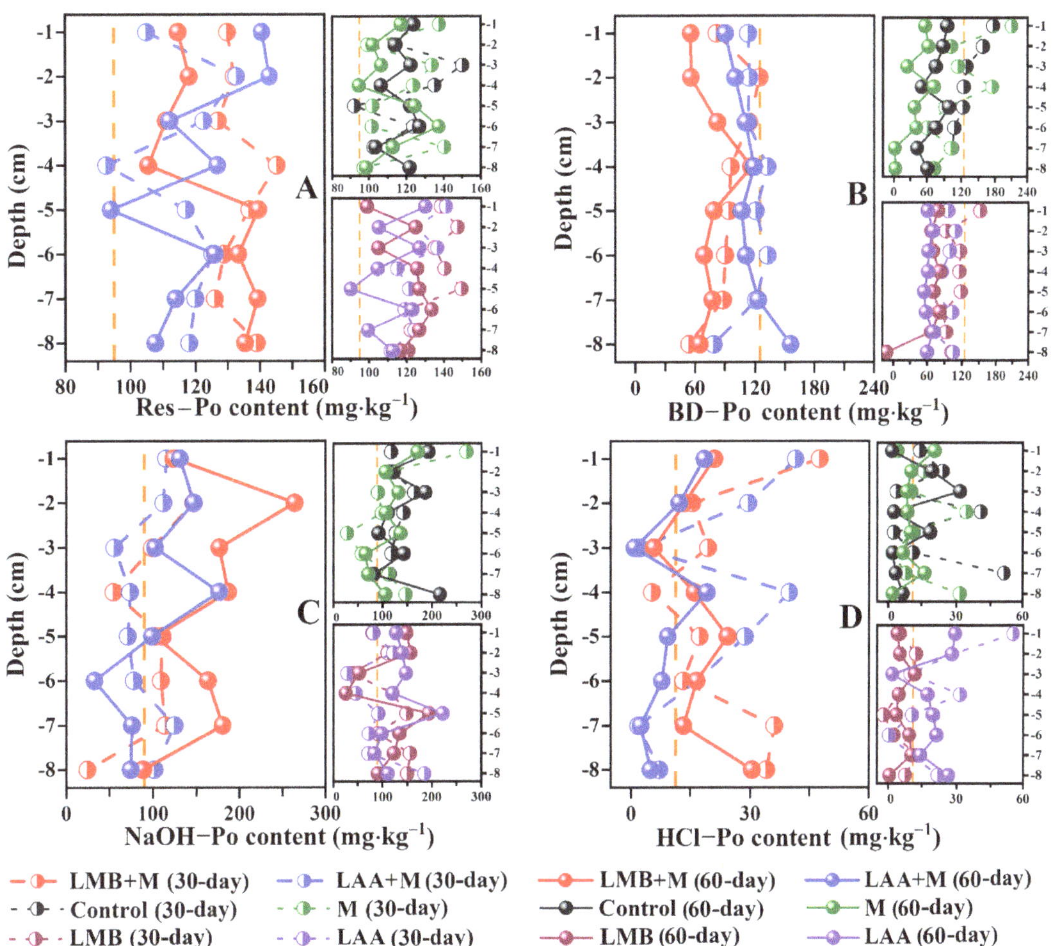

Figure 5. Vertical distribution of different P forms in sediments of the six treatment groups during the experiment. (**A–D**) represent the changes of Res−Po, BD−Po, NaOH−Po and HCl−Po contents in sediment, respectively. Each graph consists of three distinct subgraphs. The left subgraphs represent LAA+M and LMB+M groups, respectively. The upper right subgraphs represent control and M groups, respectively. The lower right subgraphs represent LAA and LMB groups, respectively. All data are triplicates. The vertical dashed line is the start P contents of the sediments.

3.4. DGT-Labile P, Fe, and S Variation

The one-dimensional distribution of DGT-labile P, Fe, and S is shown in Figure 6, with concentrations ranging from 0.01 to 0.63 mg·L⁻¹, 0.01 to 1.76 mg·L⁻¹, and 0.01 to 0.89 mg·L⁻¹, respectively. DGT-labile P, Fe, and S concentrations in the water column were about 3, 20, and 2 times higher than in the overlying water after 60 days of restoration. Overall, DGT-labile P/Fe/S concentrations in each group first increased and then stabilized or decreased with increasing depth (Figure S3). High correlations were observed for DGT-labile P concentrations and DGT-labile S concentrations along water columns, with linear correlation coefficients of >0.77 for all treatments, with the exception of the M group ($R^2 = 0.40$). The DGT-labile Fe concentrations considerably varied vertically and peaked in the top sediment layers of all experimental groups, exhibiting an increasing trend with incubation time that was most significant in the control and M groups. However, the linear correlation of DGT-labile Fe concentrations was minimal with respect to DGT-labile P and DGT-labile S concentrations (Table 2).

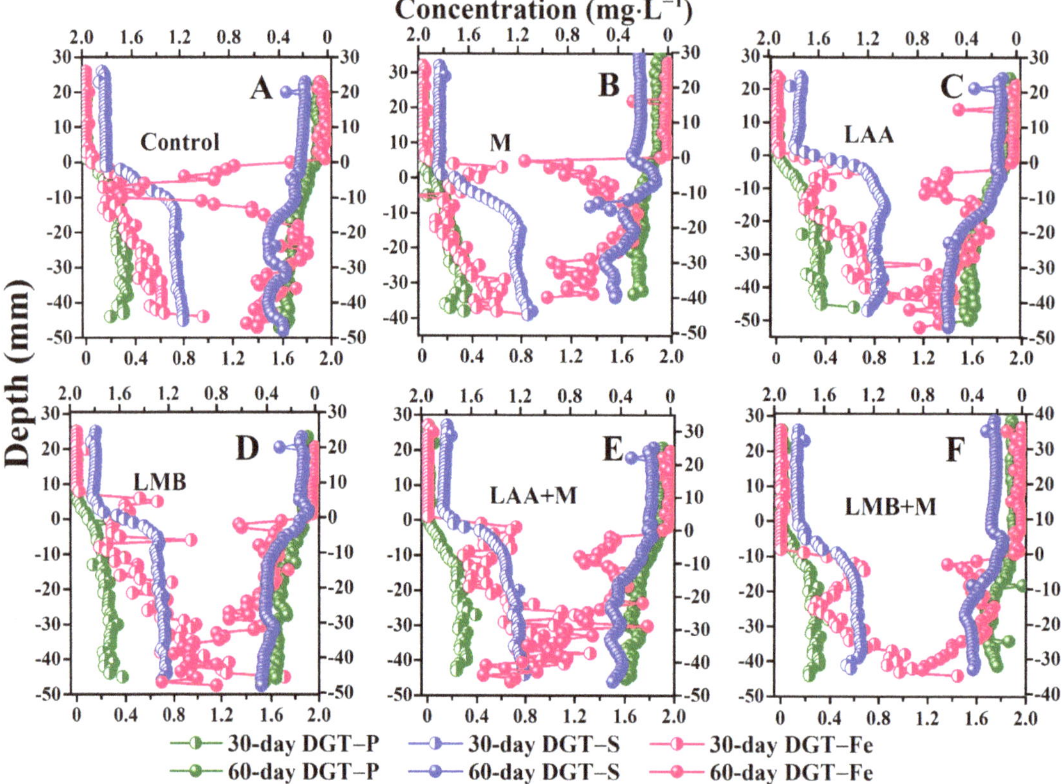

Figure 6. Vertical change of the DGT-labile P (Fe/S) concentrations in sediments of the six treatments groups during the experiment. The location of the sediment–water interface is represented by zero. (**A–F**) indicate the changes of DGT-labile P (Fe/S) concentrations in sediment of the Control, M, LAA, LMB, LAA+M, and LMB+M groups on the 30-day and 60-day, respectively.

Table 2. Relationships between DGT-labile P, DGT-labile Fe, and DGT-labile S ($p < 0.05$) in sediments of the six treatments used in the experiment.

Sample	30-Day			60-Day		
	P/Fe	P/S	Fe/S	P/Fe	P/S	Fe/S
Control	0.76	0.95	0.78	0.08	0.79	0.01
M	0.55	0.91	0.43	0.48	0.41	0.13
LAA	0.66	0.77	0.58	0.58	0.83	0.42
LMB	0.67	0.91	0.57	0.70	0.86	0.60
LAA+M	0.73	0.91	0.77	0.64	0.86	0.54
LMB+M	0.66	0.93	0.75	0.38	0.77	0.43

The internal flux of DGT-labile P at SWI was calculated based on Fick's first law of diffusion and was estimated to range from 0.28 to 0.95 mg·m⁻²·d⁻¹ during the 30 days and 60 days of restoration (Table 3). At the end of the experiment, the internal flux of DGT-labile P was lower to different degrees in all experimental groups compared with flux on the 30-day. This was particularly evident in the LAA+M group, which exhibited the lowest internal flux of 0.28 mg·m^{-2}·d^{-1} representing 60% and 44% lower levels than in the control group on the 30-day and in the LAA+M group at the beginning of the experiment, respectively. In contrast, the internal flux of DTG-labile P was 0.63 mg·m^{-2}·d^{-1}, 0.63 mg·m^{-2}·d^{-1}, 0.41 mg·m^{-2}·d^{-1}, and 0.34 mg·m^{-2}·d^{-1} in the M, LAA, LMB, and LMB+M groups, respectively.

Table 3. The apparent diffusion fluxes of P (mg·m^{-2}·d^{-1}) at sediment–water interfaces among different experimental treatments.

Time	Control	M	LAA	LMB	LAA+M	LMB+M
30-day	0.95	0.68	0.77	0.86	0.50	0.40
60-day	0.70	0.63	0.63	0.41	0.28	0.34

3.5. Macrophytes

The TP contents of macrophyte seedlings in the M, LAA+M, and LMB+M groups increased by 41%, 71%, and 57% during the incubation period, respectively (Figure 7A). The mean net height growth of seedlings was 98 cm, 58 cm, and 10 cm in the MVL, HVR, and CDL treatments, respectively (Figure 7B), with mean *RGR* values of 0.02 d^{-1}, 0.03 d^{-1}, and 0.004 d^{-1}, respectively (Figure 7C).

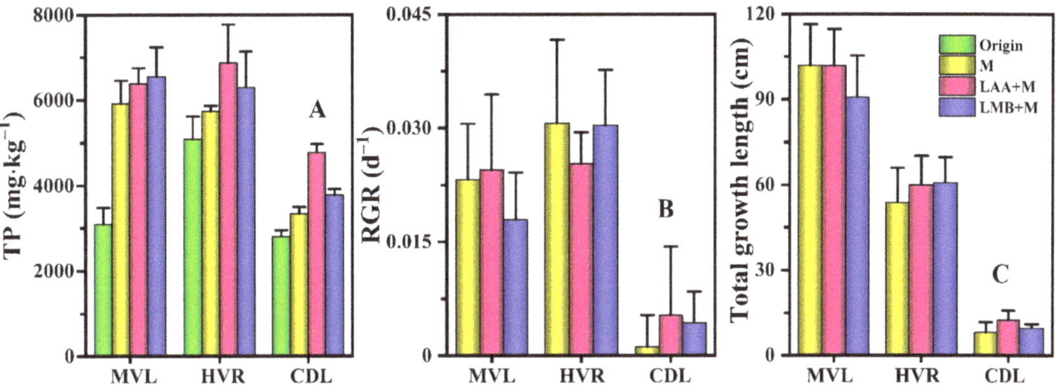

Figure 7. Macrophyte traits at the experiment. (**A**) indicates the change of TP content in different types of macrophytes. (**B,C**) represent the relative growth rate (*RGR*) and total growth length (length of macrophyte growth between the start and end of the experiment) of macrophytes in the experiments, respectively. MVL, HVR and CDL are *Myriophyllum verticillatum* L., *Hydrilla verticillata (Linn. f.) royle* and *Ceratophyllum demersum* L., respectively.

4. Discussion

4.1. Effects of LMM and Macrophytes on P Concentrations

The TP concentrations in overlying waters in the M, LAA, LMB, LAA+M, and LMB+M groups decreased by 28%, 48%, 51%, 55%, and 51% over the experimental period compared with the control group, respectively (Figure 1). Thus, combined treatment with both LMM and macrophytes led to better restoration performance of lake water P than individual treatments with LMM or macrophytes. However, the combined effects were not a simple summation of the two separate effects [3,13]. Changes in P concentrations in the overlying water indicated that the two restoration methods, whether used alone or in combination, primarily removed SRP fractions from the water column. SRP removal rates exceeded 88% in all treatment groups by the 60 days of restoration, while dissolved organophosphorus (DOP) and particulate P (PP) generally increased after stabilization of water quality in the incubators.

Combined treatment with both LMM and macrophytes effectively inhibited the release of P from sediments. Indeed, the diffusion fluxes of P in the water column after 60-day were reduced by 55% and 55% in the LAA+M group compared with the LAA and M groups, respectively, and by 16% and 46% in the LMB+M group compared with the LMB and M groups, respectively (Table 3). Consequently, both restoration methods, whether used alone or in combination, can reduce internal P loading in lakes, although combined use leads to better performance. In addition, changes in porewater P concentrations among different experimental groups indicated that the combined treatment was more effective in removing P from water columns. Between the 30 days of incubation and the end of the experiment, porewater TP concentrations decreased by 33% and 27% in the LAA+M and LMB+M groups, respectively, but increased by 34%, 13%, and 28% in the LAA, LMB, and M groups, respectively (Figure 2).

High internal P loading in the water column and high aqueous P concentrations was maintained owing to the transport and transformation of NH_4Cl–P, BD–Pi, BD–Po, and Res–Po during the experiment. The mean accumulation rates (r) of these P forms were negative for most sediments cores during the incubation period, indicating that these forms mostly entered the overlying water through mineralization, degradation, resuspension, or gradually transformed into more stable forms, such as NaOH–Pi, NaOH–Po, HCl–Pi, and HCl–Po (Table S2) [27]. The accumulation or release of different P forms in sediments cores over the incubation period was also investigated. These analyses revealed that (1) sediment accumulation of NaOH–Pi, NaOH–Po, and HCl–Pi was higher in the LAA+M and LMB+M groups than in other groups over the incubation period; (2) the release of redox-sensitive BD–P (comprising BD–Pi and BD–Po) was higher in the LAA (61.6 $mg \cdot kg^{-1}$), LMB (63.3 $mg \cdot kg^{-1}$), and M (82.2 $mg \cdot kg^{-1}$) groups than in the other treatment groups; and (3) total P accumulation was greater than total P release in the combined treatment groups, while the opposite was observed in the individual treatment groups. These phenomena may be attributed to oxygen transfer from roots to rhizosphere sediments during the growth of macrophytes, the mineralization of organophosphate, and the strong ability of La to bind P [14,29,30]. In particular, oxygen release from macrophytes roots can cause oxidation of compounds in rhizospheres such as low-valence Fe, Mn, and other metals to generate high-valence P-containing compounds, thereby leading to the continuous accumulation of P [17,31]. Further, organophosphorus mineralization also results in changes to redox conditions and pH that, in turn, affect the transport and transformation of BD–TP and NaOH–Pi in sediments [21,27]. Increased HCl–P concentrations (i.e., the sum of HCl–Pi and HCl–Po) in the sediments may be related to the generation of $LaPO_4$(s) that primarily exists as HCl–P when extracted with HCl [14,32].

4.2. Effects of LMM on Macrophytes Physiological Indices

The root systems of macrophytes are the primary area where biologically available P (BAP, comprising NH_4Cl–Pi, BD–Pi, NaOH–Pi, and HCl–Pi) is taken up from surface sediments. Thus, the reduction in sediments BAP concentrations due to the addition of LMM may negatively affect macrophyte growth [3,33]. Experimental studies have documented species-specific differences in the effects of LMM on macrophytes species [34–36], consistent with the results of this study, wherein the net height growth of the macrophytes HVR and CDL, in addition to the *RGR* of HVR, were slightly higher in the combined treatments than individual macrophytes treatments (Figure 7B,C). However, LMM addition to a 100:1 mass ratio in this study led to a low inhibitory effect on the growth of macrophytes. Differences in net height growth and the *RGR* between groups treated with individual macrophytes and combined treatments were small, likely because LMM passivation provided suitable environments but also contributed nutrients such as K and Mg that could support the growth of macrophytes (Figure 3D,F). The BAP concentrations of surface sediments on the 30 days of restoration were significantly higher in the treatments with individual macrophytes than in the LMM treatments, while the opposite was observed at the end of the experiment (Figure S1B). Further, the phosphate concentrations of macrophytes in the combined treatments were significantly higher by 24% and 11% than in the treatments with individual macrophytes, respectively (Figure 7A). Thus, insignificant passivation at the early stage of restoration led to slow macrophytes germination and growth in the lake ecosystem, while improved water quality in the middle and later stages promoted macrophytes growth, as demonstrated by differences in porewater TP concentrations between different treatment groups (Figure 2). Overall, LMM addition to a 100:1 mass ratio led to minimal negative effects on macrophytes growth.

4.3. Effects of LMM on Surface Sediments Microenvironment

Ultra-high-resolution SEM-EDS imaging and XRD spectroscopy further revealed changes in the microscopic morphologies and structures of remediating sediments in response to combined treatment with both in situ adsorption and biological forcing. SEM-EDS elemental mapping revealed that La–P minerals were predominantly present as sub-nanometre crystals in surface sediments (Figure 3B–D,F). Further, the high Si, Ca, and O elemental ratios in sediments indicated strong chemical weathering in the source area that would lead to the dissolution of phosphate-bearing minerals and phosphate release. XRD analysis also revealed that surface sediments phosphate primarily existed as taranakite, while LMM resulted in a significant decrease in the mass fraction of taranakite and $Ca(PO_3)_2$ during restoration (Table 1). Nevertheless, macrophytes enhanced the FeS_2 and $CaCO_3$ mass fractions in the surface sediments.

The concentration difference between sediments porewater and overlying water in lake systems is an important factor driving internal P release in lakes [4,17]. Due to the obvious differences between passivation materials and lake sediments in the microscopic morphology, elemental composition, and surface physical properties, the addition of LMM will change the physical and chemical properties of the surface sediments. Numerous studies have shown that the structure collapse and pore blockage caused by LMB and LAA increase the specific surface areas of the passivation materials and the adsorption capacity for phosphate during the high temperature of calcination, and the static layer formed at SWI is the key to maintaining the repair of LMM [14,32]. SEM-EDS elemental mapping revealed that LMB and LAA materials were composed of nanosheet aggregates of various sizes and contain metallic elements such as Al^{3+}, K^+, Ca^{2+}, and La^{3+} (Figure 3D,H). The modified bentonite and attapulgite have a large number of positive charges on the surface, which helps LMM to have a strong adsorption effect on P and algae [8,14]. This may be an important reason why NaOH–P and HCl–P contents different at different deposition depths in the surface sediments of the LMM treatment groups were significantly higher than that of other treatment groups. According to changes in P forms content, the redox-sensitive P forms (NH_4Cl–P and BD–P) mainly migrate to NaOH–P and HCl–P in the surface sediments, indicating that LMM can lead to the migration and transformation of redox-sensitive chemical forms at SWI. Combined with the release flux characteristics of P in the sediment profiles, it was shown that LMM-induced changes in the surface sediments microenvironment also affect the decay or release rates of P in lake systems, although it is usually limited by the stable performance of a static layer a few millimeters on top of the sediments.

4.4. P Biogeochemical Behaviour

The transport and transformation of P at SWI are primarily controlled by the stability of different P forms and various physical, geochemical, and biological processes [18,37]. The inorganic P forms of NH_4Cl–P, BD–Pi, NaOH–Pi, and HCl–Pi are considered unstable P fractions (BAP) in the modified Hupfer sequential extraction scheme and are easily released from the solid phase to the aqueous phase under certain conditions [26,38,39]. In the present study, the mean BAP to TP ratios in sediments cores after 60 days of restoration were 53%, 50%, 47%, and 51% in the LMA, LMB, LAA+M, and LMB+M groups, respectively, indicating that the P fractions in the water column are unstable when LMM are used to remediate sediments [14,32].

Evaluation of the sedimentation rates (R) of P fractions in sediments cores (Tables S3 and S4) revealed that organophosphorus mineralization is the main internal source of P in all of the LMM-treated sediments [28,40]. Although high BAP concentrations and internal P fluxes were observed in the sediments, the geochemical reactions dominated by the reduction and dissolution of P-iron oxides/hydroxides contribute little to overlying water P concentrations in macrophyte-dominated lakes because these lakes are characterized by high DO concentrations and high reduction potential (Figure S4B,D). These dynamics are further substantiated by the non-significant linear correlation between water column DGT-labile P and DGT-labile Fe during the experiment (Table 2). Further, the significant positive correlation observed between DGT-labile P and DGT-labile S was consistent with the observation that the refractory P fractions in sediments (e.g., NaOH–Po, HCl–Pi, and HCl–Po) exhibited positive release and sedimentation rates (R) since high sulfate concentrations would lead to the reductive release of some insoluble phosphates from sediments [41,42].

The concentrations of various P forms were also analyzed. The free phosphate ions released in the LMM treatment groups were primarily adsorbed by La^{3+} to form La-phosphate compounds ($LaPO_4 \cdot nH_2O$) [14,15], as confirmed by EDS spectroscopy (Figure 3B–D,F). These observations correspond to high La^{3+} concentrations in LMM and because La^{3+} exhibits stronger electronegativity than other metal ions such as Fe and Al. The mass fraction of FeS_2 in different experimental groups, the transport and transformation characteristics of P in sediments profiles, and DGT-labile Fe concentration trends in sediments cores jointly suggested that the dissimilatory reduction of Fe in macrophytes roots may be an important pathway for organic matter metabolism in surface sediments [43,44].

BD–Po and NaOH–Po are reactive organophosphate forms that can be mineralized and hydrolysis under certain conditions, leading to dissolving phosphate or small organophosphate compounds that can be easily absorbed and used by organisms [26,45]. HCl–P and Res-Po are stable forms and will be released under strongly acidic or alkaline conditions to some extent [6,46]. All forms of sediments P in each treatment group in this study, except the HCl–P and Res–Po forms, exhibited decreased concentrations across the incubation period compared with the control group (Table S5). This observation, when combined with the transport and transformation characteristics of P in the sediments (Table S2) and the XRD analyses, indicated that the active ingredient in the Supplemented Material reacted with P compounds in the sediments through ligand exchange, electrostatic interaction, and Lew is acid–base interaction to form stable La–P minerals ($LaPO_4 \cdot nH_2O$) [47,48]. Further analysis revealed that BD–Po and NaOH–Po concentrations in the 0–1 cm sediment layer were significantly lower in all treatments than in the control group. This was especially evident in the LAA+M group, where BD–Po and NaOH–Po concentrations on the 60-day were lower by 40% and 11% compared with the control, respectively, which may be related to the microbial decomposition of organophosphate. These results were consistent with the SEM-EDS imaging that revealed dentate flocs (Figure 3C,F) resembling microorganisms that have been previously reported [49,50]. Nevertheless, the short incubation period used in the present study may fail to reveal the possible inter-annual or monthly variation of macrophytes when this technique is applied in real-world scenarios. Therefore, long-term monitoring of changes in MDE lakes in response to the combined application of LMM and macrophytes is needed.

5. Conclusions

In this study, a practical approach for efficient ecological restoration is described for MDE lakes with organophosphorus-controlled internal P loading. The results indicate that mineralization of organophosphates (BD–Po, HCl–Po, and Res–Po) is an important factor for maintaining high internal P loadings and overlying water P concentrations during the experiments. The combination of LMM and macrophytes led to synergistic effects in the performance of aquatic ecological restoration compared with individual retreatments. Specifically, the combined treatments exhibited lower internal P loading than individual treatments. This was especially evident for the LAA+M group that achieved the best restoration outcome. SEM-EDS elemental mapping and XRD analysis revealed that the active ingredient in the added material reacted with sediments P forms to form stable La–P compounds ($LaPO_4 \cdot nH_2O$). LMM also enhanced the conversion rates of redox-sensitive P forms in surface sediments. Simultaneously, LMM had less effect on macrophyte traits (e.g., relative growth rate (*RGR*), length) compared with individual macrophyte groups and enhanced the absorption capacity of phosphate by macrophytes.

Supplementary Materials: The following supporting information can be downloaded at: https://www.mdpi.com/article/10.3390/w14121847/s1, Table S1. The 2θ, interlayer spacing (d) and crystal size (CS) of different experimental groups in the experiment; Table S2. The release and accumulation rates (r) of P in sediments cores during the experiment; Table S3. The fast-release and sedimentation rates (R1) of different P forms in sediments cores during the experiment; Table S4. The flow-release and sedimentation rates (R2) of different P forms in sediments cores during the experiment; Table S5. The increase or decrease rates of different P forms in the sediment cores of different treatment groups relative to the control group during the experiment; Figure S1. Vertical distribution of P content in the sediments of the different treatments during the experiment. (A–C) represent the content changes of TP, Pi, and Po in the sediments, and each graph consists of three distinct subgraphs. The left subgraphs represent LAA+M and LMB+M groups, respectively. The upper right subgraphs represent control and M groups, respectively. The lower right subgraphs represent LAA and LMB groups, respectively. All data are three replicates; Figure S2. Vertical distribution of P content in sediments of different treatments during the ex-periment. (A–D) represent the changes in the content of inorganic NH4Cl-P, BD-Pi, NaOH-Pi and HCl-Pi in the sediments, respectively. Each graph consists of three distinct subgraphs. The left subgraphs represent LAA+M and LMB+M groups, respectively. The upper right subgraphs represent control and M groups, respectively. The lower right subgraphs represent LAA and LMB groups, respectively. All data are triplicates. The vertical dashed line is the initial P content of the sediments; Figure S3. Vertical variation of DGT-labile S in sediments during the experiment. (A–F) are the changes of DGT-labile S in the sediments of the Control, M, LAA, LMB, LAA+M and LMB+M repaired on the 30-day, respectively. (A1), (B1), (C1), (D1), (E1), and (F1) are the changes of DGT-labile S in the sediments of the Control, M, LAA, LMB, LAA+M and LMB+M repaired on the 60-day, respectively; Figure S4. Water quality parameters. (A–D) respectively represent the overlying water temperature (T), dissolved oxygen (DO), pH and redox potential (ORP) changes during the experiment.

Author Contributions: W.Y., Y.H. and J.C. contributed to conception and design of the study. W.Y., J.W. (Jiaxi Wu) and H.Y. participated in field sample collection and laboratory analysis. W.Y. and J.W. (Jiaxi Wu) organized the database and statistical analysis. W.Y. wrote the first draft of the manuscript. H.Y., Y.Y., J.C., P.L., J.W. (Jingfu Wang) and D.X. guided the writing of thesis. All authors have read and agreed to the published version of the manuscript.

Funding: This study was supported by the Science and Technology Project of Guizhou Province (QianKeHe Support [2020]4Y015, [2022] general 201, [2020]4Y006, LH [2017]7344), CAS "Light of West China" Program and the Central Government Leading Local Science and Technology Development (QianKeZhongYinDi [2021]4028).

Institutional Review Board Statement: Not applicable.

Informed Consent Statement: Not applicable.

Data Availability Statement: The datasets used and/or analyzed during the current study are available from the corresponding author on reasonable request.

Acknowledgments: Thanks to the Guizhou Caohai National Nature Reserve Administration for its help in sampling. Thanks to the Guizhou Caohai Wetland Ecosystem National Positioning Observation and Research Station for supplying the experimental site.

Conflicts of Interest: The authors declare that they have no competing interests.

References

1. Smith, V.H.; Schindler, D.W. Eutrophication science: Where do we go from here? *Trends Ecol. Evol.* **2009**, *24*, 201–207. [CrossRef] [PubMed]
2. Oosterhout, F.V.; Waajen, G.; Yasseri, S.; Manzi Marinho, M.; Pessoa Noyma, N.; Mucci, M.; Douglas, G.; Lürling, M. Lanthanum in Water, Sediment, Macrophytes and chironomid larvae following application of Lanthanum modified bentonite to lake Rauwbraken (The Netherlands). *Sci. Total Environ.* **2020**, *706*, 135188. [CrossRef] [PubMed]
3. Zhang, X.M.; Zhen, W.; Jensen, H.S.; Reitzel, K.; Jeppesen, E.; Liu, Z.W. The combined effects of macrophytes (Vallisneria denseserrulata) and a lanthanum-modified bentonite on water quality of shallow eutrophic lakes: A mesocosm study. *Environ. Pollut.* **2021**, *277*, 116720. [CrossRef] [PubMed]
4. Tammeorg, O.; Nürnberg, G.; Horppila, J.; Haldna, M.; Niemistö, J. Redox-related release of phosphorus from sediments in large and shallow Lake Peipsi: Evidence from sediment studies and long-term monitoring data. *J. Great Lakes Res.* **2020**, *46*, 1595–1603. [CrossRef]
5. Recknagel, F.; Hosomi, M.; Fukushima, T.; Kong, D.S. Short- and long-term control of external and internal phosphorus loads in lakes—A scenario analysis. *Water Res.* **1995**, *29*, 1767–1779. [CrossRef]
6. Wang, Y.T.; Zhang, T.Q.; Zhao, Y.C.; Ciborowski, J.J.H.; Zhao, Y.M.; O'Halloran, I.P.; Qi, Z.M.; Tan, C.S. Characterization of sedimentary phosphorus in Lake Erie and on-site quantification of internal phosphorus loading. *Water Res.* **2021**, *188*, 116525. [CrossRef]
7. Zhang, R.Y.; Wang, L.Y.; Wu, F.C.; Song, B.A. Phosphorus speciation in the sediment profile of Lake Erhai, southwestern China: Fractionation and ^{31}P NMR. *J. Environ. Sci.* **2013**, *25*, 1124–1130. [CrossRef]
8. Waajen, G.; Pauwels, M.; Lürling, M. Effects of combined flocculant–Lanthanum modified bentonite treatment on aquatic macroinvertebrate fauna. *Water Res.* **2017**, *122*, 183–193. [CrossRef]
9. Hilt, S.; Gross, E.M. Can allelopathically active submerged macrophytes stabilise clear-water states in shallow lakes? *Basic Appl. Ecol.* **2008**, *9*, 422–432. [CrossRef]
10. Scheffer, M.; Hosper, S.H.; Meijer, M.L.; Moss, B.; Jeppesen, E. Alternative Equilibria in Shallow Lakes. *Trends Ecol. Evol.* **1993**, *8*, 275–279. [CrossRef]
11. Fathollahzadeh, H.; Kaczala, F.; Bhatnagar, A.; Hogland, W. Significance of environmental dredging on metal mobility from contaminated sediments in the Oskarshamn Harbor, Sweden. *Chemosphere* **2015**, *119*, 445–451. [CrossRef] [PubMed]
12. Yin, H.; Yang, C.; Yang, P.; Kaksonen, A.H.; Douglas, G.B. Contrasting effects and mode of dredging and in situ adsorbent amendment for the control of sediment internal phosphorus loading in eutrophic lakes. *Water Res.* **2020**, *189*, 116644. [CrossRef] [PubMed]
13. Lin, J.W.; Zhao, Y.Y.; Zhan, Y.H.; Wang, Y. Control of internal phosphorus release from sediments using magnetic lanthanum/iron-modified bentonite as active capping material. *Environ. Pollut.* **2020**, *264*, 114809. [CrossRef] [PubMed]
14. Yin, H.B.; Yang, P.; Kong, M.; Li, W. Use of lanthanum/aluminum co-modified granulated attapulgite clay as a novel phosphorus (P) sorbent to immobilize P and stabilize surface sediment in shallow eutrophic lakes. *Chem. Eng. J.* **2020**, *385*, 123395. [CrossRef]
15. Goscianska, J.; Ptaszkowska-Koniarz, M.; Frankowski, M.; Franus, M.; Panek, R.; Franus, W. Removal of phosphate from water by lanthanum-modified zeolites obtained from fly ash. *J. Colloid Interface Sci.* **2018**, *513*, 72–81. [CrossRef]
16. Xiang, S.L.; Zhou, W.B. Phosphorus forms and distribution in the sediments of Poyang Lake, China. *Int. J. Sediment Res.* **2011**, *26*, 230–238. [CrossRef]
17. Yu, W.; Yang, H.Q.; Chen, J.A.; Liao, P.; Chen, Q.; Yang, Y.Q.; Liu, Y. Organic Phosphorus Mineralization Dominates the Release of Internal Phosphorus in a Macrophyte-Dominated Eutrophication Lake. *Front. Environ. Sci.* **2022**, *9*, 812834. [CrossRef]
18. Barik, S.K.; Bramha, S.; Bastia, T.K.; Behera, D.; Kumar, M.; Mohanty, P.K.; Rath, P. Characteristics of geochemical fractions of phosphorus and its bioavailability in sediments of a largest brackish water lake, South Asia. *Ecohydrol. Hydrobiol.* **2019**, *19*, 370–382. [CrossRef]
19. Joshi, S.R.; Kukkadapu, R.K.; Burdige, D.J.; Bowden, M.E.; Sparks, D.L.; Jaisi, D.P. Organic matter remineralization predominates phosphorus cycling in the mid-Bay sediments in the Chesapeake Bay. *Environ. Sci. Technol.* **2015**, *49*, 5887–5896. [CrossRef]
20. Zhu, Y.R.; Feng, W.Y.; Liu, S.S.; He, Z.Q.; Zhao, X.L.; Liu, Y.; Guo, J.Y.; Giesy, J.P.; Wu, F.C. Bioavailability and preservation of organic phosphorus in lake sediments: Insights from enzymatic hydrolysis and ^{31}P nuclear magnetic resonance. *Chemosphere* **2018**, *211*, 50–61. [CrossRef]
21. Lv, C.W.; He, J.; Zuo, L.; Vogt, R.D.; Zhu, L.; Zhou, B.; Mohr, C.W.; Guan, R.; Wang, W.Y.; Yan, D.H. Processes and their explanatory factors governing distribution of organic phosphorous pools in lake sediments. *Chemosphere* **2016**, *145*, 125–134. [CrossRef]
22. Murphy, J.; Riley, J.P. A modified single solution method for the determination of phosphate in natural waters. *Anal. Chim. Acta* **1962**, *27*, 31–36. [CrossRef]
23. Chen, Q.; Chen, J.A.; Wang, J.F.; Guo, J.Y.; Jin, Z.F.; Yu, P.; Ma, Z. In situ, high-resolution evidence of phosphorus release from sediments controlled by the reductive dissolution of iron-bound phosphorus in a deep reservoir, southwestern China. *Sci. Total Environ.* **2019**, *666*, 39–45. [CrossRef] [PubMed]
24. Ding, S.M.; Han, C.; Wang, Y.P.; Yao, L.; Wang, Y.; Xu, D.; Sun, Q.; Williams, P.N.; Zhang, C.S. In situ, high-resolution imaging of labile phosphorus in sediments of a large eutrophic lake. *Water Res.* **2015**, *74*, 100–109. [CrossRef] [PubMed]
25. Crusius, J.; Anderson, R.F. Core Compression and Surficial Sediment Loss of Lake Sediments of High Porosity Caused by Gravity Coring. *Limnol. Oceanogr.* **1991**, *36*, 1021–1031. [CrossRef]

26. Hupfer, M.; Gfichter, R.; Giovanoli, R. Transformation of phosphorus species in settling seston and during early sediment diagenesis. *Aquat. Sci.* **1995**, *57*, 305–324. [CrossRef]
27. Rita, J.C.D.O.; Gama-Rodrigues, A.C.; Gama-Rodrigues, E.F.; Zaia, F.C.; Nunes, D.A.D. Mineralization of organic phosphorus in soil size fractions under different vegetation covers in the north of Rio de Janeiro. *Content Uploaded Emanuela Gama-Rodrigues* **2013**, *37*, 1207–1215. [CrossRef]
28. Hupfer, M.; Lewandowski, J. Retention and early diagenetic transformation of phosphorus in Lake Arendsee (Germany)-consequences for management strategies. *Arch. Hydrobiol.* **2005**, *164*, 143–167. [CrossRef]
29. Xu, Y.; Han, F.E.; Li, D.P.; Zhou, J.; Huang, Y. Transformation of internal sedimentary phosphorus fractions by point injection of CaO_2. *Chem. Eng. J.* **2018**, *343*, 408–415. [CrossRef]
30. Stephen, D.; Moss, B.; Phillips, G. Do rooted macrophytes increase sediment phosphorus release? *Hydrobiologia* **1997**, *342*, 27–34. [CrossRef]
31. Søndergaard, M.; Jensen, J.P.; Jeppesen, E. Role of sediment and internal loading of phosphorus in shallow lakes. *Hydrobiologia* **2003**, *506–509*, 135–145. [CrossRef]
32. Wang, Y.; Ding, S.M.; Wang, D.; Sun, Q.; Lin, J.; Shi, L.; Chen, M.S.; Zhang, C.S. Static layer: A key to immobilization of phosphorus in sediments amended with lanthanum modified bentonite (Phoslock®). *Chem. Eng. J.* **2017**, *325*, 49–58. [CrossRef]
33. Christiansen, N.H.; Andersen, F.Ø.; Jensen, H.S. Phosphate uptake kinetics for four species of submerged freshwater macrophytes measured by a 33P phosphate radioisotope technique. *Aquat. Bot.* **2016**, *128*, 58–67. [CrossRef]
34. Xu, Q.S.; Fu, Y.Y.; Min, H.L.; Cai, S.J.; Sha, S.; Cheng, G.Y. Laboratory assessment of uptake and toxicity of lanthanum (La) in the leaves of Hydrocharis dubia (Bl.) Backer. *Environ. Sci. Pollut. Res.* **2012**, *19*, 3950–3958. [CrossRef]
35. Wang, X.; Shi, G.X.; Xu, Q.S.; Xu, B.J.; Zhao, J. Lanthanum- and cerium-induced oxidative stress in submerged Hydrilla verticillata plants. *Russ. J. Plant Physiol.* **2007**, *54*, 693–697. [CrossRef]
36. Herrmann, H.; Nolde, J.; Berger, S.; Heise, S. Aquatic ecotoxicity of lanthanum-A review and an attempt to derive water and sediment quality criteria. *Ecotoxicol. Environ. Saf.* **2016**, *124*, 213–238. [CrossRef] [PubMed]
37. Slomp, C.P.; Mort, H.P.; Jilbert, T.; Reed, D.C.; Gustafsson, B.G.; Wolthers, M. Coupled dynamics of iron and phosphorus in sediments of an oligotrophic coastal basin and the impact of anaerobic oxidation of methane. *PLoS ONE* **2013**, *8*, e62386. [CrossRef]
38. Mu, Z.; Wang, Y.C.; Wu, J.K.; Cheng, Y.; Lu, J.; Chen, C.C.; Zhao, F.X.; Li, Y.; Hu, M.M.; Bao, Y.F. The influence of cascade reservoir construction on sediment biogenic substance cycle in Lancang River from the perspective of phosphorus fractions. *Ecol. Eng.* **2020**, *158*, 106051. [CrossRef]
39. Kwak, D.H.; Jeon, Y.T.; Duck, H.Y. Phosphorus fractionation and release characteristics of sediment in the Saemangeum Reservoir for seasonal change. *Int. J. Sediment Res.* **2018**, *33*, 250–261. [CrossRef]
40. Dittrich, M.; Chesnyuk, A.; Gudimov, A.; McCulloch, J.; Quazi, S.; Young, J.; Winter, J.; Stainsby, E.; Arhonditsis, G. Phosphorus retention in a mesotrophic lake under transient loading conditions: Insights from a sediment phosphorus binding form study. *Water Res.* **2013**, *47*, 1433–1447. [CrossRef]
41. Zak, D.; Kleeberg, A.; Hupfer, M. Sulphate-mediated phosphorus mobilization in riverine sediments at increasing sulphate concentration, River Spree, NE Germany. *Biogeochemistry* **2006**, *80*, 109–119. [CrossRef]
42. Cui, Y.; Xiao, R.; Xie, Y.; Zhang, M.X. Phosphorus fraction and phosphate sorption-release characteristics of the wetland sediments in the Yellow River Delta. *Phys. Chem. Earth* **2018**, *103*, 19–27. [CrossRef]
43. Yang, J.X.; Tam, N.F.Y.; Ye, Z.H. Root porosity, radial oxygen loss and iron plaque on roots of wetland plants in relation to zinc tolerance and accumulation. *Plant. Soil* **2013**, *374*, 815–828. [CrossRef]
44. Canfield, D.E.; Thamdrup, B.; Hansen, J.W. The anaerobic degradation of organic matter in Danish coastal sediments: Iron reduction, manganese reduction, and sulfate reduction. *Geochim. Cosmochim. Acta* **1993**, *57*, 3867–3883. [CrossRef]
45. Tiessen, H.; Stewart, J.W.B.; Cole, C.V. Pathways of Phosphorus Transformations in Soils of Differing Pedogenesis. *Soil Sci. Soc. Am. J.* **1984**, *48*, 853–858. [CrossRef]
46. Rydin, E. Potentially mobile phosphorus in Lake Erken sediment. *Water Res.* **2000**, *34*, 2037–2042. [CrossRef]
47. Zhang, L.; Zhou, Q.; Liu, J.Y.; Chang, N.; Wan, L.H.; Chen, J.H. Phosphate adsorption on lanthanum hydroxide-doped activated carbon fiber. *Chem. Eng. J.* **2012**, *185–186*, 160–167. [CrossRef]
48. Zhang, L.; Liu, Y.H.; Wang, Y.L.; Li, X.H.; Wang, Y.Y. Investigation of phosphate removal mechanisms by a lanthanum hydroxide adsorbent using p-XRD, FTIR and XPS. *Appl. Surf. Sci.* **2021**, *557*, 149838. [CrossRef]
49. Pennafirme, S.; Pereira, D.C.; Pedrosa, L.G.M.; Machado, A.S.; Silva, G.O.A.; Keim, C.N.; Lima, I.; Lopes, R.T.; Paixão, I.C.N.P.; Crapez, M.A.C. Characterization of microbial mats and halophilic virus-like particles in a eutrophic hypersaline lagoon (Vermelha Lagoon, RJ, Brazil). *Reg. Stud. Mar. Sci.* **2019**, *31*, 100769. [CrossRef]
50. Rasuk, M.C.; Fernandez, A.B.; Kurth, D.; Contreras, M.; Novoa, F.; Poire, D.; Farias, M.E. Bacterial Diversity in Microbial Mats and Sediments from the Atacama Desert. *Microb. Ecol.* **2016**, *71*, 44–56. [CrossRef]

Article

Effect of Ecosystem Degradation on the Source of Particulate Organic Matter in a Karst Lake: A Case Study of the Caohai Lake, China

Jiaxi Wu [1,2], Haiquan Yang [2,*], Wei Yu [2,3], Chao Yin [2], Yun He [2,4], Zheng Zhang [2], Dan Xu [2], Qingguang Li [1] and Jingan Chen [2]

1. College of Resources and Environmental Engineering, Guizhou University, Guiyang 550025, China; wujiaxi@mail.gyig.ac.cn (J.W.); leeqg12@163.com (Q.L.)
2. State Key Laboratory of Environmental Geochemistry, Institute of Geochemistry, Chinese Academy of Sciences, Guiyang 550081, China; yuwei@mail.gyig.ac.cn (W.Y.); yinchao@mail.gyig.ac.cn (C.Y.); heyun@mail.gyig.ac.cn (Y.H.); 1993zheng@gmail.com (Z.Z.); xudan@mail.gyig.ac.cn (D.X.); chenjingan@vip.skleg.cn (J.C.)
3. School of Geographic and Environmental Sciences, Guizhou Normal University, Guiyang 550025, China
4. College of Eco-Environmental Engineering, Guizhou Minzu University, Guiyang 550025, China
* Correspondence: yanghaiquan@vip.skleg.cn

Abstract: The cycle of biogenic elements in lakes is intimately linked with particulate organic matter (POM), which plays a critical role in ecosystem restoration and the control of eutrophication. However, little is known regarding the functionality of ecosystem degradation on the source of POM in the water of a karst lake. To fill this knowledge gap, herein we compared the temporal and spatial distribution characteristics of POM prior to and after ecosystem degradation in the karst lake Caohai Lake, located in the southwest of China, and analyzed the source of POM using a combination of carbon and nitrogen stable isotopes ($\delta^{13}C$–$\delta^{15}N$). Our results showed that the dissolved oxygen (DO) concentration and pH values decreased, and the concentrations of POM in water increased by 11% and 31% in the wet and dry seasons, respectively. The decrease in the $\delta^{13}C$ value of POM was accompanied by the increase in the $\delta^{15}N$ value of POM in the water of Caohai lake. Prior to the ecosystem's degradation, sediment resuspension (28%) and submerged macrophytes (33%) were the dominant sources of POM in lake water. In contrast, sediment resuspension (51%) was the major source of POM after the ecosystem's degradation. Environmental factors, including DO, turbidity, water depth, and water temperature, that are related to photosynthesis and sediment resuspension are the main factors controlling the spatiotemporal distribution of POM. The resuspension of sediment reduced the transparency of the water, limiting effective photosynthesis, impeding the survival of submerged macrophytes, and, consequently, deteriorating the ecosystem. We propose that the control of sediment resuspension is important for improving the water transparency that creates an appropriate habitat for the restoration of the submerged macrophyte community.

Keywords: particulate organic matter (POM); carbon and nitrogen stable isotopes; source tracing; ecosystem degradation; Caohai Lake

Citation: Wu, J.; Yang, H.; Yu, W.; Yin, C.; He, Y.; Zhang, Z.; Xu, D.; Li, Q.; Chen, J. Effect of Ecosystem Degradation on the Source of Particulate Organic Matter in a Karst Lake: A Case Study of the Caohai Lake, China. *Water* **2022**, *14*, 1867. https://doi.org/10.3390/w14121867

Academic Editors: Xiaojun Luo and Zhiguo Cao

Received: 30 April 2022
Accepted: 8 June 2022
Published: 10 June 2022

Publisher's Note: MDPI stays neutral with regard to jurisdictional claims in published maps and institutional affiliations.

1. Introduction

The transformation of clear water to turbid water in a lake is defined as degradation of a shallow lake ecosystem. In the steady state of clear water, the water body is clear, and the coverage of submerged macrophytes is high. In the steady state of turbid water, abundant phytoplankton is present, and the water quality is turbid [1]. Ecosystem degradation is the result of a comprehensive effect that comprises the influence of external driving forces (including input of exogenous nutrient loads, increase in the lake water level or destruction of lakeside belts, and biological regulation) [2]. The rapid dying-out of the submerged macrophytes causes increasing amounts of phytoplankton and, as a result, a rapid drop in the water's transparency. Additionally, the disappearance of submerged macrophytes enhances sediment resuspension, and moreover, due to the disruption of the food web, omnivorous fish must forage in the substrate, further exacerbating sediment resuspension. The combined impact of phytoplankton and sediment resuspension causes further reduction in transparency, and the lake ecosystem becomes completely transformed from a clear water steady state to a turbid water steady state [2,3].

Particulate organic matter (POM) is an important component of the cycle of biogenic elements in lakes and reservoirs and has a significant regulatory effect on lake organic carbon and inorganic carbon pools [4]. Moreover, it controls nutrients, including nitrogen (N), phosphorus (P), and heavy metals, to a large extent [5,6]. As well as the transport of organic pollutants, the identification of POM sources is the key to understanding the carbon cycle in lakes [7]. POM plays an important role in the circulation of carbon and energy within lake systems and largely controls the transport of macronutrients. Therefore, the study of the source and cycle of POM will provide a reliable basis for the prevention and, if necessary, reversal of lake ecosystem degradation [8,9].

The plateau karst lake ecosystem is very fragile and highly sensitive to environmental changes and human food disturbance [10–12]. With the urbanization of the Caohai Basin, a large amount of particulate matter and nutrients has entered Caohai Lake through surface runoff in recent years. Caohai Lake (Guizhou Province, China), a typical plateau karst lake, is facing serious environmental problems, including enhanced sedimentation, pollution of recharge water sources, and deterioration of water quality [9,13]. Since 2020, a large area of submerged macrophytes in Caohai Lake has died out and, with obvious turbid water bodies and serious water eutrophication, the aquatic ecosystem has been degraded. Currently, the ecosystem is in a critical transition stage from a steady state of clear water to a steady state of turbid water. However, few studies have focused on the source of organic matter in such types of lake particles and its impact on ecosystem degradation [14,15].

The carbon and nitrogen stable isotopes (δ^{13}C–δ^{15}N) of the endmembers of different POM sources show a significant difference. Therefore, the isotopes can trace the source and transformation process of POM, which is a reliable indicator for tracking the source of organic matter in lakes, allowing us to understand nutrient utilization and the fate of organic matter in lakes [5,11,16]. In this study, we combine the δ^{13}C–δ^{15}N isotopes and SIAR (Stable Isotope Analysis in R) isotope mixture models to (1) analyze the effect of ecosystem degradation on the POM concentrations and δ^{13}C–δ^{15}N distribution characteristics in Caohai Lake, (2) quantitatively evaluate the differences in the sources of POM in Caohai Lake prior to and after ecosystem degradation, and (3) reveal the driving factors of the source and distribution of POM in Caohai Lake.

2. Materials and Methods

2.1. Study Area

Caohai Lake (N 26°47′32″~26°52′52″, E 104°10′16″~104°20′40″) is located in the southwest of Weining County (China) at an altitude of 2170 m. The average water body temperature is 10.5 °C, the average water depth is 1.5 m, and the maximum depth is 5 m [17]. The wet season lasts annually from May to September, and the dry season lasts from October to April. The average annual precipitation in the Caohai Basin is about 951 mm, with 88% of the precipitation occurring during the wet period [18]. The lake is rich in aquatic plants, primarily *Potamogeton lucens L.*, *Myriophyllum verticillatum L.*, *Potamogeton wrightii Morong*, *Potamogeton maackianus A. Bennett*, and Najas graminea Del. Four major rivers enter the lake: the Zhong River, the Sha River, the Dongshan River, and the Maojiahaizi River, which are part of the typical plateau-karst-grass-type shallow water lake system (Figure 1).

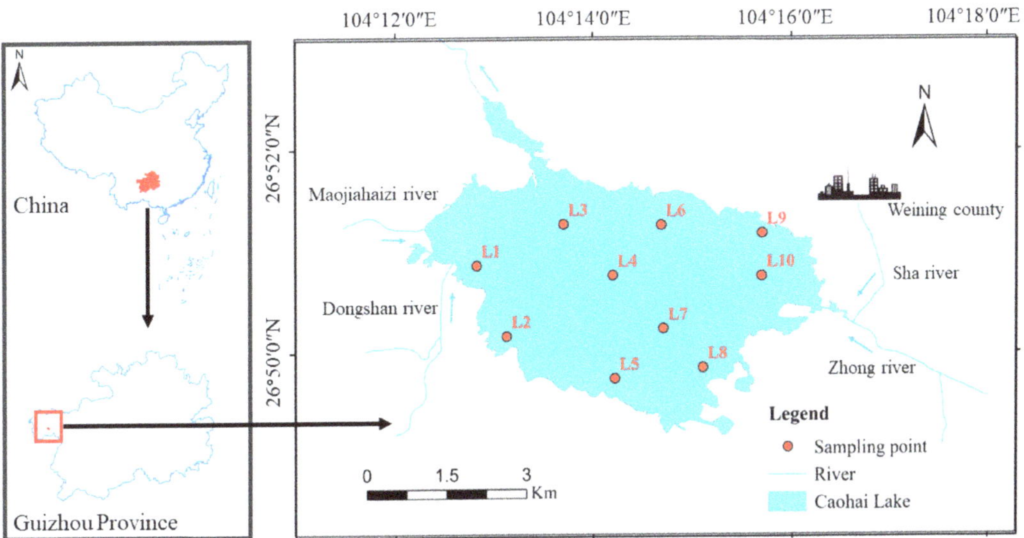

Figure 1. Distribution of sampling sites in Caohai Lake. Their positions were marked in red in the inserted maps of China and Guizhou Province.

2.2. Sample Collection and Analysis

The POM in the water of Caohai Lake was systematically sampled prior to (in 2019) and after (in 2021) ecosystem degradation. The sampling date are from April (dry season) and July (wet season), 2019 and August (wet season) and December (dry season), 2021. Ten representative sites were selected for taking the samples (L1–L10), which were collected at a depth of 0.5 m using a Niskin water collector. Water quality parameters—including temperature (T), electrical conductivity (EC), and dissolved oxygen (DO)—were measured on-site with a multi-parameter water quality monitor (YSI-6600V2). The water transparency (SD) of each sampling point in the lake area was measured with a Secchi disc. Source endmember samples of five main submerged macrophytes were collected in the center of the lake (L4) using a grab (Table S5). Surface sediment samples from the lake bottom were collected with a sediment grab sampler loaded in a crisper and dripped with saturated $HgCl_2$. The solution was poisoned and kept refrigerated.

The water samples were transferred through a silica gel tube to a stainless steel filter and percolated with a Whatman GF/F filter under positive pressure. After filtration of about 10 L of lake water per filter, we used clean tweezers to fold the filter and to place it in aluminum foil for storage at low temperature (<4 °C). After freezing, the filter membrane was freeze-dried using a freeze dryer, fumigated with 12 mol/L hydrochloric acid for 24 h to remove carbonate, subsequently dried at 60 °C, and finally stored in a desiccator. The collected fresh plant leaves were washed with ultrapure water to remove impurities, freeze-dried, grounded through a 100-mesh sieve, and packed in zip-lock bags. The surface sediment samples were centrifuged to separate the pore water of in the speed of 3800 rmp, and freeze-dried. The dried sediment samples were cleaned with an agate mortar after removing the gravel as well as animal and plant residues. After passing through a 120-mesh sieve, they were soaked in 1 mol/L hydrochloric acid for 24 h while heating to 100 °C to remove inorganic carbon. The samples were washed to neutrality, freeze-dried, ground into powder, and cooled in a desiccator. The particulate organic carbon (POC) and particulate organic nitrogen (PON) concentrations of all samples were determined with an organic element analyzer (Vario macro cube with a precision <0.5% rel); $\delta^{13}C$ and $\delta^{15}N$ values of samples were detected in all samples using a Thermo Fisher MAT-252/253 mass spectrometer with an accuracy $\leq 0.06‰$.

2.3. Isotopic Mixing Model

We used the Bayesian SIAR mixture model to quantify the contribution of different endmembers to water POM. SIAR is available to download from the packages section of the Comprehensive R Archive Network site (CRAN)—http://cran.r-project.org/ [19] (accessed on 15 July 2012). The SIAR model is expressed in Equation (1):

$$X_{ij} = \sum_{k=1}^{k} P_k\left(S_{ij} + C_{ij}\right) + \varepsilon_{ij} \tag{1}$$

$$S_{jk} \sim N(\mu_{jk}, \omega_{jk}^2) \tag{2}$$

$$c_{jk} \sim N(\lambda_{jk}, \tau_{jk}^2) \tag{3}$$

$$\varepsilon_{jk} \sim N(\sigma_j^2) \tag{4}$$

where:

X_{ij} = observed isotope value j of the consumer i;

S_{jk} = source value k on isotope j; normally distributed with mean μ_{jk} and variance ω_{jk}.

c_{jk} =TEF (sources and trophic enrichment factors) for isotope j on source k; normally distributed with mean; λ_{jk} and variance τjk^2.

P_k = dietary proportion (external sources of variation not connected to isotopic uncertainty) of source k; estimated by the model.

q_{jk} = concentration of isotope j in source k.

ε_{ij} = residual error, describing additional interobservation variance not described by the model, σ_j^2 estimated by the model.

The Bayesian paradigm enables the calculations of the uncertainty of all parameters. P and σ^2 are the most important parameters—controls for the proportional contribution and residual variance, respectively. Model fitting is hierarchical, enabling flexibility in adding further complexity [19].

2.4. Data Analysis

We used the inverse-distance-weighted method in ArcGIS 10.6 to analyze and map the concentrations of POC, PON, and $\delta^{13}C_{POC}$ and $\delta^{15}N_{PON}$ data in water prior to and after the degradation of the Caohai Lake ecosystem. We used the R 4.0.3 software to calculate the contribution of POM quantitatively and analyzed the results with Origin 2018. Significant variability between variables prior to and after ecosystem degradation was verified using the Mann–Whitney test of IBM SPSS Statistics 21.

3. Results and Discussion

3.1. Water Quality Parameters

Prior to the degradation of Caohai Lake, the average pH values of all locations in the wet season and dry season were 8.50 ± 0.18 and 9.04 ± 0.31, respectively; after the degradation, they were 8.42 ± 0.28 and 8.28 ± 0.08, respectively. The pH of the water body decreased significantly with the ecosystem's degradation ($p < 0.01$; Figure 2a). The DO (dissolved oxygen) concentrations in the wet and dry seasons prior to degradation were 8.55 ± 2.78 mg/L and 12.53 ± 1.71 mg/L, respectively; after degradation, they were 7.74 ± 0.57 mg/L and 7.23 ± 1.16 mg/L, respectively, thus indicating a significant decrease in the water DO in Caohai Lake with ecosystem degradation ($p < 0.01$; Figure 2b).

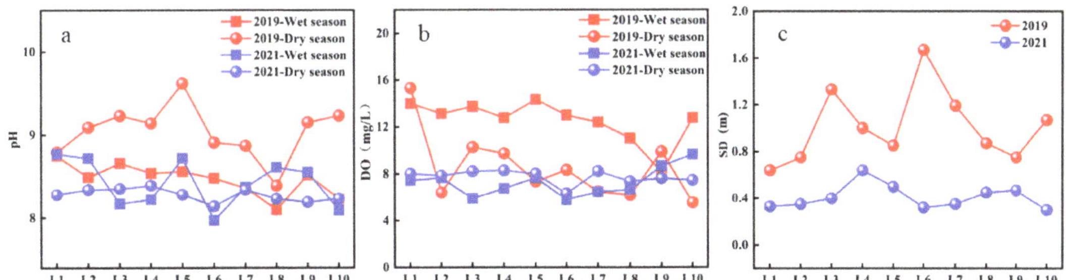

Figure 2. Distribution characteristics of environmental parameters: pH (**a**), DO (**b**), SD (**c**) in Caohai Lake (locations from L1 to L10).

After the degradation, Caohai Lake's water became turbid, with a transparency of only 0.41 m, which is a reduction of 59% compared to its pre-degradation status (Figure 2c). Weining County is located northeast of Caohai Lake. Until 2017, the sewage from Weining County was directly discharged into Caohai Lake (about 8000 tons per day), which occurred continuously [20]. The input of exogenous nutrient load caused a gradual increase in the concentration of nutrients in the water and an increase in phytoplankton production. Since March 2019, a total fishing ban has been implemented for Caohai Lake. As a consequence, the quantity of herbivorous fish in the lake has increased rapidly, which damages the submerged macrophytes community, but has also strongly disturbed the sediment, causing sediment resuspension. Due to the combined impact of increased phytoplankton production and sediment resuspension, the transparency of Caohai Lake's water became significantly reduced [2]. By 2021, submerged macrophytes had died out on a large scale, losing their fixation, and sediment resuspension has increased. In addition, due to the destruction of the food chain, omnivorous fish were forced to forage in the sediment, which further exacerbated sediment resuspension (Figure S1) [2].

High turbidity influences aquatic photosynthesis and respiration, which in turn affects the pH and DO concentrations of the lake, resulting in fluctuations in the concentrations of nutrients and other water components [21]. Prior to the degradation of the ecosystem, the water had high transparency, and the photosynthesis of aquatic plants was prosperous, causing high DO concentrations in the water. Photosynthesis was weakened after the degradation, so the consumption of free CO_2 in the water was reduced, and the free CO_2 combined with hydrogen ions in the water, thus changing the carbonate equilibrium and causing a decrease in the pH of the water (Figure 2a).

3.2. Spatiotemporal Distribution Characteristics of POM Concentrations in Water Prior to and after Ecosystem Degradation

After ecosystem degradation, the POC concentrations in the water of Caohai Lake increased by 11% and 31% in the wet and dry seasons, respectively, the concentrations of POC increased to 1.67 ± 0.67 mg/L and 1.28 ± 0.15 mg/L after degeneration. The concentrations of PON remained almost unchanged. Prior to the degradation, the PON concentrations were 0.30 ± 0.14 mg/L and 0.22 ± 0.10 mg/L in the wet and dry seasons, respectively, and after degradation, the PON concentrations were 0.26 ± 0.09 mg/L and 0.25 ± 0.04 mg/L, respectively (Figures 3 and 4). The overall POM concentrations decrease from the east towards the west and are higher during the wet season than during the dry season (Figures 3 and 4). The seasonal difference is significant ($p < 0.01$; Figures 3 and 4).

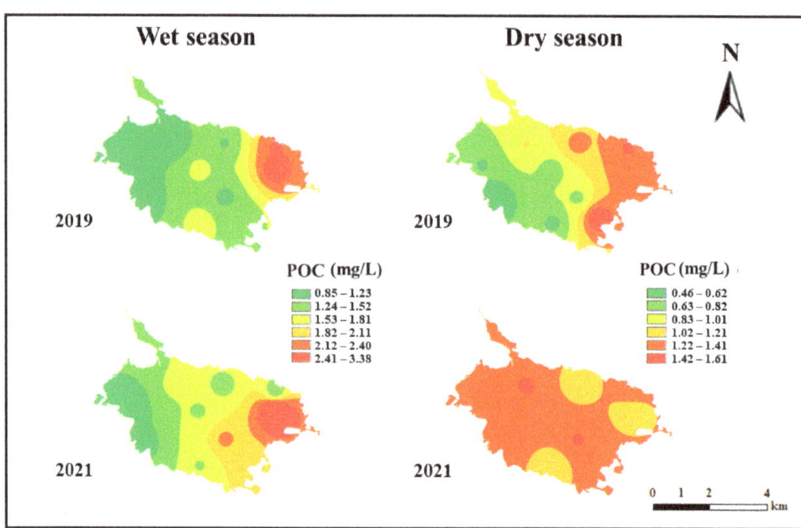

Figure 3. The temporal and spatial distribution of the concentrations of POC in wet season and dry season of Caohai Lake.

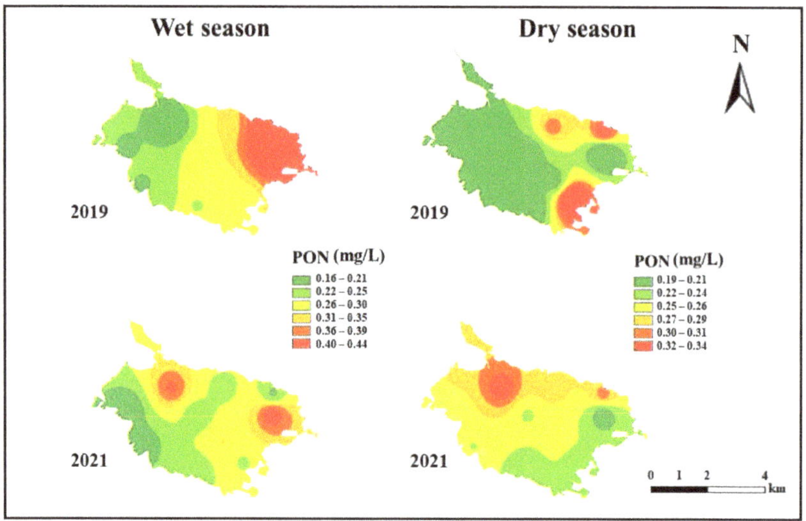

Figure 4. The temporal and spatial distribution of the concentrations of PON in wet season and dry seasons of Caohai Lake.

The strong sunlight in the wet season (from May to September) causes the growth of aquatic plants and algae that increase the endogenous contribution of the lake [22]. In addition, due the heavy rainfall in the wet season, the particles on both sides of the river entering the lake are drained into the river channel and enter the lake with the surface runoff, thus enhancing the external input in the wet season. Therefore, the concentrations of POM are significantly higher in the wet season than in the dry season ($p < 0.01$; Figures 3 and 4). Related to the discharge from Weining County, the POC concentrations of the sampling point L10 at the eastern tributary of the lake (with a flow rate of 225 L/s, which is the highest among the four main rivers) prior to and after ecosystem degradation during the wet season are very high—2.69 mg/L and 3.38 mg/L, respectively—which are 40% and

125% higher than in the center of the lake (Figure 3). These data show that terrigenous input has an important impact on the composition of the lake POM in this area.

The endogenous sources of POM in the Caohai Lake are mainly sediment resuspension and submerged macrophytes. After degradation of the ecosystem, the immobilization of submerged macrophytes gets lost, and sediment resuspension increases. The inferred release caused the increases of 11% and 31% in the POC concentrations in water of the Caohai Lake in the wet and dry periods, respectively (Figure 3). After the degradation, the concentrations of POC and PON increased in the dry season of the western outlet, which may be due to the deterioration of the water quality (decreased DO, increased pH, and water transparency less than 0.5 m), which promoted the growth of algae.

3.3. Spatiotemporal Distribution Characteristics of POM Carbon and Nitrogen Isotopes in Lake Water Prior to and after Ecosystem Degradation

The $\delta^{13}C_{POC}$ value in the water of Caohai Lake after the degradation is lower than prior to the degradation. The $\delta^{13}C_{POC}$ value in the water of Caohai Lake in wet season and dry season prior to the degradation is $-21.6 \pm 2.7‰$ and $-24.2 \pm 1.7‰$, respectively, and $-23.9 \pm 3.2‰$ and $-25.1 \pm 0.6‰$ after the degradation (Figure 5). The $\delta^{13}C_{POC}$ value in the water of Caohai Lake is generally higher than the average $\delta^{13}C_{POC}$ value of -29.7 ‰ for global lakes [16]. The karst basin has a strong weathering effect; the HCO_3^- enters the water, which leads to the DIC increases in the water. The $\delta^{13}C_{DIC}$ of the water, with a range from $-15.7‰$ to $3.7‰$, eventually lead to the higher $\delta^{13}C_{POC}$ value in the water of Caohai Lake (Table S6). Prior to ecosystem degradation, the average values of $\delta^{15}N_{PON}$ in the water of Caohai Lake in the wet and dry seasons are $3.3 \pm 1.5‰$ and $3.4 \pm 1.5‰$, respectively, and $3.7 \pm 2.3‰$ and $0.1 \pm 0.6‰$ after degradation (Figure 6). The $\delta^{13}C_{POC}$ value in water of Caohai Lake is higher in the wet season than in the dry season and lower in the east than in the west. During the dry season, the $\delta^{15}N_{PON}$ value in the water of Caohai Lake became significantly lower after ecosystem degradation (Figure 6; $p < 0.01$).

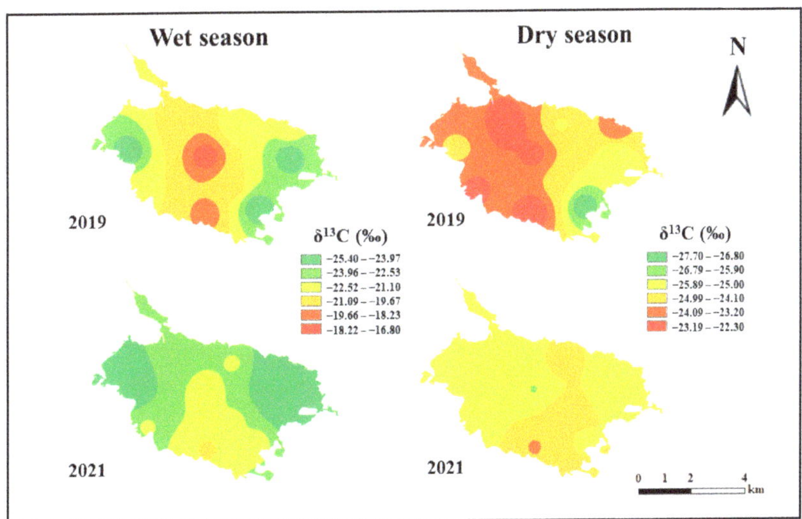

Figure 5. Temporal and spatial variation of carbon stable isotope in wet season and dry season of Caohai Lake.

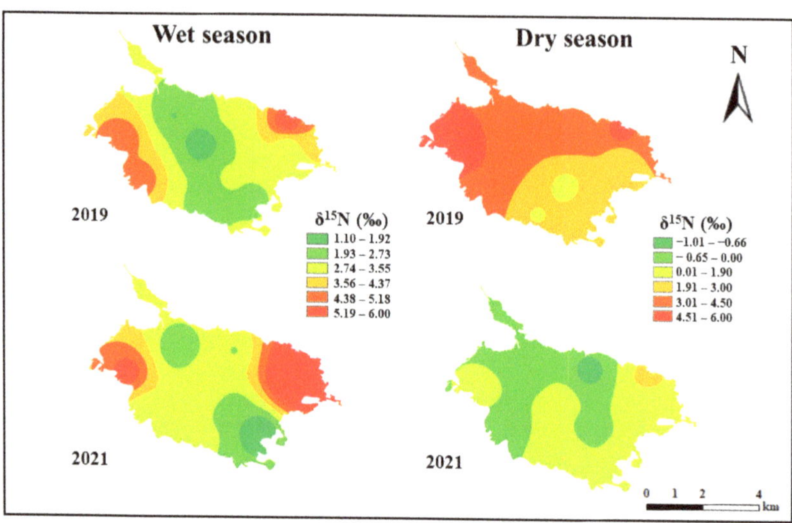

Figure 6. Temporal and spatial variation of nitrogen stable isotope in wet season and dry season of Caohai Lake.

Due to temperature and light intensity, photosynthesis of aquatic plants is intense in the wet season, leading to the fixation of atmospheric CO_2 into the water [21]. The $\delta^{13}C$ value of atmospheric CO_2 is higher than the $\delta^{13}C$ of the water and of bottom biodegradation. Organic matter preferentially releases lighter ^{12}C during the degradation process, making the $\delta^{13}C$ value of POM in the wet season higher than that in the dry season [23,24]. POM in dry seasons contains more detritus, which reduces water transparency and promotes the growth of microzooplankton. This results in higher $\delta^{15}N$ values in dry seasons compared to wet seasons [5]. The $\delta^{13}C$ value of terrigenous C3 plants is lower than that of sediments and submerged macrophytes, and the $\delta^{15}N$ value is higher (Table 1, Figure 7). Under the influence of exogenous input (such as plant debris and soil organic matter), the values of $\delta^{13}C_{POC}$ and $\delta^{15}N_{PON}$ were lower and higher, respectively, in nearshore area than those of in offshore area during the wet seasons prior to and after degradation (Figure 6; Tables S1 and S2).

Prior to the ecosystem's degradation, submerged macrophytes and sediment resuspension were the main source of POM in the water of Caohai Lake. However, sediment resuspension dominated as a source of POM after degeneration. The change in sources caused variation in the isotopic composition in water of Caohai Lake: lower $\delta^{13}C_{POC}$ value in the water regardless of wet season or dry season after degeneration (Figure 7). In addition, after the ecosystem's degradation, the $\delta^{15}N_{PON}$ value in water of Caohai Lake in the dry season became significantly lower than that in the wet season, which may be related to the absorption of nitrate by phytoplankton and isotopic fractionation under the condition of light limitation. Low temperature further aggravated the extinction of aquatic plants after degradation. At the same time, diatoms preferentially utilize ^{14}N in nitrate under low light intensity, resulting in a lower $\delta^{15}N_{PON}$ value of the water in the dry season [25,26].

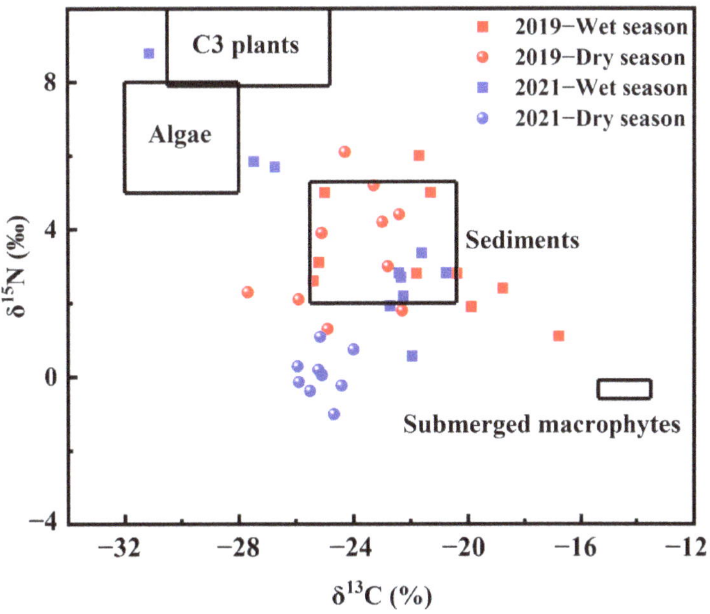

Figure 7. Mixing plots of the δ^{15}N and δ^{13}C values of the four different POM sources during the different periods (2019−wet season, 2019−dry season, 2021−wet season, 2021−dry season) of the Caohai Lake. The different POM sources were sediments, submerged macrophytes, algae, and C3 plants.

Table 1. The δ^{13}C and δ^{15}N (‰) of different sources to POM in water of Caohai Lake.

End Member	δ^{13}C		δ^{15}N		References
	Range of δ^{13}C	Mean $\pm$ SD	Range of δ^{15}N	Mean $\pm$ SD	
Sediment (n = 10)	−20.4 to −25.5	−23.3 $\pm$ 1.7	2.0 to 5.3	3.3 $\pm$ 1.1	this study
Submerged macrophytes (n = 5)	−13.5 to −15.4	−14.9 $\pm$ 0.7	−0.6 to −0.1	−0.1 $\pm$ 0.4	this study
Algae	−32.0 to −28.0	−30.0 $\pm$ 2.0	5.0 to 8.0	6.5 $\pm$ 1.5	[21,27]
C3 plants	−30.5 to −24.8	−27.6 $\pm$ 2.0	7.9 to 12.8	10.3 $\pm$ 2.0	[28,29]

3.4. POM Source Characteristics and Their Drivers

Exogenous input (C3 plants) and endogenous production (sediment resuspension, phytoplankton, and submerged plants) are the main sources of POM in Caohai Lake (Figure 7). We measured the δ^{13}C and δ^{15}N values of the endmembers of sediments and submerged macrophytes to determine the range of variation of the stable isotope values of sediments and submerged macrophytes. Citing data of previous studies on carbon and nitrogen stable isotopes of terrestrial plants and lake algae [21,27–29], our data document the following variation (Table 1, Figure 7): The range of δ^{13}C of terrigenous C3 plants is −30.5 to −24.8‰, with an average value of −27.6 $\pm$ 2.0‰. The range of δ^{15}N of C3 plants is 7.9 to 12.8‰, with an average value of 10.3 $\pm$ 2.0‰, and the range of δ^{13}C of lake algae is −32.0 to −28.0‰, with an average value of −30 $\pm$ 2‰. The range of δ^{15}N of lake algae is 5.0 to 8.0‰, and the average value is 6.5 $\pm$ 1.5‰.

The SIAR model shows that from the wet season of 2019 to the dry season of 2021, the contribution percentages of sediment resuspension increased from 25% to 71% (Figure 8).

The contribution percentages of submerged macrophytes decreased from 41% to 12%. The changes in the contribution percentages of algae do not show any obvious systematic trend or variation, and the mean value was 22%. The contribution percentages of C3 plants were lower, with a mean value of 12% (Figure 8).

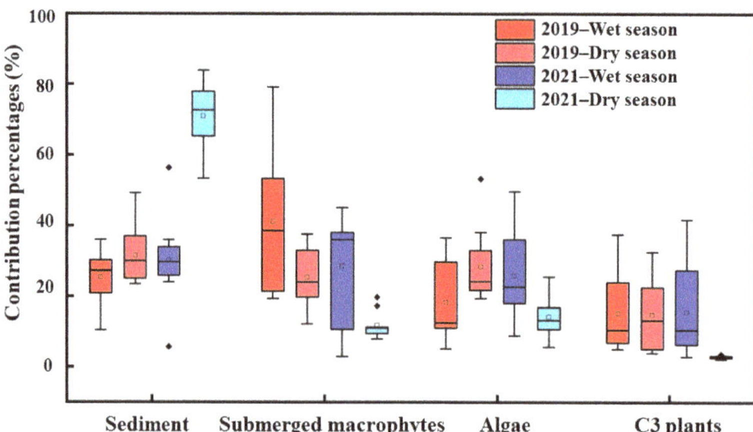

Figure 8. Relative contribution percentages of sediment, submerged macrophytes, algae, and C3 plants to POM in water of Caohai Lake. Red, light red, blue, and light blue represent the 2019−wet season, 2019−dry season, 2021−wet season, and 2021−dry season, respectively.

The dominant role of endogenous POM has been documented in several plateau karst lakes [5,11,16]. Our results show that the average contribution percentages of endogenous POM (including algae, submerged macrophytes, and sediment resuspension) in Caohai Lake can reach to 91% (Figure 8, Table S4). Compared with large deep-water lakes, such as Tai Lake, the material exchange between shallow-water lakes and sediments is stronger, and sediment resuspension contributes more to the POM, suggesting that shallow-water lakes are more susceptible to sediment resuspension than deep-water lakes [4]. However, the contribution percentages of algae in Caohai Lake are comparably small, and its contribution percentages are much lower than those of Tai Lake (the maximum contribution rates of algae is 57%), showing that the growth of algae in lakes with lush aquatic plants is significantly inhibited [30]. Compared with the floodplain lake Poyang Lake, the sediments in Caohai Lake are the result of multivariate mixing, whereas those of Poyang Lake are predominantly derived from soil erosion, indicating that large karst lakes have more complex carbon cycle patterns than other lakes [14].

Although the POM of Caohai Lake is generally dominated by endogenous sources, the influence of land sources is striking in some parts of the lake. Terrestrial organic carbon is mainly composed of terrigenous plant debris and organic matter. Physical erosion is generally the main controlling factor for POC concentrations in rivers. Soil erosion causes land degradation and migration of soil particles, related POC, and nutrients, thus affecting the global carbon cycle [31]. Four major rivers discharge into Caohai Lake at the eastern and western lakeshores. During the strong rainfall in the wet season, terrigenous soil debris enters the Caohai Lake through surface runoff, which affects the POM composition of Caohai Lake's water in the eastern and western parts of the lake. The contribution rates of submerged macrophytes are lower in the east and west than in the center of the lake (see Figure S1). On the contrary, the contribution of terrigenous sources in the nearshore lake area (L1, L2, L9) is significant, and the maximum value reaches 41% of the total POM.

As a typical karst lake system, Caohai Lake is highly vulnerable and sensitive to environmental changes and human disturbances [11]. Since 2013, the water level of the Caohai Lake has gradually increased, and since 2015, the water level has remained on a high value (more than 1 m higher than in 2012). In addition, the water level fluctuation amplitude has decreased, and the hydrological rhythm has changed significantly. The continuously high water level, reduced water level variation, and poor water mobility in Caohai Lake caused the extinction of emergent plants in the nearshore area, the inhibition of seed germination of submerged macrophytes, the degradation of aquatic vegetation, the weakening of wave-dissipating effects, and the reduction of water self-purification [32–34]. Moreover, in March 2019, a comprehensive fishing ban was implemented for Caohai Lake. As a consequence, the number of herbivorous and omnivorous fish (including carp, crucian carp, and wheat ear fish) in the lake area has increased rapidly. This further strongly disturbed the sediment, causing the resuspension of particulate matter, thus inducing water turbidity. As a consequence, photosynthesis of submerged macrophytes became limited and gradually vanished [35–37].

The source to POM in the water of Caohai Lake changed significantly after ecosystem degradation (Figures 2, 8 and S1). The POM in the water of Caohai Lake was mainly controlled by sediment resuspension and submerged macrophytes prior to the degradation. In the wet season, the temperature was high and the photosynthesis of submerged macrophytes was intense. The average contribution percentages were higher than that of the sediments (41%)—as high as 79% in the center of the lake. The contribution (25%) in the dry season was slightly lower than that of sediment, further confirming that environmental factors related to photosynthesis and sediment resuspension are the main factors affecting the contribution of POM source in karst lakes (Figure 2). After the degradation, the POM of Caohai Lake's water gradually changed and became controlled by sediment resuspension. Due to the low temperature, inhibiting the photosynthesis of aquatic plants, and the minor rainfall, which reduced external input, the average contribution of sediment resuspension during the dry season reached 71%.

Significant seasonal differences in POM contribution percentages indicate the direct effects of water movement on air temperature and rainfall caused by seasonal changes (Figure 8; $p < 0.01$) [23]. Turbidity has an important indirect effect on phytoplankton photosynthesis through its impact on the availability of sunlight in the water. The DO is controlled by water−vapor exchange, water photosynthesis and respiration, mineral oxidation, and water mixing that occur in aquatic ecosystems, as well as by cycling and other important indicators of physicochemical processes [38]. Moreover, aquatic photosynthesis is an important source of DO [39]. The pH value of water is an important factor in the eutrophication of the lake. It is the result of the comprehensive effect of various factors, such as geological background, climate conditions, exogenous input, etc., and is closely related to algae growth [21]. After the degradation, change in the carbonate equilibrium caused a decrease in the pH in Caohai Lake. In addition, nutrient levels (i.e., N and P) are crucial for the primary production and growth of phytoplankton. We detected for Caohai Lake that algae have higher contribution percentages at the sites with higher PON concentrations (L9, L10) (Table S1). During the wet season, the contribution percentages of algae attained a maximum of 50% in the east after the system was degraded (Figure 8), further confirming the important contribution of primary production to POM [40]. Fish, wind, and wave disturbances can cause sediment resuspension, which has a major impact on POM. Our study shows that the contribution percentages of sediments continue to increase after ecosystem degradation, and the average value can reach 71% in the dry season after ecosystem degradation (Figure 8). The spatial distribution characteristics of POM and the differences among the influences of various driving factors further illustrate the complexity of the carbon cycle process in large karst lakes, such as Caohai Lake.

We have detected that the ecosystem degradation in Caohai Lake is related to the following main factors: sediment resuspension caused by wind and waves, sediment resuspension caused by fish foraging, reduced transparency impeding the survival of submerged macrophytes, increased nutrient concentration in the water, and the growth of large quantities of phytoplankton in submerged water. These factors lead to a reduction in the transparency of the water (Figure 2) and, as a consequence, the effective photosynthesis of the submerged macrophytes is largely inhibited (Figure 8). These processes improve the stability of the turbid water steady state. Therefore, we suggest controlling the sedimentation of the water particles and the resuspension of the sediment. This strategy will improve the transparency of the water and will create an appropriate habitat for the restoration of submerged macrophyte communities. Based on this strategy, the submerged macrophyte community will be reconstructed, and the food web will be optimized and regulated to restore the aquatic ecosystem of Caohai Lake [41].

4. Conclusions

We used carbon and nitrogen stable isotopes (δ^{13}C$-\delta^{15}$N) and SIAR mixed model to quantitatively assess the source of POM in the water of karst Caohai Lake prior to and after ecosystem degradation and to explore the driving factors of POM. Our results document a trend of the spatial distribution of the POM concentrations in the water of Caohai Lake, with high values in the east moving towards low values in the west, which is related to the influence of terrigenous input and rainfall distribution. Moreover, our data show that the distribution characteristics of the POM concentrations in the wet season are larger than those in the dry season. The isotope characteristics show low δ^{13}C values near the shore and high values near the lake's center. In contrast, the δ^{15}N values are high near the shore and low in the lake's center. After the degradation of the ecosystem, the DO concentrations and pH value in the water of Caohai Lake decreased due to the extinction of a large quantity of submerged macrophytes. Moreover, POM concentrations increased by 11% and 31% in the wet season and the dry season, respectively. Prior to the ecosystem's degradation, sediment resuspension (28%) and submerged macrophytes (33%) were the major sources of POM, while sediment resuspension (51%) dominated the source of POM after the ecosystem's degradation. Environmental parameters (including DO, turbidity, water depth, and water temperature) that are related to photosynthesis and sediment resuspension are the main factors driving the spatiotemporal distribution of POM. Sediment resuspension reduces water transparency, thereby impeding the survival of submerged macrophytes. Growth of large quantities of phytoplankton is related to increased nutrient concentrations in water. Finally, the reduced water transparency induced by these factors caused the ecosystem's degradation.

Supplementary Materials: The following supporting information can be downloaded at: https://www.mdpi.com/article/10.3390/w14121867/s1, Figure S1. Photos of Caohai Lake before and after the ecosystem degradation between 2019 and 2021. Figure S2. Contribution percentages (%) of sediment, submerged macrophytes, algae and C3 plants to POM in water of Caohai Lake. W and D represents the wet season and dry season, respectively. Table S1. Hydrochemical parameters of the lake water (T, DO, EC, and pH), concentrations of POC and PON, δ13CPOC and δ15NPON in Caohai Lake of wet season and dry season in 2019. Table S2. Hydrochemical parameters of the lake water (T, DO, EC, and pH), concentrations of POC and PON, δ13CPOC and δ15NPON in Caohai Lake of wet season and dry season in 2021. Table S3. Relative contribution percentages (%) of sediment, submerged macrophytes, algae and C3 plants to POM in water of Caohai Lake in 2019. Table S4. Relative contribution percentages (%) of sediment, submerged macrophytes, algae and C3 plants to POM in water of Caohai Lake in 2021. Table S5. The δ13C and δ15N (‰) of sediment (n = 10) and submerged macrophytes (n = 5) of Caohai Lake. Table S6. The δ13CDIC (%) in water of Caohai Lake of wet season and dry season in 2019.

Author Contributions: J.W., H.Y. and Z.Z. contributed to conception and design of the study. J.W. and Z.Z. organized the database and statistical analysis. J.W. wrote the first draft of the manuscript. Y.H., C.Y., H.Y., Q.L., D.X., J.C. and W.Y. guided the writing of the thesis. All authors have read and agreed to the published version of the manuscript.

Funding: This research was supported by the National Natural Science Foundation of China (No. 41807394), the CAS "Light of West China" Program, the Science and Technology Project of Guizhou Province (QianKeHe Support [2022]general 201, [2020]4Y015), and the Central Government Leading Local Science and Technology Development (QianKeZhongYinDi [2021]4028).

Institutional Review Board Statement: Not applicable.

Informed Consent Statement: Not applicable.

Data Availability Statement: The datasets used and/or analyzed during the current study are available from the corresponding author on reasonable request.

Conflicts of Interest: The authors declare that they have no competing interest.

References

1. Scheffer, M. Multiplicity of stable states in freshwater systems. In *Biomanipulation Tool for Water Management*; Springer: Berlin/Heidelberg, Germany, 1990; pp. 475–486. [CrossRef]
2. Scheffer, M.; Carpenter, S.R. Catastrophic regime shifts in ecosystems: Linking theory to observation. *Trends Ecol. Evol.* **2003**, *18*, 648–656. [CrossRef]
3. Scheffer, M.; van Nes, E.H. Mechanisms for marine regime shifts: Can we use lakes as microcosms for oceans? *Prog. Oceanogr.* **2004**, *60*, 303–319. [CrossRef]
4. Chen, J.; Yang, H.; Zeng, Y.; Guo, J.; Song, Y.; Ding, W. Combined use of radiocarbon and stable carbon isotope to constrain the sources and cycling of particulate organic carbon in a large freshwater lake, China. *Sci. Total Environ.* **2018**, *625*, 27–38. [CrossRef]
5. Kumar, S.; Finlay, J.C.; Sterner, R.W. Isotopic composition of nitrogen in suspended particulate matter of Lake Superior: Implications for nutrient cycling and organic matter transformation. *Biogeochemistry* **2010**, *103*, 1–14. [CrossRef]
6. Kaiser, D.; Unger, D.; Qiu, G. Particulate organic matter dynamics in coastal systems of the northern Beibu Gulf. *Cont. Shelf Res.* **2014**, *82*, 99–118. [CrossRef]
7. Gawade, L.; Krishna, M.; Sarma, V.; Hemalatha, K.; Rao, Y.V. Spatio-temporal variability in the sources of particulate organic carbon and nitrogen in a tropical Godavari estuary. *Estuar. Coast. Shelf Sci.* **2018**, *215*, 20–29. [CrossRef]
8. Gu, B.; Schelske, C.L.; Waters, M.N. Patterns and controls of seasonal variability of carbon stable isotopes of particulate organic matter in lakes. *Oecologia* **2011**, *165*, 1083–1094. [CrossRef]
9. Hu, J.; Long, Y.; Zhou, W.; Zhu, C.; Yang, Q.; Zhou, S.; Wu, P. Influence of different land use types on hydrochemistry and heavy metals in surface water in the lakeshore zone of the Caohai wetland, China. *Environ. Pollut.* **2020**, *267*, 115454. [CrossRef]
10. Dai, X.; Zhou, Y.; Ma, W.; Zhou, L. Influence of spatial variation in land-use patterns and topography on water quality of the rivers inflowing to Fuxian Lake, a large deep lake in the plateau of southwestern China. *Ecol. Eng.* **2017**, *99*, 417–428. [CrossRef]
11. Chen, Q.; Ni, Z.; Wang, S.; Guo, Y.; Liu, S. Climate change and human activities reduced the burial efficiency of nitrogen and phosphorus in sediment from Dianchi Lake, China. *J. Clean Prod.* **2020**, *274*, 122839. [CrossRef]
12. Duan, L.; Zhang, H.; Chang, F.; Li, D.; Liu, Q.; Zhang, X.; Liu, F.; Zhang, Y. Isotopic constraints on sources of organic matter in surface sediments from two north–south oriented lakes of the Yunnan Plateau, Southwest China. *J. Soils Sediments* **2022**, *22*, 1597–1608. [CrossRef]
13. Cao, X.; Yang, S.; Wu, P.; Liu, S.; Liao, J. Coupling stable isotopes to evaluate sources and transformations of nitrate in groundwater and inflowing rivers around the Caohai karst wetland, Southwest China. *Environ. Sci. Pollut. Res.* **2021**, *28*, 45826–45839. [CrossRef] [PubMed]
14. Wang, S.; Gao, Y.; Jia, J.; Lu, Y.; Sun, K.; Ha, X.; Li, Z.; Deng, W. Vertically stratified water source characteristics and associated driving mechanisms of particulate organic carbon in a large floodplain lake system. *Water Res.* **2021**, *209*, 117963. [CrossRef] [PubMed]
15. Fernandes, D.; Wu, Y.; Shirodkar, P.V.; Pradhan, U.K.; Zhang, J. Sources and implications of particulate organic matter from a small tropical river—Zuari River, India. *Acta Oceanol. Sin.* **2020**, *39*, 18–32. [CrossRef]
16. Liu, D.; Chen, G.; Li, Y.; Gu, B. Global pattern of carbon stable isotopes of suspended particulate organic matter in lakes. *Limnology* **2012**, *13*, 253–260. [CrossRef]
17. Cao, X.; Wu, P.; Han, Z.; Zhang, S.; Tu, H. Sources, Spatial Distribution, and Seasonal Variation of Major Ions in the Caohai Wetland Catchment, Southwest China. *Wetlands* **2016**, *36*, 1069–1085. [CrossRef]
18. Zhu, Z.; Chen, J.; Zeng, Y. Paleotemperature variations at Lake Caohai, southwestern China, during the past 500 years: Evidence from combined δ 18O analysis of cellulose and carbonates. *Sci. China Earth Sci.* **2014**, *57*, 1245–1253. [CrossRef]
19. Parnell, A.; Inger, R.; Bearhop, S.; Jackson, A.L. Source partitioning using stable isotopes: Coping with too much variation. *PLoS ONE* **2010**, *5*, e9672. [CrossRef]
20. Long, Y.; Jiang, J.; Hu, X.; Hu, J.; Ren, C.; Zhou, S. The response of microbial community structure and sediment properties to anthropogenic activities in Caohai wetland sediments. *Ecotoxicol. Environ. Saf.* **2021**, *211*, 111936. [CrossRef]
21. Thornton, S.; McManus, J. Application of Organic Carbon and Nitrogen Stable Isotope and C/N Ratios as Source Indicators of Organic Matter Provenance in Estuarine Systems: Evidence from the Tay Estuary, Scotland, Estuarine, Coastal and Shelf Science. *Estuar. Coast. Shelf Sci.* **1994**, *38*, 219–233. [CrossRef]
22. Edje, B.O.; Ishaque, A.B.; Chigbu, P. Spatial and Temporal Patterns of δ^{13}C and δ^{15}N of Suspended Particulate Organic Matter in Maryland Coastal Bays, USA. *Water* **2020**, *12*, 2345. [CrossRef]
23. Huang, S.; Pu, J.; Li, J.; Zhang, T.; Cao, J.; Pan, M. Sources, variations, and flux of settling particulate organic matter in a subtropical karst reservoir in Southwest. China. *J. Hydrol.* **2020**, *586*, 124882. [CrossRef]
24. Yi, Y.; Zhong, J.; Bao, H.; Mostofa, K.M.; Xu, S.; Xiao, H.-Y.; Li, S.-L. The impacts of reservoirs on the sources and transport of riverine organic carbon in the karst area: A multi-tracer study. *Water Res.* **2021**, *194*, 116933. [CrossRef] [PubMed]

25. Saino, T.; Hattori, A. 15N natural abundance in oceanic suspended particulate matter. *Nature* **1980**, *283*, 752–754. [CrossRef]
26. Kumar, S.; Ramesh, R.; Bhosle, N.B.; Sardesai, S.; Sheshshayee, M.S. Natural isotopic composition of nitrogen in suspended particulate matter in the Bay of Bengal. *Biogeosciences* **2004**, *1*, 63–70. [CrossRef]
27. Ye, F.; Guo, W.; Shi, Z.; Jia, G.; Wei, G. Seasonal dynamics of particulate organic matter and its response to flooding in the Pearl River Estuary, China, revealed by stable isotope (δ^{13}C and δ^{15}N) analyses. *J. Geophys. Res. Oceans* **2017**, *122*, 6835–6856. [CrossRef]
28. Cloern, J.E. Stable carbon and nitrogen isotope composition of aquatic and terrestrial plants of the San Francisco Bay estuarine system. *Limnol. Oceanogr.* **2002**, *47*, 713–729. [CrossRef]
29. Meyers, P. Preservation of elemental and isotopic source identification of sedimentary organic matter. *Chem. Geol.* **1994**, *114*, 289–302. [CrossRef]
30. Xu, J.; Lyu, H.; Xu, X.; Li, Y.; Li, Z.; Lei, S.; Bi, S.; Mu, M.; Du, C.; Zeng, S. Dual stable isotope tracing the source and composition of POM during algae blooms in a large and shallow eutrophic lake: All contributions from algae? *Ecol. Indic.* **2019**, *102*, 599–607. [CrossRef]
31. Gomes, T.F.; Van de Broek, M.; Govers, G.; Silva, R.W.; Moraes, J.M.; Camargo, P.B.; Mazzi, E.A.; Martinelli, L.A. Runoff, soil loss, and sources of particulate organic carbon delivered to streams by sugarcane and riparian areas: An isotopic approach. *Catena* **2019**, *181*, 104083. [CrossRef]
32. Barko, J.W.; Gunnison, D.; Carpenter, S.R. Sediment interactions with submersed macrophyte growth and community dynamics. *Aquat. Bot.* **1991**, *41*, 41–65. [CrossRef]
33. Asaeda, T.; Fujino, T.; Manatunge, J. Morphological adaptations of emergent plants to water flow: A case study withTypha angustifolia, Zizania latifoliaandPhragmites australis. *Freshw. Biol.* **2005**, *50*, 1991–2001. [CrossRef]
34. Jeppesen, E.; Søndergaard, M.; Søndergaard, M.; Christoffersen, K. (Eds.) *The Structuring Role of Submerged Macrophytes in Lakes*; Springer Science & Business Media: Berlin/Heidelberg, Germany, 2012; p. 131. ISBN 0-387-98284.
35. Zhang, Y.; Shi, K.; Zhang, Y.; Moreno-Madriñán, M.J.; Zhu, G.; Zhou, Y.; Yao, X. Long-term change of total suspended matter in a deep-valley reservoir with HJ-1A/B: Implications for reservoir management. *Environ. Sci. Pollut. Res.* **2018**, *26*, 3041–3054. [CrossRef] [PubMed]
36. He, H.; Han, Y.; Li, Q.; Jeppesen, E.; Li, K.; Yu, J.; Liu, Z. Crucian Carp (Carassius carassius) Strongly Affect C/N/P Stoichiometry of Suspended Particulate Matter in Shallow Warm Water Eutrophic Lakes. *Water* **2019**, *11*, 524. [CrossRef]
37. Han, Y.; Li, Q.; He, H.; Gu, J.; Wu, Z.; Huang, X.; Zou, X.; Zhang, Y.; Li, K. Effect of juvenile omni-benthivorous fish (Carassius carassius) disturbance on the efficiency of lanthanum-modified bentonite (LMB) for eutrophication control: A mesocosm study. *Environ. Sci. Pollut. Res. Int.* **2021**, *28*, 21779–21788. [CrossRef]
38. Mader, M.; Schmidt, C.; van Geldern, R.; Barth, J.A. Dissolved oxygen in water and its stable isotope effects: A review. *Chem. Geol.* **2017**, *473*, 10–21. [CrossRef]
39. Paerl, H.W. Controlling cyanobacterial harmful blooms in freshwater ecosystems. *Microb Biotechnol.* **2017**, *10*, 1106–1110. [CrossRef]
40. Zheng, X.; Como, S.; Magni, P.; Huang, L. Spatiotemporal variation in environmental features and elemental/isotopic composition of organic matter sources and primary producers in the Yundang Lagoon (Xiamen, China). *Environ. Sci. Pollut. Res.* **2019**, *26*, 13126–13137. [CrossRef]
41. Zhang, X.; Liu, Z.; Jeppesen, E.; Taylor, W.D.; Rudstam, L.G. Effects of benthic-feeding common carp and filter-feeding silver carp on benthic-pelagic coupling: Implications for shallow lake management. *Ecol. Eng.* **2016**, *88*, 256–264. [CrossRef]

 *water*

Article

Influence of Cascade Hydropower Development on Water Quality in the Middle Jinsha River on the Upper Reach of the Yangtze River

Tianbao Xu [1], Fengqin Chang [2,*], Xiaorong He [3], Qingrui Yang [4] and Wei Ma [4,*]

1   Power China Kunming Engineering Co., Ltd., Kunming 650051, China; xutianbao00@163.com
2   Institute for Ecological Research and Pollution Control of Plateau Lakes, School of Ecology and Environmental Science, Yunnan University, Kunming 650500, China
3   Environmental Monitoring Center of Middle Reaches of Jinsha River, Kunming 650034, China; xiaorong-he@chd.com
4   China Institute of Water Resources and Hydropower Research, Beijing 100038, China; lzyqr2004@163.com
*   Correspondence: changfq@ynu.edu.cn (F.C.); mawei@iwhr.com (W.M.)

Abstract: In recent decades, there has been unprecedented development of hydropower in China, especially in the Yangtze River Basin, which has changed the hydrological and hydraulic conditions of natural rivers and has an impact on water quality. However, the spatial—temporal extent, factors, and the reasons behind the influence of cascade hydropower development are not clear. The six hydropower stations on the main course of the middle reach of the Jinsha River in Yunnan and Sichuan Provinces have been in joint operation for seven years, and the impact of cascade hydropower development on water quality has begun to appear. In this paper, in order to accurately determine the causal relationship between cascade hydropower development and water quality changes on the middle reaches of the Jinsha River and their trends using regression discontinuity analysis, we collected monitoring data on water quality from 2004 to 2019. The results show that cascade hydropower development on the middle reach of the Jinsha River led to a decrease in TP concentration in that section of the river and an increase in the concentration of COD_{Mn} and NH_3-N. Furthermore, increase in sedimentation following the impoundment of cascade hydropower development is the main reason for the decrease in TP concentration, and the regional economic and social development driven by cascade hydropower development are external sources of the increase in the concentration of COD_{Mn} and NH_3-N. In addition, influenced by rainfall, the concentrations of COD_{Mn} and TP are higher in the rainy season and lower in the dry season, which is directly related to the input of non-point-source pollutants in the basin during the former. This study established a model to accurately judge the causal relationship between cascade hydropower development and water quality changes in the basin, which was then used to assess the impact of cascade hydropower development on water quality. Our results provide a basis for the formulation and implementation of a water quality protection plan for the middle reach of the Jinsha River and can also provide a basis for the development of cascade hydropower in other river basins.

Keywords: middle reach of the Jinsha River; cascade hydropower development; water quality; regression discontinuity analysis

Citation: Xu, T.; Chang, F.; He, X.; Yang, Q.; Ma, W. Influence of Cascade Hydropower Development on Water Quality in the Middle Jinsha River on the Upper Reach of the Yangtze River. *Water* **2022**, *14*, 1943. https://doi.org/10.3390/w14121943

Academic Editor: Aonghus McNabola

Received: 15 May 2022
Accepted: 14 June 2022
Published: 16 June 2022

Publisher's Note: MDPI stays neutral with regard to jurisdictional claims in published maps and institutional affiliations.

1. Introduction

China's potential reserves and technically exploitable hydropower resources rank first in the world. By the end of 2019, China's total installed hydropower capacity reached 356 million kW, accounting for 17.7% of installed power capacity, and the average annual hydropower generation capacity was 1.15 terawatt hours (TW·h), accounting for 16% of the total power generation [1]. At the basin scale, the development and construction of hydropower stations change natural river channels into cascade hydropower stations, which can change the hydrological and hydraulic conditions of natural rivers, destroy river continuity, and block material transmission and energy flow between upstream and downstream. Pollutants from river confluences are retained in the reservoir area, which has an impact on water quality, such as reduction of water self-purification capacity, increased water salinity, stratification of water temperature in the reservoir area, and water eutrophication [2–5]. After cascade hydropower development, increasing water depth and decreased flow velocity can cause reduction of the reoxygenation coefficient of the water body. Meanwhile, water quality may also be strongly influenced by agriculture activities within the catchment area, besides water quality deterioration perhaps due to contaminated sediment flux releases [6–9].

Numerous researchers have studied water quality changes since the impoundment of reservoirs. Winton, R et al. [10] sampled a large suite of biogeochemical water quality parameters in central and southern Zambia in 2018 and 2019 to characterize seasonal changes in water quality in response to large hydropower dams. Alvarez Xana et al. [11] investigated the environmental effects of four small hydropower plants in northwestern Spain. Tomczyk Pawel and Wiatkowski Miroslaw [12] studied the effects of hydropower plants on the physicochemical parameters of the Bystrzyca River in Poland. Peng Chunlan et al. [13] analyzed inter-annual and intra-annual variation characteristics of water quality indicators before and after the impoundment of the Three Gorges Reservoir. Xue Lianfang et al. [14] compared water quality of the reservoir before and after construction of the cascade hydropower system in the main course of the Hongshui River. Ba Chongzhen et al. [15] used SPSS software to analyze water quality monitoring data of the Manwan and Dachaoshan hydropower stations on the main course of the Lancang River. There are obvious differences in different watersheds because of different geographical location, climate factors, and so on. In addition to the characteristics of randomness, water quality is also affected by factors such as air and water temperature, rainfall, and sediment content [16]. Therefore, directly analyzing the results of water quality monitoring may cause the results to be inconsistent with the actual situation. To solve this problem, some scholars construct mathematical models to conduct research. Huang Yue et al. [17] used the differential autoregressive integrated moving average (ARIMA) model to predict the quality of the inflow and outflow after the Three Gorges Project began operating. Both Hu Guohua et al. [18] and Wu Jianing et al. [19] applied the seasonal Kendall method to analyze trends in water quality in the Xiaolangdi Reservoir of the Yellow River and the lower reach of the Dongjiang River. However, these mathematical models only examine water quality trends and cannot accurately analyze the causal relationship between water quality changes and hydropower development.

Regression discontinuity analysis was developed by Thistlethwaite and Campbell in 1960, which is the best tool if we care about causal effects. As a statistical method, this approach can efficiently analyze data and is widely used in effect assessment and causality relationships in many fields. Though it did not attract much attention when it was first proposed, regression discontinuity analysis has been widely applied since it was theoretically proven by Hahn et al. in 2010 [20–23]. This method is widely used not only in sociology and economics [24–26] but also in ecology and environment studies. By the regression discontinuity method, Ying et al. [27] investigated the spatiotemporal patterns of the global warming, and Auffhammer et al. [28] studied the effects of gasoline content regulation on air quality. Judith et al. [29] demonstrated the use of regression discontinuity as a statistical technique to model ecological thresholds. Referring to the effectiveness of the "River Chief System" policy of Wanhua Li et al. [30], we adopt the regression discontinuity method to estimate the effect of cascade hydropower development.

In addition to regression discontinuity analysis, the following methods are often used to analyze water quality: Kendall method, PCA (principal component analysis), and ANOVA (variance analysis). Among them, Kendall method has obvious advantages in water concentration prediction, PCA helps to identify the main influencing factors, and ANOVA is mainly used to analyze what factors affect water quality. Compared with the previous three methods, regression discontinuity analysis is more suitable for analyzing the impact of a policy on environment. However, the result of the regression discontinuity analysis is the average causal effect at the discontinuity point, which cannot accurately simulate the water quality.

Currently, the six hydropower stations on the main course of the middle reach of the Jinsha River have been in joint operation for seven years, and the influence of cascade hydropower development in the basin on water quality has begun to appear. Therefore, our research will achieve the following objectives: Firstly, water quality changes on the middle reach of the Jinsha River will be analyzed. Secondly, the causal relation between water quality and cascade hydropower development will be found. Thirdly, a theoretical basis for water quality protection will be provided.

2. Data and Model

2.1. Study Site

The Jinsha River, on the upper reaches of the Yangtze River, is divided into three sections: the upper reach above Shigu, the middle reach from Shigu to the junction of the Yalong River estuary, and the lower reach from the confluence of the Yalong River to Yibin. The middle reach of the Jinsha River has a total length of 563.5 km, a drop of 837.9 m, and an average gradient of 1.49‰. There are eight planned hydropower stations, including Longpan, Liangjiaren, Liyuan, Ahai, Jin'anqiao, Longkaikou, Ludila, and Guanyin [31]. At present, except for the Longpan and Liangjiaren reaches, which are still in the program comparison stage, the remaining six hydropower stations have ready been impounded and are operating (Table 1, Figure 1).

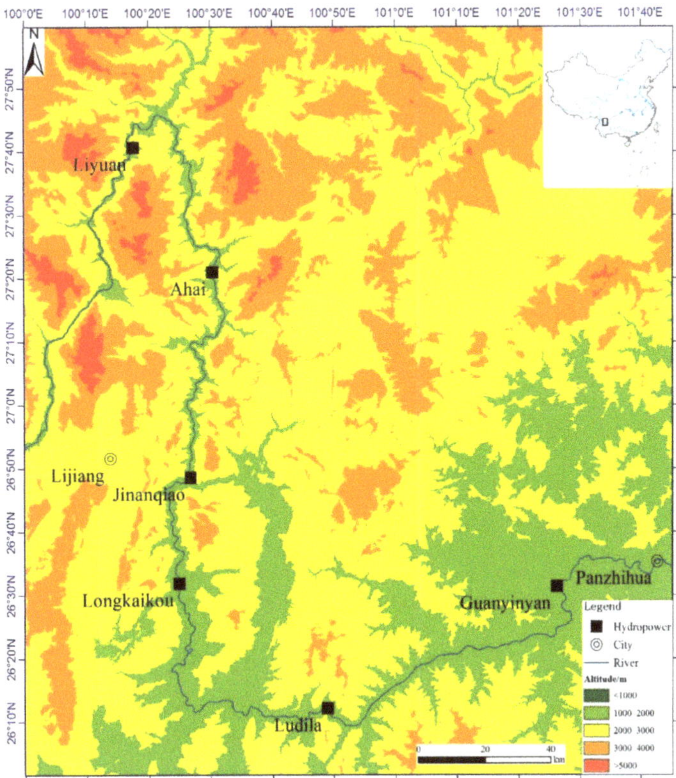

Figure 1. Distribution map of hydropower stations on the main course of the middle reach of the Jinsha River (DEM data from the website https://www.gscloud.cn/ and accessed on 8 November 2021).

Table 1. Constructed hydropower stations on the main course of the middle reach of the Jinsha River.

Name	Date of Impoundment	Total Storage (10^8 m^3)	Regulation Capacity	Watershed Area (km^2)	Generation Capacity (MW)	Normal Water Level (m)
Liyuan	November 2014	7.3	Daily	1533	2400	1618
Ahai	December 2011	8.1	Daily	15,328	2000	1504
Jin'anqiao	November 2010	8.5	Weekly	1977	2400	1418
Longkaikou	November 2012	5.1	Daily	2637	1800	1298
Ludila	May 2013	15.5	Daily	7351	2160	1223
Guanyinyan	November 2014	20.7	Weekly	9172	3000	1134

2.2. Water Quality Indicators and Data Sources

In China, chemical oxygen demand and ammonia–nitrogen, which are the most important water quality indicators, are the total control index of water pollutants. Total phosphorus is an index of eutrophication in lakes and reservoirs. In general, increase of the total phosphorus concentration means increased risk of eutrophication in lakes and reservoirs. Therefore, the three water quality indicators, COD_{Mn}, NH_3-N, and TP, were selected as the representative indicators of the study reaches. The years of these monitoring data are 2004 to 2019. There are six monitoring sections: the Liyuan, Ahai, Jin'anqiao, Longkaikou, Ludila, and Guanyinyan dam sites. The monitoring data before operation of the hydropower station began derived mainly from the Environmental Impact Report of that hydropower station, which were discontinuous and were generated about three times a year. The monitoring data after operations began came mainly from the Investigation Report on Environmental Protection Acceptance of Completion of the hydropower station and routine monitoring data from the cascade hydropower station, which were generated about once a month. (Table 2)

Table 2. Overview of water quality monitoring data.

Name	Before Impoundment		After Impoundment	
	Years of the Data	Quantities of the Data	Years of the Data	Quantities of the Data
Liyuan	2007–2014	25	2015–2018	36
Ahai	2004–2011	25	2015–2019	44
Jin'anqiao	2004–2005	11	2016–2019	43
Longkaikou	2005–2006	11	2013–2018	40
Ludila	2005–2013	36	2013–2019	63
Guanyinyan	2005	5	2017–2019	30

The temperature monitoring data in 2004–2019 were from the Lijiang Meteorological Station, which were monitored daily and then averaged monthly. The Lijiang Meteorological Station is near the middle reach of the Jinsha River. The data were obtained from the U.S. National Oceanic and Atmospheric Administration (NOAA) and the National Center for Environmental Information (NCEI) (NOAA—National Centers for Environmental Information, https://www.ncei.noaa.gov accessed on 10 June 2022).

2.3. Methods and Models

The regression discontinuity method assumes that there is a treatment variable, if the running variable is greater than a certain critical value, then the treatment variable is 1, otherwise it is 0. In this study, hydropower development (D_t) was the treatment variable, and time (t) was the running variable. Since the impact of hydropower development on the hydrological and hydraulic conditions of the river starts from the impoundment of the hydropower station, the time (t_0) of the impoundment of the hydropower station was used as a certain critical value to study. When $t < t_0$, D_t is 0, indicating that the power station does not store water; when $t \geq t_0$, D_t equals to 1, indicating that the power station has stored water (Equation (2)). D_t is a dimensionless variable. Based on the above methods of regression discontinuity, the model constructed in this study is shown in Equation (1).

$$C_t = \mu + \alpha D_t + \sum_{i=1}^{k} \beta_i (t - t_0)^i + D_t \sum_{i=1}^{k} \gamma_i (t - t_0)^i + \sum_{j=1}^{l} \delta_j X_{jt} + \varepsilon_t \tag{1}$$

$$D_t = \begin{cases} 0, (t < t_0) \\ 1, (t \geq t_0) \end{cases} \tag{2}$$

where C_t is the result variable, which refers to the concentration of pollutants such as COD_{Mn}, NH_3-N, and TP of the hydropower station in month t, t is the running variable, which is time in this study, and the unit is month; t_0 is the time of the impoundment of the power station; $t - t_0$ indicates how many months the current time is from the time of the impoundment of the power station. Both β_i and γ_i are the correlation coefficient of the running variable. D_t is the treatment variable, and the subscript t represents the time. α is the correlation coefficient of the treatment variable. X_{jt} is other covariate, δ_j is the correlation coefficient of other covariates. ε_t is random error, and μ is a constant term.

The regression discontinuity method focuses on the regression coefficient α of the treatment variable. When α is negative, it represents a negative correlation between hydropower development and pollutant concentration, and when α is positive, it represents a positive correlation between hydropower development and pollutant concentration.

Water quality is affected not only by hydropower development but also by other factors such as water temperature, rainfall, sediment content, etc., among which water temperature has an important impact on water quality. Therefore, at higher water temperatures, most chemical reactions and bacteriological processes progress faster, which would change the water's self-clarification ability [32–37]. The water temperature of the research area is affected by hydropower development and air temperature. Because the effect of hydropower development is reflected in the D_t variable, air temperature was selected as a covariate in the regression discontinuity model used in this study, which also reflects seasonality.

In this study, the regression discontinuity procedure was as follows: Firstly, the data were sorted out and classified into pre-hydropower development and post-hydropower development. Secondly, a figure of the result variable with the treatment variable was plotted to help us analyze whether there was a discontinuity point. Thirdly, the data were regressed on both sides of the discontinuity point by the Stata software.

3. Results

3.1. Monitoring Results of Water Quality Indicators

In China, COD_{Mn}, NH_3-N, and TP have different water quality standards. The current surface water standard in China is "Environmental Quality Standard for Surface Water" (GB3838-2002) [38], which divides water quality into five categories: Class I, Class II, Class III, Class IV, and Class V. According to the Functional Zoning of Surface Water and Water Environment in Yunnan Province (2010–2020) [39], the water environment functions in the middle reach of the Jinsha River include industrial water, agricultural water, and drinking water, and the water quality standards of Class II are implemented, in which TP adopts Lake Standards (Table 3). The Stata software (Version 15.1) was used for analysis.

Table 3. Limit value of water quality standard for Class II.

Variable	COD_{Mn}	NH_3-N	TP
Water quality standard limit (mg/L)	≤ 4	≤ 0.5	≤ 0.025

This study collected and analyzed the water quality monitoring data for six hydropower stations on the middle reach of the Jinsha River before and after impoundment. (Table 4, Figure 2). From the perspective of water quality indicators, most of the COD_{Mn} and NH_3-N concentrations of the six hydropower stations met the standard, and only part of the Ludila and Guanyinyan Hydropower Stations exceeded the standard by small amounts. The TP concentration of all six hydropower stations exceeded the standard ranging from 3.3% to 100%, and the maximum exceeding multiple was 0.2–8.6 times.

Table 4. Water quality monitoring results before and after water impoundment at hydropower stations.

Hydropower Station	Variable	Before Impoundment			After Impoundment		
		Ave. (mg/L)	Exceeding Ratio	Max. Exceeding Multiple	Ave. (mg/L)	Exceeding Ratio	Max. Exceeding Multiple
Liyuan	COD_{Mn}	1.1	0.0%	0	1.3	0.0%	0
	NH_3-N	0.12	0.0%	0	0.13	0.0%	0
	TP	0.014	8.0%	0.8	0.014	8.3%	1.1
Ahai	COD_{Mn}	0.9	0.0%	0	1.3	0.0%	0
	NH_3-N	0.15	0.0%	0	0.14	0.0%	0
	TP	0.1	76.0%	8.6	0.015	4.5%	0.2
Jin'anqiao	COD_{Mn}	0.9	0.0%	0	1.1	0.0%	0
	NH_3-N	0.17	0.0%	0	0.12	0.0%	0
	TP	0.035	100.0%	0.6	0.035	41.9%	6.8
Longkaikou	COD_{Mn}	3.1	0.0%	0	3.5	0.0%	0
	NH_3-N	0.26	0.0%	0	0.27	0.0%	0
	TP	0.072	81.8%	4.6	0.025	35.0%	0.2
Ludila	COD_{Mn}	2.2	8.3%	0.1	2.5	0.0%	0
	NH_3-N	0.2	0.0%	0	0.21	0.0%	0
	TP	0.043	97.2%	2.5	0.025	49.2%	0.5
Guanyinyan	COD_{Mn}	3.4	40.0%	0.2	1.2	0.0%	0
	NH_3-N	0.44	40.0%	0.4	0.12	0.0%	0
	TP	0.1	80.0%	5.2	0.015	3.3%	0.2

According to time distribution, the average COD_{Mn} concentration of each hydropower station before impoundment was 0.9–3.4 mg/L, and the average COD_{Mn} concentration of each hydropower station after impoundment was 1.1–3.5 mg/L; the average NH_3-N concentration of each hydropower station before impoundment was 0.12–0.44 mg/L, the average NH_3-N concentration of each hydropower station after impoundment was 0.12–0.27 mg/L; the average TP concentration of each hydropower station before impoundment was 0.014–0.1 mg/L, and the average TP concentration of each hydropower station was 0.014–0.035 mg/L. The average COD_{Mn} concentration at Liyuan, Ahai, Jin'anqiao, Longkoukou, and Ludila increased after impoundment; the average COD_{Mn} concentration at Guanyinyan decreased after impoundment. The average NH_3-N concentration increased at Liyuan, Longkoukou, and Ludila increased after impoundment; the average NH_3-N concentration decreased after impoundment at Ahai, Jin'anqiao, and Guanyinyan. The average TP concentration at Ahai, Longkaikou, Jin'anqiao, Ludila, and Guanyinyan decreased after impoundment; the average TP concentration at Liyuan and Jin'anqiao remained unchanged. Overall, after impoundment, the average concentrations of COD_{Mn} and NH_3-N at different power stations increased or decreased, but the average concentration of TP remained unchanged or decreased.

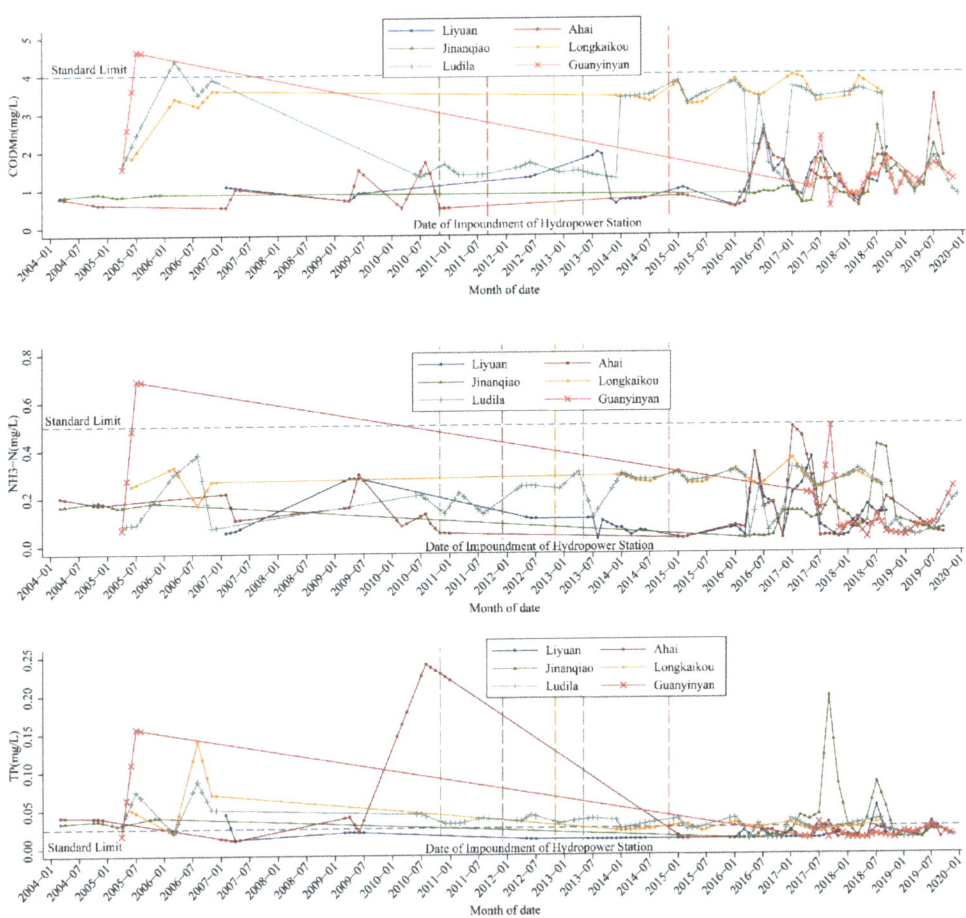

Figure 2. Water quality monitoring results at six hydropower stations on the middle reach of the Jinsha River.

Spatially, the three water quality indicators generally showed a trend of gradual deterioration from upstream to downstream. Before impoundment, the average concentration of COD_{Mn} at Liyuan Hydropower Station was 1.1 mg/L, the average concentration of NH_3-N was 0.12 mg/L, and the average concentration of TP was 0.014 mg/L. The average concentration of COD_{Mn} at Guanyinyan Hydropower Station was 3.4 mg/L, the average NH_3-N concentration was 0.44 mg/L, and the average TP concentration was 0.1 mg/L. After impoundment, the average concentration of COD_{Mn} at Liyuan Hydropower Station was 1.3 mg/L, the average concentration of NH_3-N was 0.13 mg/L, and the average concentration of TP was 0.014 mg/L; while the average concentration of COD_{Mn} at Longkou Hydropower Station was 3.5 mg/L, the average NH_3-N concentration was 0.27 mg/L, and the average TP concentration was 0.25 mg/L (Figure 3).

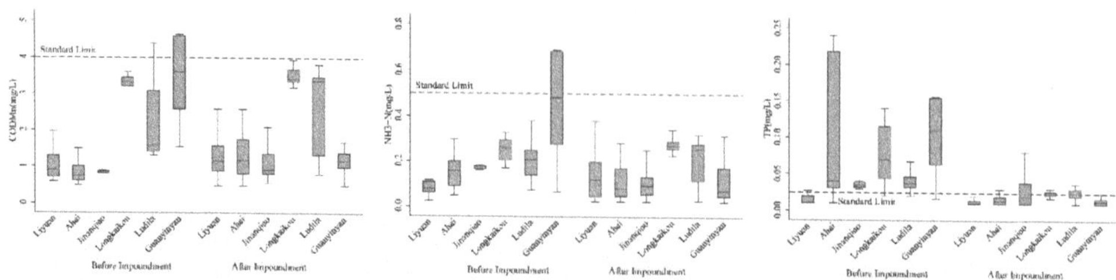

Figure 3. Boxplot of water quality indicators before and after water impoundment at six hydropower stations on the Jinsha River.

3.2. Seasonal Variations in Water Quality Indicators

According to monthly changes in pollutant concentrations at each cascade hydropower station after the hydropower development along the main course of the Jinsha River (Figure 4), the COD_{Mn} and TP concentrations had obvious seasonal variation at Liyuan, Ahai, Jin'anqiao, and Guanyinyan. Seasonal variation of COD_{Mn} and TP concentrations was not obvious at Longkaikou and Ludila. Specifically, as for the hydropower stations at Liyuan, Ahai, Jin'anqiao, and Guanyinyan, the COD_{Mn} concentration was generally higher from May to November and lower from December to April of the following year; the TP concentration was generally higher from June to October and lower from November to May of the following year. The NH_3-N concentration at Liyuan was higher from April to September and lower from October to March of the following year. The seasonal distribution characteristics of NH_3-N concentrations showed no obvious trend at the other five hydropower stations.

3.3. Correlation Analysis

Statistical analysis of the correlation of variables (Table 5) showed that the correlation coefficient between the treatment variables of hydropower development and the TP index was significantly negative. We preliminarily conclude that hydropower development is negatively correlated with TP concentration. The correlation coefficient between the treatment variables of hydropower development and COD_{Mn} index was significantly positive. We tentatively conclude that hydropower development is positively correlated with COD_{Mn} concentration. The correlation coefficient between the treatment variables of hydropower development and NH_3-N was negative but not significant, so it is impossible to judge directly whether hydropower development caused the apparent change in NH_3-N concentration. Certainly, the direct correlation analysis here was just a preliminary study, and the final result depended on the regression discontinuity analysis.

Table 5. Analysis of the correlation coefficient between treatment variables of hydropower development and water quality indexes.

Variable	Concentration			Air Temperature	Hydropower Development
	COD_{Mn}	NH_3-N	TP		
COD_{Mn}	1				
NH_3-N	0.5944 ***	1			
TP	0.0891	0.1267 *	1		
Air temperature	0.1175 *	0.0386	0.064	1	
Hydropower development	0.1026 *	−0.0598	−0.3855 ***	−0.0164	1

Note: *** and * are significant at the level of 0.1% and 5%, respectively.

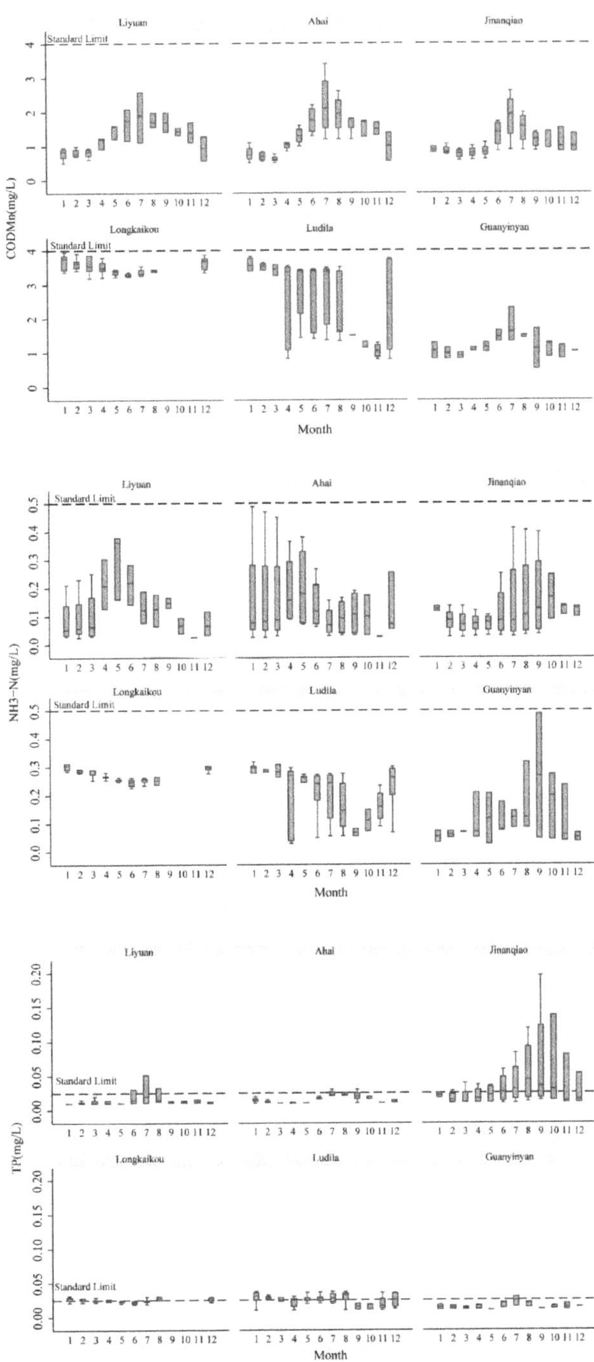

Figure 4. Boxplot of monthly changes in pollutant concentrations at each cascade hydropower station after hydropower development on the middle reach of the Jinsha River.

3.4. Regression Discontinuity

In the graph of regression discontinuity (Figure 5), the horizontal ordinate is the running variable value, which represents the value of the monitoring time subtracted from the impoundment time of the hydropower station. Therefore, the values on the left side of the 0 point on the horizontal ordinate are monitoring data obtained before hydropower development, and the values on the right side of the 0 point are monitoring data obtained after hydropower development. The ordinate of the graph is the concentration of COD_{Mn}, NH_3-N, and TP. It can be seen from Figure 5 that all three indicators of COD_{Mn}, NH_3-N, and TP have an obvious discontinuity point at 0 point. According to the construction assumption of the regression discontinuity model, if the running variable is continuous but the result variable jumps at the discontinuity point, it can be concluded that the treatment variable caused the result variable to jump. The running variable time in this study was itself continuous. The result variable was the concentration of COD_{Mn}, NH_3-N, and TP, and the treatment variable was hydropower development. Therefore, the discontinuity point can indicate that hydropower development led to a marked increase in the concentrations of COD_{Mn}, NH_3-N, and TP, proving that cascade hydropower development on the middle reach of the Jinsha River has had a significant impact on water quality.

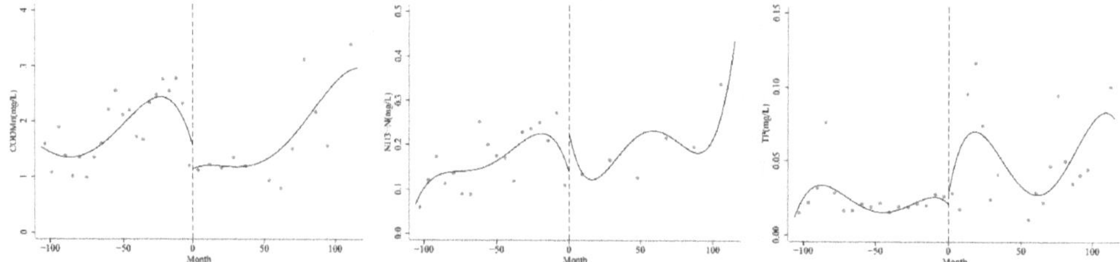

Figure 5. Regression discontinuity analysis on the discontinuity point. Note: the points are the monitoring data, and the lines are modeled using regression discontinuity.

All the estimation results were significant (Table 6), and the concentrations of COD_{Mn}, NH_3-N, and TP all jumped at the discontinuity point. In order to exclude the influence of other covariates on water quality, the air temperature was chosen as a covariate to use in regression discontinuity analysis, then all the results were still significant. The concentration of the three indicators still jumped at the discontinuity points after adding covariates. It can be concluded that hydropower development has a significant impact on the concentrations of COD_{Mn}, NH_3-N, and TP. Regardless of whether covariates were added or not, the correlation coefficients corresponding to TP were negative, while the correlation coefficients corresponding to COD_{Mn} and NH_3-N were positive. This is consistent with the results of variable correlation analysis.

Table 6. Regression discontinuity estimation results.

Estimate	COD_{Mn}	NH_3-N	TP
No covariates added	0.883 ***	0.172 **	−0.033 *
	(0.233)	(0.062)	(0.016)
Covariates added	0.934 ***	0.179 **	−0.034 *
	(0.218)	(0.062)	(0.016)

Note: ***, **, and * are significant at the level of 0.1%, 1%, and 5%, respectively. The standard error in brackets passed the robustness test and independence test.

Through analysis of the monitoring date, hydropower development led to a decrease of 0.034 mg/L in the TP concentration. Although the absolute value of the influence of hydropower development on TP concentration was small, because the standard limit of TP in the research river section was only 0.025 mg/L, the impact of hydropower development on TP concentration was as high as 1.36 times the standard limit value. Therefore, we can conclude that cascade hydropower development on the middle reach of the Jinsha River has greatly influenced the concentration of TP and that it has a strong negative correlation with the concentration of TP.

Similarly, hydropower development led to an increase in the concentration of COD_{Mn} by 0.934 mg/L and of NH_3-N by 0.179 mg/L. The standard limit of COD_{Mn} in the research river section was 4 mg/L, and the standard limit of NH_3-N was 0.5 mg/L. The impact of hydropower development on the COD_{Mn} concentration was 0.23 times the COD_{Mn} standard limit, and the impact on the NH_3-N concentration was 0.36 times the NH_3-N standard limit. Therefore, it can be said that cascade hydropower development on the middle reach of the Jinsha River has a weak positive correlation with the COD_{Mn} and NH_3-N concentrations. Through analysis of the absolute value of the correlation coefficient, it was determined that hydropower development on the middle reach of the Jinsha River has slightly increased the concentration of COD_{Mn} and NH_3-N limit. The maximum concentrations of COD_{Mn} and NH_3-N were 3.5 and 0.44 mg/L, respectively, while the standard limits of COD_{Mn} and NH_3-N were 4 and 0.5 mg/L. Thus, there is a risk that the above two indicators will exceed the standard values in the future.

4. Discussion

4.1. Influence of Cascade Hydropower Development on Water Quality

This study focuses on the impact of cascade hydropower development on water quality and uses discontinuity regression to analyze water quality monitoring data before and after the construction of six hydropower stations on the middle reach of the Jinsha River. The results of regression discontinuity show that cascade hydropower development on the middle reach of the Jinsha River had different impacts on three water quality indicators, including COD_{Mn}, NH_3-N, and TP. Hydropower development had a strong negative correlation with TP concentration, and a weak positive correlation with concentrations of COD_{Mn} and NH_3-N.

Among cascade hydropower stations located on the middle reach of the Jinsha River, four are daily adjustment hydropower stations and two are weekly adjustment stations (Table 1). Therefore, flow and water volume processes show little change, but water level and flow velocity changed greatly after cascade hydropower development on the middle reach of the Jinsha River (Figure 6). The water levels were 1020 to 1503 m before cascade hydropower development and 1132 to 1618 m after cascade hydropower development. The flow velocities were 1.53 to 3.39 m/s before cascade hydropower development and 0.03 to 0.08 m/s after cascade hydropower development. According to research conducted by Li Jinxiu et al. [40], increasing water depth and decreased flow velocity can cause reduction of the reoxygenation coefficient of the water body, which will be unfavorable for the attenuation of pollutant concentrations in the river. When sedimentation increases, pollutants settle together with the sediment, which is beneficial to the reduction of pollutant concentrations in the river. Ultimately, the concentration of pollutants increases or decreases depending on the predominant mechanism.

In addition to changes in water flow, hydropower development also affects pollution sources in the reservoir area, and pollution sources will also lead to changes in water pollutant concentrations. Firstly, after the cascade hydropower development of the Jinsha River, the original inhabitants of the reservoir area were relocated and resettled (Figure 7). Complete sewage treatment and water and soil conservation measures have been undertaken in the resettlement areas. Compared with the previous extensive production and lifestyle, pollutants entering the river will be reduced. Secondly, hydropower development will drive economic and social development of the reservoir area, which in turn will increase the discharge of pollutants. The middle reach of the Jinsha River includes thirteen counties in five prefectures in the Sichuan and Yunnan provinces. Before the development of hydropower, economic development in this rural area was relatively slow, and there were essentially no major industrial pollution sources except for Panzhihua City in southern Sichuan, which is downstream of the Guanyinyan Hydropower Station. Therefore, the pollution in the reservoir area mainly derived from agricultural non-point sources and domestic pollution sources before the development of hydropower. However, after the development of hydropower, the regional economy has become well developed. According to the research of He Xiaorong et al. [41], after the development of cascade hydropower on the middle reach of the Jinsha River, the main sources of pollutants in the river now are tourism, shipping, domestic sources, and agricultural non-point sources. Finally, after the cascade reservoirs were impounded on the Jinsha River, residual pollutants in submerged cultivated land and forests have been gradually released (Figure 7), also leading to increased pollutants in the water body [42]. In summary, although pollution prevention and control measures in the submerged area improved during the construction of hydropower stations, the total amount of pollutants flowing into the middle reach of the Jinsha River after cascade hydropower development is generally on the rise due to the influences of hydropower development on the regional economy and society.

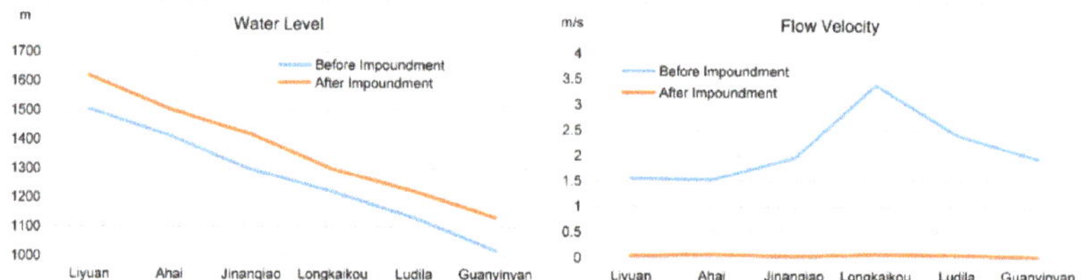

Figure 6. Changes in water level in front of dams and velocity in reservoirs after cascade hydropower development on the middle reaches of the Jinsha River.

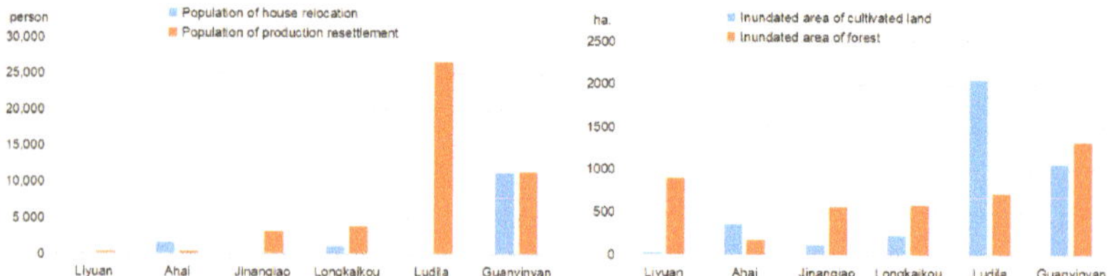

Figure 7. Number of resettled populations and inundated land areas for cascade hydropower development on the middle reach of Jinsha River.

According to correlation coefficient analysis of the regression discontinuity, cascade hydropower development on the middle reach of the Jinsha River has a strong negative correlation with TP concentration. On the Jinsha's middle reach, although the total amount of TP pollutants discharged into the river increased, at the same time, the degradation capacity of the water body decreased and the TP concentration decreased, due to the deposition of TP with a large amount of sediment after the impoundment of the hydropower station. This result is consistent with the research of Lou Baofeng et al. [43] on the concentration of TP with sediment deposition before and after the impoundment of the Three Gorges Reservoir, who noted that the decrease in TP concentration can reach 61–65%. Therefore, we believe that increased sedimentation caused by cascade hydropower development on the middle reach of the Jinsha River is the principal reason for the noted decrease in TP concentration.

Furthermore, according to correlation coefficient analysis of the regression discontinuity, cascade hydropower development on the middle reach of the Jinsha River has a weak positive correlation with concentrations of COD_{Mn} and NH_3-N. According to the previous analysis, the total amount of COD_{Mn} and NH_3-N flowing into rivers experiencing regional economic and social development increases after hydropower development. However, different from the case of TP, the effect of COD_{Mn} and NH_3-N with sediment settlement is not so obvious. From the results, we conclude that hydropower development led to increased COD_{Mn} and NH_3-N concentrations. Therefore, we believe that the regional economic and social development brought by cascade hydropower development on the middle reach of the Jinsha River is the principal source of increased COD_{Mn} and NH_3-N concentrations.

4.2. Seasonal Variations

According to monthly variations in water quality indicators, at four of the six hydropower stations on the middle reach of the Jinsha River, COD_{Mn} concentrations are higher from May to November and lower from December to April of the following year, and TP concentrations are higher from June to October and lower from November to May of the following year, which may be the result of rainfall. The uneven distribution of water resources on the middle reach of the Jinsha River, affected by the plateau monsoon regime with distinct rainy and dry seasons must also be borne in mind since water resources vary greatly throughout the year. During the rainy season (May to October), heavy rains are frequent, and pollutants enter water bodies with rainfall runoff and eventually sink into the main course of the Jinsha River, resulting in increased concentrations of pollutants. This result is consistent with the research of Lou Ba Chongzhen et al. on the main course of the Lancang River.

Seasonal variation of COD_{Mn} and TP concentrations is not obvious at Longkaikou and Ludila. The inundated land areas of Ludila are the largest from Figure 7, so the release of pollutants from the submerged land is the important factor affecting the water quality. The total storage of Longkaikou is the smallest from Table 1, so the water quality is easily affected by residual pollutants in the submerged land.

4.3. Water Quality Affected by Hydropower Development in Other Watersheds

In general, the water quality change is a complicated process after cascade hydropower development, and there are obvious differences between different watersheds. Peng Chunlan et al. [13] and Lou Baofeng et al. [43] all suggest that increased sedimentation after the impoundment of the Three Gorges Reservoir caused a decrease in TP concentration. Xue Lianfang et al. [14] believes the self-purification capacity of the water body was improved because of reservoir construction in the Hongshui River; Tomczyk Pawel and Wiatkowski Miroslaw's [12] research result was that hydropower plants affect the physicochemical parameters of the water. In these studies, hydropower development had positive effects on water quality. However, there are some different results in other studies: research on water quality change of Dadu River showed that untreated domestic sewage around the watersheds was the main reason for water quality deterioration, while hydropower development played little role in water quality [44]. Alvarez Xana et al. [11] also think that the presence of the hydropower plants did not significantly influence the physical and chemical characteristics of the water. After Li Jinpeng et al. [45] analyzed water quality monitoring data before and after the construction of the Xiaowan Dam on the Lancang River, they argued that water quality in the reservoir area gradually worsened relative to that of natural rivers after impoundment. Winton, R et al. [10] anticipate that with agricultural intensification, urbanization, and future hydropower development in the Zambezi River Basin in Southern Africa, the number and extent of these hotspots of water quality degradation will grow in response.

The above research results show that there are watersheds with degraded water quality after hydropower development, and watersheds with improved water quality after hydropower development (Table 7). By comparing our research with the work of other scholars, we believe that the different causes of water quality changes should be investigated according to the characteristics of the watershed.

Table 7. Overview of water quality affected by hydropower development in other watersheds.

Study Area	Water Quality Indicators	Change of Water Quality	Authors	Publication Year
Yangtze River in China	TP	Improved	Lou Baofeng et al. [43]	2011
Hongshui River in China	TN, BOD$_5$, TP	Improved	Xue Lianfang et al. [14]	2013
Yangtze River in China	TP, DO, SS, heavy metal	Improved	Peng Chunlan et al. [13]	2016
Lancang River in China	Goodnight–Whitley index	Degraded	Li Jinpeng et al. [45]	2018
Lerez, Umia, Ulla and Mandeo rivers in northwestern Spain	pH, DO, conductivity, water temperature	Unchanged	Alvarez Xana et al. [11]	2020
Dadu River in China	pH, COD$_{Mn}$, NH$_3$-N, DO, BOD$_5$, heavy metal	Unchanged	Gao Jian et al. [44]	2021
Zambezi River Basin in southern Africa	pH, TP, DO, TN, SS, chlorophyll, and so on	Degraded	Winton, R et al. [10]	2021
Bystrzyca River in Poland	DO, conductivity, TP	Improved	Tomczyk Pawel and Wiatkowski Miroslaw [12]	2021

4.4. Suggestions on Water Quality Protection Measures

According to the main reasons cited for changes in the concentrations of COD$_{Mn}$, NH$_3$-N, and TP pollutants, we propose three suggestions for water quality protection on the middle reach of the Jinsha River.

Firstly, local governments should take steps to control pollution sources. The main reasons for increases in COD$_{Mn}$ and NH$_3$-N concentrations is elevated pollutant emissions caused by economic and social development of the middle reach of the Jinsha River. While developing the economy of the reservoir area, local governments should also effectively control pollutants to increase production without escalating water pollution.

Secondly, the operating division of each hydropower station should carry out sediment transportation through reasonable scheduling. The main reason for decreased TP concentrations is enhanced sedimentation after the reservoir was impounded. TP on the middle reach of the Jinsha River is not degraded but deposited on the bottom of the reservoirs along with sediment. Therefore, the operating division of each hydropower station should transport more sediment downstream through reasonable dispatch of the cascade hydropower station, in order to mitigate phosphorus transport [46,47].

Thirdly, both local governments and operating divisions of hydropower stations should augment water quality monitoring on the middle reach of the Jinsha River. Due to the deposition of TP with sediment, a large quantity of TP pollutants accumulates at the bottom of each cascade hydropower station reservoir on the middle reach of the Jinsha River. Combined with the current situation of TP exceeding set standards during some periods, monitoring TP pollutants will be an important task for water quality conservation at cascade hydropower stations on the middle reach of the Jinsha River in the future. In addition, the maximum concentration of COD$_{Mn}$ and NH$_3$-N are 3.5 mg/L (the standard limit: 4 mg/L) and 0.44 mg/L (the standard limit: 0.5 mg/L), so there is a risk that the COD$_{Mn}$ and NH$_3$-N will exceed the standard limit in the future, and the monitoring of them should also be emphasized.

5. Conclusions

In this paper, we chose the main course of the middle reach of the Jinsha River as the subject of our research to analyze water quality monitoring data before and after construction of hydropower stations, using the regression discontinuity method. The results lead us to the following conclusions:

1. The concentration of COD$_{Mn}$ and NH$_3$-N at each hydropower station meet standards most of the time, but the concentration of TP exceeds standards sometimes. From

the perspective of spatial distribution, concentrations of COD_{Mn}, NH_3-N, and TP generally showed a trend toward gradual worsening from upstream to downstream.

2. After cascade hydropower development on the middle reach of the Jinsha River, water quality changed significantly, including decreased TP concentration and increased concentrations of COD_{Mn} and NH_3-N. Mechanisms influencing the concentration of different pollutants vary. After reservoir impoundment, the intensified sedimentation effect leads to decreased TP concentrations, while the economic and social development driven by cascade hydropower development leads to increasing concentrations of COD_{Mn} and NH_3-N.

3. The water quality of most hydropower stations on the middle reach of the Jinsha River reflects obvious seasonal characteristics. Since the study area is characterized by marked rainy and dry seasons, rainfall affects water quality during different seasons. In the rainy season, concentrations of COD_{Mn} and TP are higher, and they are concomitantly lower during the dry season. This is directly related to non-point-source pollution in reservoir catchments. It is very important that the nutrient input of non-point source pollution is controlled for controlling the input of P in reservoirs.

Based on these research results, we suggest actions to enhance water quality protection in the middle reach of the Jinsha River, including strengthening the control of pollution sources in the reservoir area, carrying out sediment transportation through reservoir scheduling, and strengthening the water quality monitoring. Shortcomings of this study include the fact that our analyzes and interpretations are based only on water quality monitoring data from 2004 to 2019. The collection of additional, more complete, data in the future will help improve our model and potentially modify the results.

Author Contributions: Conceptualization, T.X. and F.C.; methodology, T.X.; software, T.X. and Q.Y.; validation, W.M.; formal analysis, T.X.; resources, X.H.; writing—original draft preparation, T.X.; writing—review and editing, F.C.; supervision, W.M.; project administration, T.X.; funding acquisition, Q.Y. All authors have read and agreed to the published version of the manuscript.

Funding: This study was supported by a grant from the National Natural Science Foundation of China (No. 51609263).

Data Availability Statement: Restrictions apply to the availability of these data. Data were obtained from the Environmental Monitoring Center of Middle Reaches of Jinsha River and are available from the authors with the permission of the Environmental Monitoring Center of Middle Reaches of Jinsha River.

Acknowledgments: Many thanks to Hucai Zhang for his kind invitation and careful comments and suggestions that have improved the earlier draft greatly.

Conflicts of Interest: The authors declare no conflict of interest.

Abbreviations

COD_{Mn}	Chemical oxygen demand (potassium permanganate index)
NH_3-N	Ammonia–nitrogen
TP	Total phosphorus
SS	Suspended solids
DO	Dissolved oxygen
pH	Pondus hydrogenii

References

1. Gang, L.; Li, P. Research on the construction of safety emergency platform for cascade hydropower stations in the basin. *Water Power* **2021**, *47*, 104–109. [CrossRef]
2. Chen, K.-Q.; Chang, Z.-N.; Cao, X.-H. Thinking highly of environment and developing hydropower orderly. *China Water Power Electrif.* **2010**, *31*, 13–17. [CrossRef]
3. Bao-Zhu, H.; Lei-Lei, G.; Na, W. Analysis on reservoir construction upon ecological environment. *J. Zhejiang Univ. Water Resour. Electr. Power* **2008**, *20*, 41–43. [CrossRef]

4. Tianbao, X.; Jing, P.; Chong, L. Influence of Gezhouba Dam on the eco-hydrological characteristics in the middle reach of the Yangtze River. *Resour. Environ. Yangtze Basin* **2007**, *16*, 72–75. [CrossRef]

5. Ran, X.B.; Yu, Z.G.; Yao, Q.Z.; Chen, H.T.; Mi, T.Z.; Yao, P. Advances in nutrient retention of dams on river. *J. Lake Sci.* **2009**, *21*, 614–622. [CrossRef]

6. Kuriqi, A.; Pinheiro, A.N.; Sordo-Ward, A.; Bejarano, M.D.; Garrote, L. Ecological impacts of run-of-river hydropower plants—Current status and future prospects on the brink of energy transition. *Renew. Sustain. Energy Rev.* **2021**, *142*, 110833. [CrossRef]

7. Kuriqi, A.; Pinheiro, A.N.; Sordo-Ward, A.; Garrote, L. Influence of hydrologically based environmental flow methods on flow alteration and energy production in a run-of-river hydropower plant. *J. Clean. Prod.* **2019**, *232*, 1028–1042. [CrossRef]

8. Młyński, D.; Wałęga, A.; Kuriqi, A. Influence of meteorological drought on environmental flows in mountain catchments. *Ecol. Indic.* **2021**, *133*, 108460. [CrossRef]

9. Kuriqi, A.; Pinheiro, A.N.; Sordo-Ward, A.; Garrote, L. Water-energy-ecosystem nexus Balancing competing interests at a run-of-river hydropower plant coupling a hydrologic–ecohydraulic approach. *Energy Convers. Manag.* **2020**, *223*, 113267. [CrossRef]

10. Winton, R.S.; Teodoru, C.R.; Calamita, E.; Kleinschroth, F.; Banda, K.; Nyambe, I.; Wehrli, B. Anthropogenic influences on Zambian water quality: Hydropower and land-use change. *Environ. Sci. Processes Impacts* **2021**, *23*, 981–994. [CrossRef]

11. Alvarez, X.; Valero, E.; Torre-Rodríguez, N.D.; Acuna-Alonso, C. Influence of small hydroelectric power stations on river water quality. *Water* **2020**, *12*, 312. [CrossRef]

12. Tomczyk, P.; Wiatkowski, M. The effects of hydropower plants on the physicochemical parameters of the Bystrzyca River in Poland. *Energies* **2021**, *14*, 2075. [CrossRef]

13. Peng, C.-I.; Jiang, Y.-J.; Zhang, X.-Y. Discussion on water quality changes and causes in reservoir area and downstream of Three Gorges Reservoir before and after impoundment. *Express Water Resour. Hydropower Inf.* **2016**, *37*, 35–40. [CrossRef]

14. Lianfang, X.; Hongbin, G.; Bin, W. Study on cumulative effects of water quality by hydroelectric cascade development in Hongshui River. *Water Power* **2013**, *39*, 9–12. [CrossRef]

15. Aijun, B.C.; Ying, P. Study on the spatiotemporal differences in the reservoir of cascade hydropower of Lancang River. *Environ. Sci. Surv.* **2015**, *34*, 35–40. [CrossRef]

16. Xinyue, Z.; Peiming, M.; Qianhong, G.; Haitao, Y.; Bao, Q. Spatial-temporal variations of water quality in upstream and downstream of Three Gorges Dam. *J. Lake Sci.* **2019**, *31*, 633–645. [CrossRef]

17. Yue, H.; Zhilin, H.; Wenfa, X.; Lixion, Z.; Liang, M. Water quality evaluation and prediction of the reservoir inflow and outflow after the Three Gorges Project operation. *Environ. Pollut. Control.* **2019**, *41*, 211–215. [CrossRef]

18. Hu, G.; Wu, H.; Zhang, Z. Trend analysis on water quality of Xiaolangdi Reservoir of Yellow River. *Areal Res. Dev.* **2003**, *22*, 86–88. [CrossRef]

19. Wu, J.; Chen, M.; Yuan, R.; Li, H.; Liu, L. Application of seasonal Kendall method in downstream water quality trend analysis for Dongjiang River. *Guangxi Water Resour. Hydropower Eng.* **2017**, *3*, 14–18. [CrossRef]

20. Xie, Q.; Xue, X.; Fu, M. A literature review of the application of regression discontinuity design. *Rev. Econ. Manag.* **2019**, *35*, 69–79. [CrossRef]

21. Thistlethwaite, D.L.; Campbell, D.T. Regression-discontinuity analysis: An alternative to the ex-post facto experiment. *J. Educ. Psychol.* **1960**, *51*, 309–317. [CrossRef]

22. Hahn, J.; Todd, P.; Van der Klaauw, W. Identification and estimation of treatment effects with a regression-discontinuity design. *Econometrica* **2001**, *69*, 201–209. [CrossRef]

23. Yu, J.; Wang, C. The Rise of a New Quasi-Random Experiment Method—Breakpoint Regression and Application in Economics. *Econ. Perspect.* **2011**, 125–131. Available online: http://www.jjxdt.org/Admin/UploadFile/20150323002/2016-08-25/Issue/14 yb0ltf.pdf (accessed on 26 February 2022).

24. Card, D.; Giuliano, L. Can tracking raise the test scores of high—Ability minority students? *Am. Econ. Rev.* **2016**, *106*, 2783–2816. [CrossRef]

25. Malamud, O.; Popeleches, C. Home computer use and the development of human capital. *Q. J. Econ.* **2011**, *126*, 987–1027. [CrossRef] [PubMed]

26. Pinotti, P. Clicking on Heaven's Door: The effect of immigrant legalization on crime. *Am. Econ. Rev.* **2017**, *107*, 138–168. [CrossRef]

27. Ying, L.; Shen, Z.; Piao, S. The recent hiatus in global warming of the land surface: Scale-dependent breakpoint occurrences in space and time. *Geophys. Res. Lett.* **2015**, *42*, 6471–6478. [CrossRef]

28. Auffhammer, M.; Kellogg, R. Clearing the air? The effects of gasoline content regulation on air quality. *Am. Econ. Rev.* **2011**, *101*, 2687–2722. [CrossRef]

29. Toms, J.D.; Lesperance, M.L. Piecewise regression: A tool for identifying ecological thresholds. *Ecology* **2003**, *84*, 2034–2041. [CrossRef]

30. Wanhua, L.; Yaodong, Z.; Zhijia, D. The effectiveness of "River Chief System" policy: An empirical study based on environmental monitoring samples of China. *Water* **2021**, *13*, 1988. [CrossRef]

31. Hydro China Co., Ltd. Report on environmental impact Assessment and countermeasures of hydropower cascade development planning in mid-reach of Jinsha River. 2009.

32. Kang, J. Climate change characteristics and its influence on hydrology and water quality in Wuliangsuhai Basin in recent 50 years. *Hebei Univ. Eng.* **2016**, *10*, 1–97. [CrossRef]

33. Tianyu, L.; Lei, W.; Guohu, M.; Weihua, G. Numerical simulation for impacts of hydrodynamic conditions on algae growth in Chongqing Section of Jialing River, China. *Ecol. Model.* **2010**, *222*, 112–119. [CrossRef]
34. Whitehead, P.G.; Wilby, R.L.; Battarbee, R.W.; Kernan, M.; Wade, A.J. A review of the potential impacts of climate change on surface water quality. *Hydrol. Sci. J.* **2009**, *54*, 101–123. [CrossRef]
35. Hassan, H.; Hanaki, K.; Matsuo, T. A modeling approach to simulate impact of climate change in lake water quality: Phytoplankton growth rate assessment. *Water Sci. Technol.* **1998**, *37*, 177–185. [CrossRef]
36. Gómez-Martínez, G.; Galiano, L.; Rubio, T.; Prado-López, C.; Redolat, D.; Paradinas Blázquez, C.; Gaitán, E.; Pedro-Monzonís, M.; Ferriz-Sánchez, S.; Añó Soto, M.; et al. Effects of climate change on water quality in the Jucar River Basin (Spain). *Water* **2020**, *13*, 2424. [CrossRef]
37. Zhang, Z.; Wang, X.; Ma, W.; Zhang, J.; Pan, R.; Yang, R. The effects of global warming on purification processes of Tongzhou section of Beiyun river. *China Environ. Sci.* **2017**, *37*, 730–739. [CrossRef]
38. Ministry of Ecology and Environment of the People's Republic of China. Environmental Quality Standards for Surface Water (GB3838-2002). Available online: http://www.mee.gov.cn/ywgz/fgbz/bz/bzwb/shjbh/shjzlbz/200206/t20020601_66497.shtml (accessed on 26 February 2022).
39. Department of Ecology and Environment of Yunnan Province. Surface Water Environmental Functional Zoning of Yunnan Province (2010–2020). Available online: http://sthjt.yn.gov.cn/zwxx/zfwj/gsgg/201804/t20180417_178523.html (accessed on 26 February 2022).
40. Li, J.X.; Liao, W.G.; Huang, Z. Numerical simulation of water quality for the Three Gorges Project Reservoir. *J. Hydraul. Eng.* **2002**, *12*, 7–10. [CrossRef]
41. He, X.R.; Wang, X.Y.; He, W. Study on comprehensive water pollution control measures for reservoir areas along mid-reach of Jinshajiang River. *Water Resour. Hydropower Eng.* **2015**, *46*, 71–76. [CrossRef]
42. Wei, W.E.; Tao, L.I.; Lu, H.A. Analysis of influence of water environment on development of hydropower cascade downstream of the Hanjiang River. *J. Environ. Eng. Technol.* **2016**, *6*, 259–265. [CrossRef]
43. Lou, B.; Yin, S.; Mu, H.; Zhang, X. Comparison of total phosphorus concentration of Yangtze River within the Three Gorges Reservoir before and after impoundment. *J. Lake Sci.* **2011**, *23*, 863–867. [CrossRef]
44. Gao, J.; Yao, F.; Lei, X. Study on water quality evaluation and water quality distribution characteristics of main stream of Daduhe River under background of cascade hydropower development. *Water Resour. Hydropower Eng.* **2021**, *52*, 133–145. [CrossRef]
45. Li, J.; Peng, M.; Dong, S.; Li, C.; Wang, L. Assessment of benthic macro invertebrate assemblages and water quality in Xiaowan Reservoir before and after dam operation, Lancang River. *Res. Environ. Sci.* **2018**, *31*, 1900–1908. [CrossRef]
46. Han, C.; Zheng, B.; Qin, Y.; Ma, Y.; Yang, C.; Liu, Z.; Cao, W.; Chi, M. Impact of upstream river inputs and reservoir operation on phosphorus fractions in water-particulate phases in the Three Gorges Reservoir. *Sci. Total Environ.* **2018**, *610*, 1546–1556. [CrossRef] [PubMed]
47. Wang, D.; Wu, X. Analysis of phosphorus transport in the Three Gorges Reservoir. *J. Hydraul. Eng.* **2021**, *52*, 885–895. [CrossRef]

Article

Spatial and Temporal Distribution Characteristics of Nutrient Elements and Heavy Metals in Surface Water of Tibet, China and Their Pollution Assessment

Jiarui Hong [1,2], Jing Zhang [1,2,*], Yongyu Song [1,2] and Xin Cao [1,2]

1   Beijing Laboratory of Water Resources Security, Beijing 100048, China
2   Key Laboratory of 3D Information Acquisition and Application of Ministry of Education, Capital Normal University, Beijing 100048, China
*   Correspondence: 5607@cnu.edu.cn

Abstract: In the context of global climate change, the ecological environment of Tibet has been gaining attention given its unique geographical and fragile nature. In this study, to understand the pollution status of the surface water of Tibet, China, we collected monthly data of 12 indicators from 41 cross-sectional monitoring sites in 2021 and analyzed the spatial and temporal variations of nutrients and heavy metal elements, water quality conditions, and pollutant sources in surface water. All 12 polluting elements, except lead (Pb), had significant seasonal variations, but the magnitude of the differences was very small. Spatially, nutrient elements were relatively concentrated in the agricultural and pastoral development areas in central and northern Tibet. In general, the water quality in most parts of Tibet was found to be good, and the water quality of 41 monitoring sections belonged to Class I water standard as per the entropy method–fuzzy evaluation method. The study used a multivariate statistical method to analyze the sources of pollution factors. The principal component analysis method identified four principal components. The results of this study can provide a scientific basis for pollution prevention and control in the Tibet Autonomous Region, and contribute to further research on water ecology.

Keywords: nutrient elements; heavy metal elements; spatiotemporal characteristics; entropy method-fuzzy evaluation method; principal component analysis

Citation: Hong, J.; Zhang, J.; Song, Y.; Cao, X. Spatial and Temporal Distribution Characteristics of Nutrient Elements and Heavy Metals in Surface Water of Tibet, China and Their Pollution Assessment. *Water* **2022**, *14*, 3664. https://doi.org/10.3390/w14223664

Academic Editors: Hucai Zhang, Jingan Chen and G.D. Haffner

Received: 21 October 2022
Accepted: 11 November 2022
Published: 14 November 2022

Publisher's Note: MDPI stays neutral with regard to jurisdictional claims in published maps and institutional affiliations.

1. Introduction

Tibet is the main part of the Qinghai–Tibet Plateau in China with the highest and largest mountain systems in the world, such as the Himalayas, Nyingchi Tanggula, and Karakoram. It is also the main snow distribution area in China, and its glacial meltwater is the source of river runoff recharge for more than 10 countries in Asia [1]. With the global climate change and rapid development of the Tibetan society and economy, the exploitation of natural resources and development of secondary and tertiary industries have gradually accelerated, considerably impacting the management and utilization of local water resources. Therefore, research on the security of water resources in Tibet is of great scientific research value and practical significance for the harmonious coexistence of man and nature, and the maintenance of ecological security is a prerequisite for development.

The deterioration of aquatic environments owing to urbanization and economic development is becoming a serious problem [2]. In recent years, domestic studies on heavy metal pollution in surface water have mainly focused on areas with developed water systems or industrial bases. He et al. collected exposure concentration data of ten typical heavy metals from eight watersheds in China; they assessed and compared the ecological risks in the water bodies of each watershed, and mining was identified to be the main cause of heavy metal pollution [3]. Zhu et al. [4] conducted a study on the distribution characteristics of heavy metals in the sediments of major water systems and concluded that heavy metal pollution gradually developed into complex pollution of multiple elements due to the development of recent industries. For the western region, especially the Qinghai–Tibet Plateau, which is a unique geographical unit in the world, there are few studies on its surface water pollution and the concentration of nutrients and heavy metals in the local surface water. Based on the analysis of surface water samples collected from the Gongga mountain area, He et al. [5] found that the mass concentrations of heavy metal elements and nutrients in surface water generally show an increase from west to east under the influence of topography and monsoon. Chen et al. [6] investigated the enrichment and migration characteristics of heavy metal elements in polyunsaturated copper ores and found that the distribution of heavy metal elements was remarkable influenced by the minerals. Yang et al. [7] analyzed the enrichment characteristics of organochlorine pesticides and heavy metals in fish from remote mountain lakes and the Lhasa River on the Tibetan Plateau and concluded the plateau to be a regional pollutant convergence area due to long-range atmospheric transport and topographic cold traps. To date, the relationship among water quality, organic matter pollution, and heavy metal pollution in Tibet has not been studied thoroughly. Moreover, information on the distribution of surface water pollution and water quality evaluation lacks the scope of large watersheds, and most studies were mainly conducted in small watersheds.

While connecting and allocating rivers, lakes and other water resources for human use, there are more and more studies on water quality security and ecological impact [8]. Polluted and low-quality water resources will damage natural ecosystems and endanger human and animal health [9]. At present, arsenic (As) is a ubiquitous toxic substance in Tibetan rivers, and its compounds have teratogenic and carcinogenic effects [10,11]. Zhang et al. [12]. observed a very high concentration of dissolved As (1130–9760 μg/L) in the hot springs in the upper reaches of the Yarlung Zangbo River. Wang et al. [13] conducted a health risk assessment in the Yamdrok-tso basin, southern Tibetan Plateau, and found that when residents were exposed to As in the lake water through oral and skin channels for a long time, there was a potential danger in this area, and the contribution rate of cancer risk was 97.31%. Zhang et al. [14] found that changes in the nutrient structure and biodiversity of the Lhasa River forced the number of maladapted fish to decline. Based on the fully covered water quality data of Tibet, the non-parametric Kruskal-Wallis test is used to show the seasonal and regional changes in nutrient elements and heavy metals in 2021. According to the pollution situation in different regions and the health threat caused by local pollution, researchers can propose some adaptive treatment measures to alleviate the pollution.

Human activities are slowly affecting Tibet. So far, there has been no comprehensive understanding of water quality in Tibet. It is very necessary to carry out water quality assessment for its surface water. Analytical hierarchy process (AHP) and fuzzy AHP are usually used, but it is difficult and uncertain to determine model parameters (such as variable weights) under actual conditions, which leads to inaccurate water quality risk assessment [15]. Entropy method-fuzzy evaluation method combines the objectivity of entropy method with the fuzziness of fuzzy evaluation method [16–18]. This method can not only use the inherent information of the original data, but also takes into account the judgment of the practical experience of the evaluators to obtain the importance coefficient of each pollution element. It makes up for the shortcomings of the two methods, provides a more accurate and objective water quality assessment and determines river pollution risk indicators [18]. However, this method has never been used in the water quality assessment of Tibetan rivers, and its applicability is unknown.

In the context of global climate change, due to the unique geographical features and fragile ecological environment of Tibet, ecological and environmental problems are gaining attention, and it is important to study the local water resource situation. With the acceleration of urbanization and industrialization, the pollution of water resources in Tibet has become more serious and the seasonal spatial-temporal distribution of pollutant elements has changed. The impact on water resources, hydrological processes, social development and economy of downstream China will become more complex. Reviewing the literature, it was easy to realize that previous studies mostly focused on small watersheds or cities in Tibet, and until now there has been no comprehensive exploration in whole Tibet yet. Therefore, this study analyzed the spatial and temporal distribution of surface water pollution, for the first time, using data from 41 monitoring sites in Tibet from a large watershed perspective to better protect the water resources in Tibet. The study objectives were (1) select the non-parametric Kruskal–Wallis test combined with box line and spatial distribution maps to analyze the seasonal and inter-basin variability of elements, (2) combine the entropy value and fuzzy evaluation methods to develop a more objective evaluation of water quality at the cross-sectional monitoring sites, and (3) use Pearson correlation and principal component analyses to determine the source of the polluting elements.

2. Materials and Methods

2.1. Study Area Overview

The Tibet Autonomous Region is located in the southwest of the Qinghai–Tibet Plateau at 26°50′–36°53′ N and 78°25′–99°06′ E, with an average altitude >4000 m. It covers an area of 1,228,400 km^2, accounting for approximately 1/8 of the total area of China, and is known as the roof of the world [1]. The region's local annual average temperature is low, the temperature difference between day and night is high, solar radiation is strong, average sunshine time is long, and natural resources are extremely rich. Tibet is one of the provinces with the largest number of rivers and most concentrated swampy lakes in China, with more than 20 rivers with watershed areas >10,000 km^2 and 819 lakes with areas >1 km^2.

The natural ecosystem types (grasslands, forests, and wetlands) of Tibet account for more than 90% of the total area of the region and are the main ecosystems of the region. The arable land in Tibet is mainly concentrated in the middle valley of the Yarlung Zangbo River, the valley of "three rivers" in eastern Tibet, and in the lower valley of the Niyang River in the Linzhi region of southeastern Tibet. With the growth of population and the improvement of urbanization, the construction lands have increased slightly.

Major tectonic blocks of Tibet are bound by the Yarlung Zangbo river fault, which divides the northern Tibetan Plateau and southern Tibetan Mountains. The geological structure of Tibet can be divided into five units: the Himalayan, Lhasa–Bomi, Tanggula, northern Tibet and Xinjiang adjacent area, and Chengdu area linear fold systems. The mountain range, river course, and rock stability in Tibet are varyingly affected by the fracture zone activities. Tibet is located in the Mediterranean–Tethyan metallogenic belt, which makes the local mineral resources extremely rich, and the development of mineral resource research has become an important industrial project in Tibet.

As an important ecological barrier in East Asia, Tibet is of national macro-strategic significance for studying the local ecological environment. In this study, data from 41 cross-sectional monitoring points in seven municipal administrative units—Lhasa, Shigatse, Chengdu, Shannan, Linzhi, Nagqu, and Ali—in the Tibet Autonomous Region were collected. The specific locations of monitoring points are shown in Figure 1.

2.2. Data Collection

The water quality data were obtained from the surface water quality data released by the Ministry of Ecology and Environment [19]. Monthly water quality monitoring data from 2021.01 to 2021.12 were collected by using python 3.8 crawler. Every 4 h, the automatic detection station of surface water quality will conduct the whole process of automatic collection, processing, analysis and data transmission. We used Dissolved Oxygen (DO), Permanganate Value (PV), Chemical Oxygen Demand (COD), Ammonia Nitrogen (NH$_3$-N), Total Phosphorous (TP), and Total Nitrogen (TN) as nutrient monitoring indicators and copper (Cu), zinc (Zn), As, cadmium (Cd), chromium (Cr), and Pb as heavy metal monitoring indexes. The study collected data from 41 cross-sectional monitoring points from seven cities, namely Shannan, Lhasa, Shigatse, Chengdu, Linzhi, Nagqu, and

Ali and divided them into 15 groups in terms of the watershed to facilitate inter-regional significance analysis (Table 1).

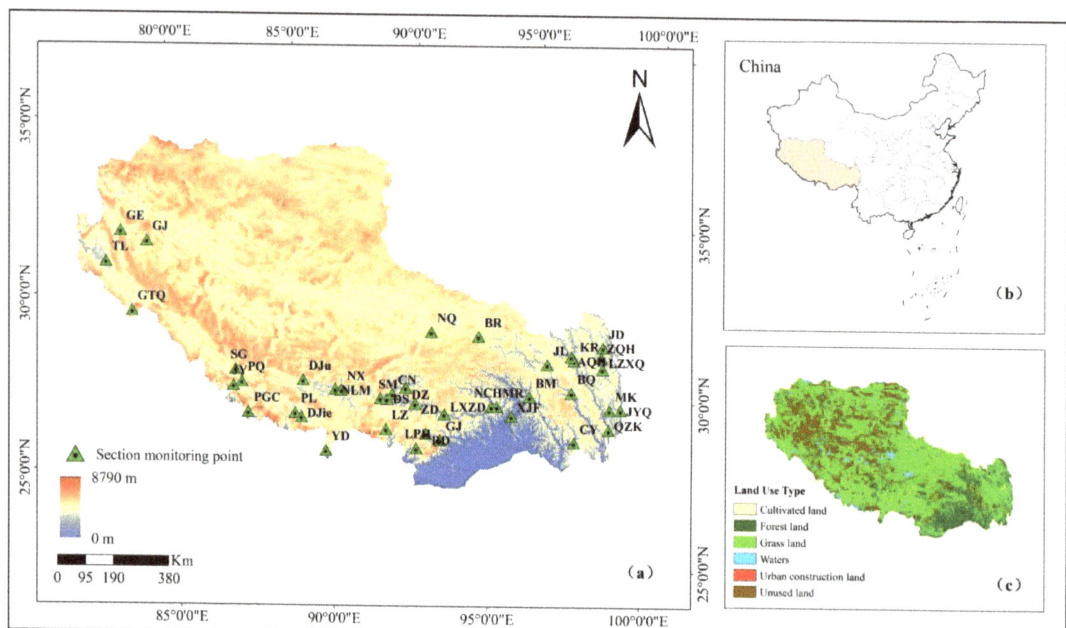

Figure 1. (**a**) Specific location map of monitoring sections in Tibet, China; (**b**) Geographic location map; (**c**) Land use map of the study area.

2.3. Analytical Method of Chemical Indicators

In the monitoring work of the National Surface Water Environmental Quality Monitoring Network, the dissolved oxygen in the basic items of the surface water environmental quality standards is measured on site, and the other 11 monitoring items are analyzed in a laboratory.

Precision control: at least 10% of the parallel double samples shall be measured for each batch of samples ($\leq$20). When the number of samples is less than 10, one parallel double sample shall be measured. According to the sample content, the relative deviation of the parallel double sample determination results should be controlled.

Accuracy control: for each batch of samples ($\leq$20), at least one certified standard sample or matrix spiked recovery sample shall be measured, and the measured value of certified standard sample shall be within the allowable range. Then the spiking recovery rate can be determined according to the sample content.

Elemental concentrations were quantified using external calibration standards. The Table 2 is for specific monitoring and analysis method standards and quality control standards.

Table 1. Basic information of monitoring sections.

Basin	River Basin Abbreviation	Name of Section Monitoring Point	Abbreviation of Section Monitoring Point
Yarlung Zangbo River	YLZBJ	Dongsa	DS
		Zedang	ZD
		Bianxiong	BX
		Daju	DJ
		Nyalam	NCH
		Pengqu	PKC
		Saga	SG
		Xinjiefang Bridge	BB
		Bomi	BM
		Lang County Zhongda	LX
		Nianchu River	LZ
		Miru	MR
		Caina	CN
		Dazi	DZ
		Semai	SM
Xiba Xiaqu	XBXQ	Ridang	RD
		Gongjue	JY
Langpo River	LPH	Langpo River	LP
Luozhaqu	LZXQ	Linzhi	ZM
Peng Qu	PQ	Pulan	PQ
		dingjie	DJ
Gyirong River	JLZB	Jiayu	JL
Poqu River	BQH	Peikucuo	NLM
Yadong River	YDH	Yadong	YD
Nujiang River	NJ	Biru	BR
		Naqu	NQ
		baxiu	BX
		Jilong	LL
Jinsha River	JSJ	Mangkang	MK
		Zaqu River Ruyi	ZQ
		Downstream of Loza xiongqu	GJ
		Jiangda	JD
		Jiayu Bridge	GTQ
Lantsang River	LCJ	Karuo	KR
		Oron River	AQ
		Quzika	KM
Lohit River	CYH	Chayu	CY
Shiquan River	SGZB	Gar	GE
		Geji	GJ
Kongque River	MJZB	Gangtuo bridge on Jinsha River	PL
Xiangquan River	LQZB	Tuolin	TL

Table 2. Analysis method and quality control standard of monitoring indicators.

Monitoring Items	Analytical Methods	Method Source	Accuracy	Precision	Recovery Rate of Spiking
DO	Electrode method	GB/T 7489-1987	-	±0.3	-
PV	Acid process	GB/T 11914-1989	±5%	±20%	-
COD	Dichromate process	HJ 505-2009	±10%	±10%	-
NH_3-N	Nessler's reagent spectrophotometry	HJ 535-2009	±10%	±15%	80–120%
TP	Ammonium molybdate spectrophotometry	GB/T 11893-1989	±10%	±10%	80–120%
TN	Alkaline potassium persulfate digestion ultraviolet spectrophotometry	HJ 636-2012	±5%	±10%	90–110%
Cu	Inductively coupled plasma mass spectrometry	HJ 700-2014	±10%	±20%	70–130%
Zn	Inductively coupled plasma mass spectrometry	HJ 700-2014	±10%	±20%	70–130%
As	Inductively coupled plasma mass spectrometry	HJ 700-2014	±10%	±10%	80–120%
Cd	Inductively coupled plasma mass spectrometry	HJ 700-2014	±10%	±20%	70–130%
Cr	Diphenylcarbazide spectrophotometry	GB/T 7467-1987	±5%	±5%	90–110%
Pb	Inductively coupled plasma mass spectrometry	HJ 700-2014	±10%	±20%	70–130%

2.4. Method

2.4.1. Entropy Value Method-Water Quality Fuzzy Evaluation

The concept of entropy originates from thermodynamics and is a measure of the uncertainty of system state. The entropy method is a relatively objective evaluation method. The higher the entropy value of the information, the more balanced the structure of the system, and the smaller the error [20,21]. Therefore, the weights can be derived from the calculated entropy values. With m monitoring sections and n monitoring indicators, the original data matrix $X = (x_{ij})_{m \times n}$ is formed. Given the differences in the scale, order of magnitude, and positive and negative indicators in the matrix, it is necessary to normalize the original data matrix X [22]. The greater the value of the indicator, the better the water quality, i.e., the positive indicator calculation method. On the other hand, the smaller the value of the indicator, the lower the water quality, i.e., a negative indicator calculation method.

In the entropy value method, the calculation steps are as follows [23]:

$$\begin{cases} Y_{ij} = \dfrac{X_{ij}}{\sqrt{\sum_{i=1}^{m} X_{ij}^2}} \text{ Positive index} \\ Y_{ij} = \dfrac{X_{ij}}{\sqrt{\sum_{i=1}^{m} X_{ij}^2}} \text{ Negative index} \end{cases} \tag{1}$$

$$e_j = -\frac{1}{\ln m} \sum_{i=1}^{m} Y_{ij} \times \ln Y_{ij} \tag{2}$$

$$d_j = 1 - e_j \tag{3}$$

$$w_i = \frac{d_i}{\sum_{j=1}^{n} d_j} \tag{4}$$

where e_j denotes the information entropy of the indicator, d_j denotes the information entropy redundancy, and w_i denotes the indicator weights.

Fuzzy evaluation is based on the fuzzy mathematical affiliation theory. The above two methods are combined to develop a comprehensive evaluation method where the qualitative evaluation changes into quantitative evaluation, with the advantages of clear results and systematic evaluation. This method can better quantify water quality ratings. The steps are as follows: (1) determine the evaluation object factor set and evaluation set and (2) establish the affiliation degree function and fuzzy matrix R.

In the fuzzy evaluation method, the calculation steps are as follows [24]:

$$\text{When } j = 1, \ r_{ij} = \begin{cases} 0, & T_i \geq S_{i(j+1)} \\ \frac{S_{i(j+1)} - T_i}{S_{i(j+1)} - S_{ij}} & S_{ij} < T_i < S_{i(j+1)} \\ 1, & T_i \leq S_{ij} \end{cases} \tag{5}$$

$$\text{When } 1 < j < n, \ r_{ij} = \begin{cases} 0, & T_i \geq S_{i(j+1)} \\ \frac{T_i - S_{i(j-1)}}{S_{ij} - S_{i(j-1)}} & S_{i(j-1)} < T_i < S_{ij} \\ \frac{S_{i(j+1)} - T_i}{S_{i(j+1)} - S_{ij}} & S_{ij} < T_i < S_{i(j+1)} \end{cases} \tag{6}$$

$$\text{When } j = n, \ r_{ij} = \begin{cases} 0, & T_i \leq S_{i(j-1)} \\ \frac{T_i - S_{i(j-1)}}{S_{ij} - S_{i(j-1)}} & S_{i(j-1)} < T_i < S_{ij} \\ 1, & T_i \geq S_{ij} \end{cases} \tag{7}$$

$$R = \begin{bmatrix} r_{11} & \cdots & r_{1n} \\ \vdots & \ddots & \vdots \\ r_{m1} & \cdots & r_{mn} \end{bmatrix} \tag{8}$$

where T_i denotes the actual measured value of element I, S_{ij} denotes the corresponding standard value of level j for element I, and r_{ij} denotes the affiliation degree of element i to level j.

(3) derive the weights of the entropy value method W $(w_1, w_2, \cdots, w_i)$, and (4) construct the fuzzy evaluation results matrix B. Water quality evaluation results were analyzed according to the principle of maximum affiliation.

$$B = W \times R = (b_1, b_2, \cdots, b_n) \tag{9}$$

2.4.2. Principal Component Analysis Method

The Pearson product-moment correlation matrix was used to analyze the correlation between the elements. Principal component analysis (PCA) was performed through varimax rotation, which used small independent variables to explain the variance of interrelated large datasets, conducive to the analysis of PCA results [25].

The original data matrix X = $(x_{ij})_{m \times n}$ was standardized to obtain the standardized data matrix R = $(z_{ij})_{m \times n}$, followed by the Kaiser–Meyer–Olkin (KMO) measure of sampling adequacy and Bartlett's test of sphericity. Subsequently, the eigenvalue λ_i of R was calculated as the variance. The eigenvalues were ranked from the largest to the smallest, and the variance and cumulative variance contribution rates were derived. The eigenvectors corresponding to the eigenvalues were used to transform the normalized data into the principal component F.

The number of principal components was determined by the cumulative contribution of the variance or the magnitude of the eigenvalues.

$$Z_{ij} = \frac{x_{ij} - \overline{x_j}}{\sigma_j} \tag{10}$$

where x_{ij} is the measured value of the indicator, $\overline{x_j}$ is the mean value of the indicator, σ_j is the standard deviation of the jth indicator term, and Z_{ij} is the standardized value.

2.4.3. Non-Parametric Kruskal-Wallis Test

Kruskal-Wallis test was a test method for non-parametric multi-sample comparison in statistics. It was used for the comparison of multiple continuous independent samples. It did not require the normality of the overall probability distribution [26].

Suppose there were m mutually independent simple random samples $(X_1, \ldots, X_{ni})$ $(i = 1, \ldots, m)$. We arranged all of the observations of each sample into a column in increasing order, and R_i $(i = 1, \ldots, m)$ represented the sum of the ranks of n_i observations $X_1, \ldots, X_{ni}$ of the ith sample in this arrangement. The calculation statistics were as follows:

$$K = \frac{12}{N(N+1)} \sum_{i=1}^{m} \frac{R_i^2}{n_i} - 3(N+1) \tag{11}$$

If each sample had r identical data, let t_1 $(i = 1, \ldots, r)$ be the number of occurrences of the ith public observation of each sample in all N observations, then calculated the following correction statistics

$$K' = \frac{N(N^2 - 1)}{\sum_{i=1}^{r}(t_i^3 - t_i)} K \tag{12}$$

When N was sufficiently large, H and H' approximately obey the distribution, and the degree of freedom $v = m - 1$.

3. Results

3.1. Time Distribution

By testing the normal distribution of the elements, it was found that the data did not conform to a normal distribution. Therefore, the non-parametric Kruskal–Wallis test was chosen to analyze the inter-seasonal and inter-basin variabilities and to produce box plots to observe the degree of dispersion of the data (shown in Table 3). The results showed that only the test for Pb had a p-value > 0.05 (0.2774). This indicates that there were no significant inter-seasonal fluctuations and Pb concentration may be influenced by natural biogeochemical processes [27]. The p-values of all the remaining elements were less than 0.05, implying that they were significantly different from one season to another.

Table 3. Results of Kruskal–Wallis test.

	Season			Group		
	Statistical Value	p-Value	Cohen's f Value	Statistical Value	p-Value	Cohen's f Value
DO	27.152	0.000 ***	0.063	14.001	0.45	0.137
PV	18.658	0.000 ***	0.058	39.224	0.000 ***	0.193
COD	42.351	0.000 ***	0.078	12.479	0.568	0.091
NH$_3$-N	9.196	0.027 **	0.034	15.804	0.325	0.122
TP	55.82	0.000 ***	0.072	18.853	0.171	0.122
TN	8.459	0.037 **	0.021	37.595	0.001 ***	0.119
Cu	31.225	0.000 ***	0.031	35.385	0.001 ***	0.071
Zn	22.848	0.000 ***	0.044	21.154	0.098 *	0.108
As	9.803	0.020 **	0.013	54.074	0.000 ***	0.309
Cd	56.313	0.000 ***	0.068	7.22	0.926	0.081
Cr	34.244	0.000 ***	0.075	9.953	0.766	0.096
Pb	3.882	0.274	0.031	32.461	0.003 ***	0.201

Note: Correlation coefficient between water quality indicators and physicochemical parameters of water was significant at: *** $p < 0.01$, ** $p < 0.05$, * $p < 0.1$.

According to Figure 2, the COD, PV, TP, NH$_3$-N, and Cr contents in the surface water during different seasons were in the following order: winter > autumn > spring > summer. Combined with the monthly average precipitation in Figure 3, it is speculated that an increase in summer precipitation may lead to an increase in runoff, resulting in dilution [28]. The DO, TN, and As contents in the surface water during the different seasons were in the following order: winter < autumn < spring < summer. In winter, the rivers in Tibet are icebound, which hinders the gaseous exchange between the atmosphere and rivers, resulting in low DO content. The temperature changes in all four seasons also affect the chemical activity of the elements. In spring, snowmelt starts, and an increase in heavy metal concentrations in the surface water may be influenced by atmospheric deposition [29]. The Zn content in summer and autumn was significantly higher than in winter and spring. This is because, due to the high temperature in summer, Zn is more easily released from sediments [28,30]. In addition, as Cohen's f values of all elements are <0.1, it can be inferred that the magnitude of the difference between the four seasons is very small.

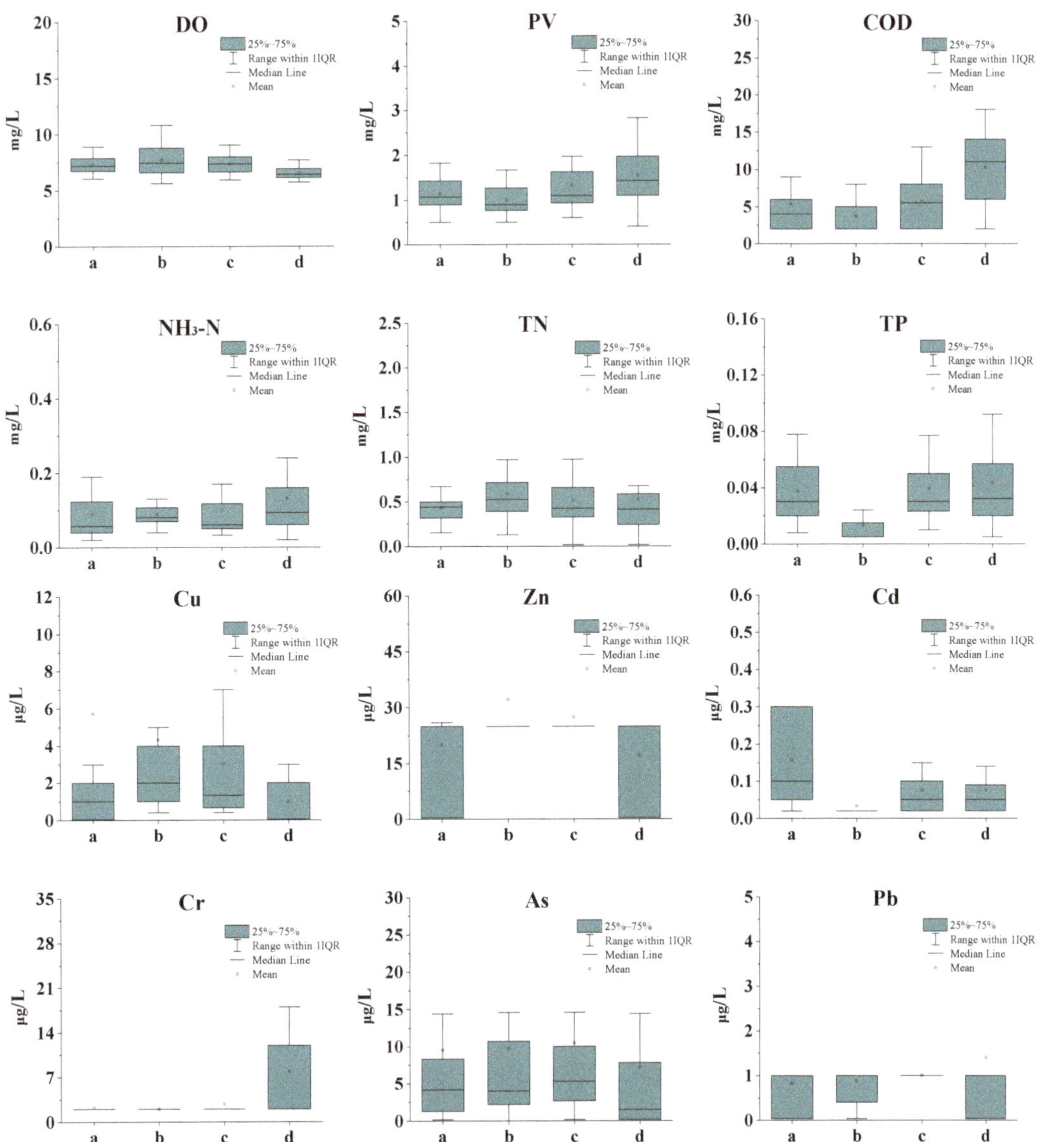

Figure 2. Heavy metal box diagram (a—spring; b—summer; c—autumn; d—winter).

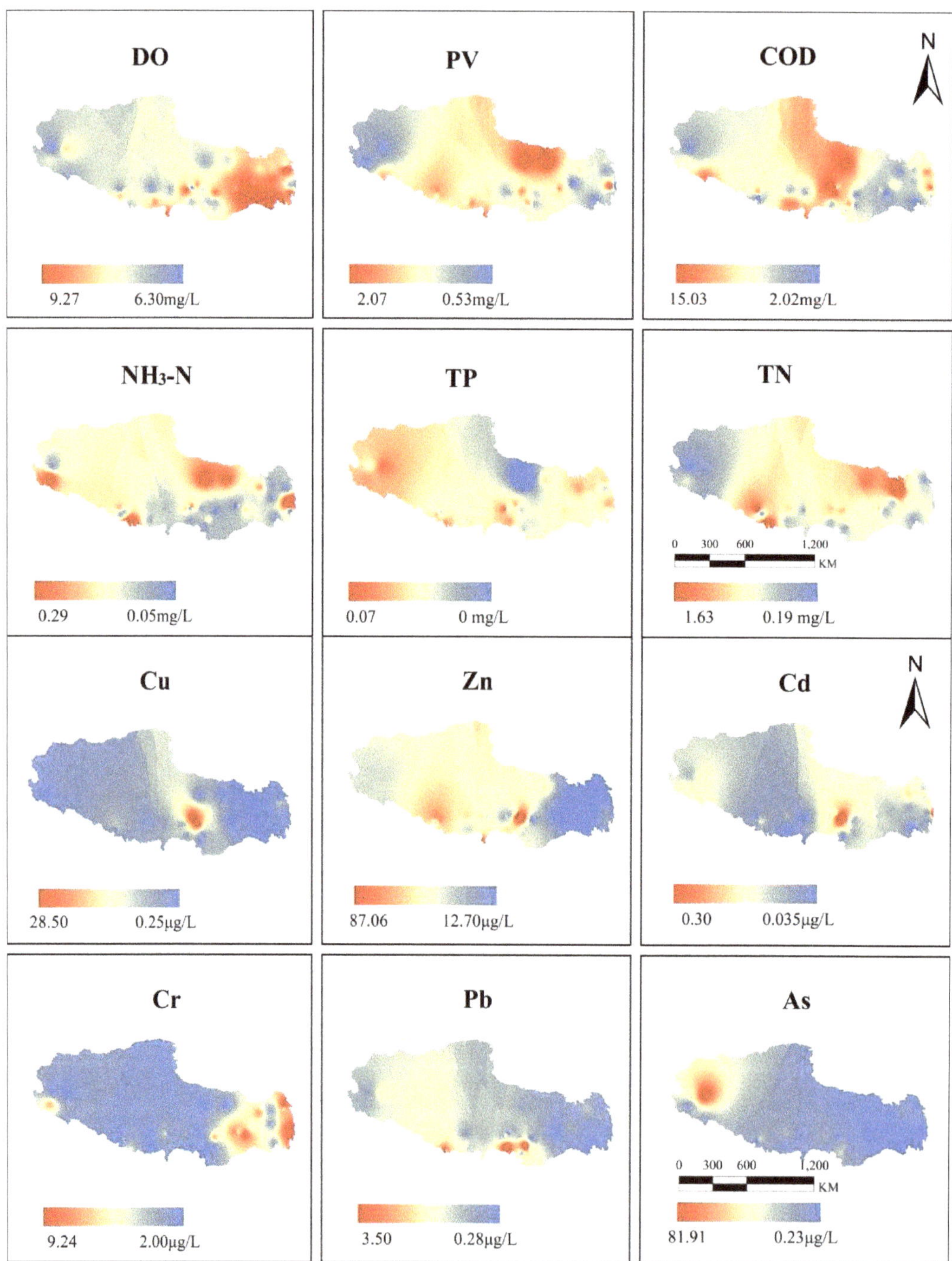

Figure 3. Spatial and temporal distribution of nutrient and heavy metal elements in Tibet, in 2021.

3.2. Spatial Distribution

According to the non-parametric Kruskal–Wallis test, the variables DO, COD, NH_3-N, TP, Cd, and Cr with p-values > 0.05 were statistically insignificant. This indicates that there are no significant differences between watersheds, and the magnitude of differences is small. The p-values for PV, TN, Cu, As, Zn, and Pb were <0.05, indicating that there are significant differences between watersheds. This implies that the sources of these elements are relatively complex. Cohen's f values of the elements were used to infer the magnitude of inter-basin variation, and the inverse distance weighted (IDW) interpolation method of ArcMap 10.2 (Environment System Research Institute, ESRI, Redlands, CA, USA) was used to investigate the internal similarity between the basins and analyze the spatial distribution of the basins.

The spatial distribution patterns of NH_3-N, PV, COD, and TN were relatively similar (Figure 3) with almost overlapping high-content areas and poor water quality in the central and northern parts due to agricultural fertilizers and pesticides in the central Tibetan valley and biological emissions from livestock farming in northern Tibet. Conversely, TP distribution was lower in the central and northern parts and highest in the western and eastern parts. High TP content in the west may be related to the phosphorus content of the soil-forming parent material, mainly from soil erosion and rock weathering [31]. The eastern part is more urbanized; therefore, the phosphorus present in water bodies may originate from urban wastewater [32]. The DO content increases from west to east. The eutrophication of water bodies depletes DO, affecting the metabolism of aquatic organisms and resulting in the deterioration of the water environment [33]. Cohen's f value of As was 0.309, indicating the largest variation between watersheds, with content in the range of 0.23–81.91 µg/L. The high-content area is in the western Bangongcuo-Nujiang collision zone. The spatial distribution trend of As is consistent with the analysis result of Zhang et al.'s field sampling in the Yarlung Zangbo River in 2017–2018. Their conclusion is that the upper reaches of Yarlung Zangbo River have comparatively high levels of dissolved As (4.7–81.6 µg/L), while the tributaries of the lower reaches have relatively low levels (0.11–1.3 µg/L) [12]. The Pb content was high in the south of the study area, and the spatial pattern demonstrated a decreasing trend from south to north. The highest content of Zn appears in the south-central direction close to the Tibetan Gondian mineralization zone [34]. The spatial distributions of Cu and Cd were relatively similar. High contents were found in areas with more developed non-ferrous metals. The development of large cities with high population density is relatively fast in the central part. Mineral development and human activities are the main causes of heavy metal enrichment [34].

Further, the significant changes between regions are diluted by water [35], and the complex compounds combined with heavy metals and organics result in the reduction in contents downstream [36].

Collectively, the high contents of nutrient elements, under the influence of chemical fertilizers and pesticides and biological emissions, are mainly concentrated in the agricultural and animal husbandry development areas in central and northern Tibet. Areas with high heavy metal content are mainly those with rich mineral resources, more developed cities, or frequent geological activities.

3.3. Entropy Value Method-Ater Quality Fuzzy Evaluation

The application of the entropy method to determine the weights can weaken the association between samples, eliminate human interference, and make the evaluation results more scientific and accurate [37,38]. From the weight information matrix in Table 4, Cu has the largest index weight, accounting for 19.3%, and TP has the smallest index weight, accounting for only 0.8%. The set of weights can be represented as W {0.0036, 0.056, 0.121, 0.008, 0.051, 0.17, 0.013, 0.17, 0.081, 0.193, 0.082, 0.019}.

Table 4. Results of weight calculation using the entropy method.

Term	Information Entropy, e	Information Utility Value, d	Weight
PV	0.965	0.035	0.036
COD	0.945	0.055	0.056
NH_3-N	0.881	0.119	0.121
TP	0.992	0.008	0.008
Pb	0.95	0.05	0.051
As	0.833	0.167	0.17
Cd	0.988	0.012	0.013
Cr	0.834	0.166	0.17
TN	0.92	0.08	0.081
Cu	0.811	0.189	0.193
Zn	0.92	0.08	0.082
DO	0.981	0.019	0.019

The water quality fuzzy evaluation was performed according to the water quality standard of the Environmental Quality Standards in Surface Water (GB3838-2002) [39]. Forty-one monitoring sections were selected as water quality evaluation sets. DO, PV, COD, NH_3-N, TP, TN, Cu, Zn, As, Cd, Cr, and Pb were selected as evaluation factors, and the evaluation categories were I, II, III, IV, and V. Subsequently, the water quality category of the monitoring section was determined according to the maximum affiliation [40].

According to Table 5, the nutrient and heavy metal contents in the surface water in Tibet are low, and most areas have a higher content than at the level of surface water category III. The water quality of 41 cross-sections in the Tibet Autonomous Region was evaluated as category I. Monitoring point JYQ comes under categories IV and V, and the percentage of water quality IV at monitoring point GJ reaches 10.9%, which indicates that both sections are slightly affected by pollutants [41]. The Exceedance factor of the GJ monitoring point is As, and according to the previous analysis, there are significant seasonal and regional differences in As contents. The As content in water is mainly controlled by lithology, and the shale that is rich in As is widely distributed in Qinghai-Tibet Plateau [42–44].

In general, the water quality in most areas of Tibet is relatively good as per the national sanitary standards for drinking water; some indicators even exceed the standards for drinking mineral water in China. Local exceedance factors are mainly affected by agricultural production and geological and geothermal activities.

3.4. Analysis of Pollution Sources

The correlation between the concentrations of water quality indicators can help to analyze the interactions among indicators in the water body, as well as the possibility of homology among indicators or the relationship between migration and transformation processes. Table 6 shows the correlation between the physicochemical parameters of water and water quality elements. DO, TN, Zn, As, and Pb showed significant negative correlations with water temperature (WT). With an increase in WT, the increase in biological activity reduces the nutrient element index, and the adsorption of metal elements by sediment significantly reduces their content. Similar correlations have been observed in the source area of the Yellow River and the Yarlung Zangbo River on the Qinghai Tibet Plateau [45]. Additionally, DO, TN, Zn, Pb, and NH_3-N also showed significant negative correlations with precipitation. The increase in runoff due to the increase in summer precipitation, results in a dilution effect [25]. The water with low concentration of these elements diluted the river water. This conclusion is consistent with that of the previous studies. On the contrary, Cd has a significant positive correlation with WT and precipitation. Due to the high content of Cd pollutants in aerosols and surface soil of the watershed, affected by the increase in precipitation, Cd was transported to the river channel. In the case of Cd and Cu, pH had significant negative correlations, indicating that Cd and Cu are more easily dissolved in weakly acidic conditions [46]. Alkaline river water may facilitate the uptake and oxidation of dissolved heavy metals [47]. DO, NH_3-N, TN showed significant positive correlation with pH, indicating that the local water quality was good. The coupling effect of pH and dissolved oxygen in the water column of Erhai Lake on nitrogen release shows that when the water quality starts to deteriorate, the pH value shows an upward trend (8.48–8.87), while the DO concentration shows a downward trend (7.42–6.61 mg/L), and the TN concentration increases significantly [48]. COD and Zn had a significant positive correlation with turbidity. Conductivity was significantly correlated only with NH_3-N. The relatively weak correlation between Cr in heavy metals and the physicochemical parameters of water implies that it is more influenced by human activities. The results of the correlation analysis can provide information regarding the sources of nutrients and metal elements, which is helpful in the subsequent determination of the sources of major pollutants.

Table 5. Results of entropy method–fuzzy evaluation for water quality.

Section	I	II	III	IV	V	Evaluation Results	Exceedance Factor
ZD	0.914	0.063	0.023	0	0	I	-
YD	0.944	0.056	0	0	0	I	-
TL	0.922	0.078	0	0	0	I	-
SM	0.914	0.074	0.011	0	0	I	-
SG	0.905	0.014	0.077	0.004	0	I	TN
RD	0.901	0.090	0.001	0	0	I	-
GTQ	0.936	0.064	0	0	0	I	-
PQ	0.961	0.039	0	0	0	I	-
PGC	0.982	0.018	0	0	0	I	-
NLM	0.872	0.047	0.024	0.057	0	I	TN
NQ	0.820	0.083	0.015	0	0	I	-
MR	0.919	0.060	0.021	0	0	I	-
QZK	0.923	0.077	0	0	0	I	-
MK	0.859	0.141	0	0	0	I	-
LZ	0.933	0.067	0	0	0	I	-
JL	0.926	0.074	0	0	0	I	-
NCH	0.975	0.025	0	0	0	I	-
LPH	0.968	0.032	0	0	0	I	-
LXZD	0.925	0.075	0	0	0	I	-
KR	0.915	0.074	0.011	0	0	I	-
JYQ	0.913	0.006	0	0.060	0.021	I	TN
JD	0.988	0.012	0	0	0	I	-
GJ	0.940	0.060	0	0	0	I	-
JY	0.909	0.091	0.001	0	0	I	-
LZXQ	0.954	0.046	0	0	0	I	-
GJ	0.811	0.019	0.061	0.109	0	I	As
GE	0.984	0.016	0	0	0	I	-
DS	0.918	0.067	0.015	0	0	I	-
DJie	0.947	0.053	0	0	0	I	-
PL	0.964	0.036	0	0	0	I	-
DZ	0.913	0.085	0.002	0	0	I	-
DJu	0.904	0.081	0.015	0	0	I	-
AQH	0.925	0.075	0	0	0	I	-
ZQH	0.934	0.066	0	0	0	I	-
CY	0.980	0.020	0	0	0	I	-
CN	0.847	0.127	0.026	0	0	I	-
BM	0.961	0.039	0	0	0	I	-
NX	0.906	0.079	0.015	0	0	I	-
BR	0.910	0.043	0.048	0	0	I	-
XJF	0.965	0.035	0	0	0	I	-
BQ	0.933	0.067	0	0	0	I	-

Table 6. Correlation between water quality indicators and physicochemical parameters of water.

Index	DO	PV	COD	NH$_3$-N	TP	TN	Cu	Zn	As	Cd	Cr	Pb
WT	−0.386 **	−0.069	0.051	−0.183	0.06	−0.228 *	0.023	−0.404 **	−0.225 *	0.297 **	0.066	−0.226 *
Precipitation	−0.441 **	−0.066	0.012	−0.341 **	−0.005	−0.332 **	0.011	−0.312 **	−0.126	0.453 **	0.005	−0.22 *
pH	0.234 *	0.087	0.06	0.282 **	−0.053	0.255 *	−0.198 *	0.115	−0.063	−0.332 **	0.018	0.129
Turbidity	0.064	−0.051	0.411 **	0.061	−0.154	−0.062	0.015	0.286 **	0.086	0.128	−0.096	0.177
Conductivity	−0.077	0.127	0.021	0.327 **	−0.15	0.076	−0.056	−0.098	−0.055	−0.021	−0.02	−0.029

Note: Correlation coefficient between water quality indicators and physicochemical parameters of water was significant at: ** $p < 0.01$, * $p < 0.05$.

As shown in Table 7, PV–NH$_3$-N–TP, Zn–Cu–Cd, and Zn–Pb were significantly positively correlated ($p < 0.01$). Table 7 shows a significant negative correlation between Cr and Zn. However, there were significant correlations between Cr and PV, COD, NH$_3$-N, TP, and TN, implying that the source of Cr is relatively similar to that of the nutrient elements. The correlations between As and all the other elements were weak, indicating that their sources might be specific.

Table 7. Results of correlation analysis between water quality indicators.

Index	DO	PV	COD	NH$_3$-N	TP	TN	Cu	Zn	As	Cd	Cr	Pb
DO	1											
PV	−0.2	1										
COD	−0.329 **	0.275 **	1									
NH$_3$-N	−0.25 **	0.459 **	0.291 **	1								
TP	−0.01	0.258 **	0.28 **	0.385 **	1							
TN	−0.03	0.3 **	0.023	0.364 **	0.217 **	1						
Cu	0.063	0.012	−0.030	−0.055	−0.030	0.002	1					
Zn	−0.017	−0.023	0.047	−0.035	−0.189 *	0.063	0.367 **	1				
As	−0.067	−0.211 **	−0.049	−0.115 *	0.025	−0.179 *	−0.013	0.068	1			
Cd	−0.055	0.046	0.077	−0.062	0.163 *	−0.061	0.422 **	0.267 **	0.035	1		
Cr	−0.14 *	0.42 **	0.318 **	0.443 **	0.398 **	0.25 **	−0.107	−0.25 **	−0.174 *	0.074	1	
Pb	−0.149 *	−0.008	0.306 **	−0.051	0.067	−0.045	0.035	0.230 **	0.043	0.072	−0.093	1

Note: Correlation coefficient between nutrient elements and heavy metal elements was significant at: ** $p < 0.01$, * $p < 0.05$.

PCA was conducted to determine the sources of nutrients and heavy metals, and the results are shown in Table 8. The data were first subjected to the KMO and Bartlett's tests to analyze the feasibility of PCA. KMO was calculated as 0.603. The *p*-value of Bartlett's spherical test was <0.01, showing significance at the level. The variables were correlated, indicating that the PCA was valid. Figure 4 shows the explanation of variance by the normalized rotation obtained using the maximum variance method for nutrients and heavy metals elements [49]. Four effective principal components were identified according to the Kaiser criterion, with eigenvalues > 1 [50], which accounted for up to 56.08% of the contribution.

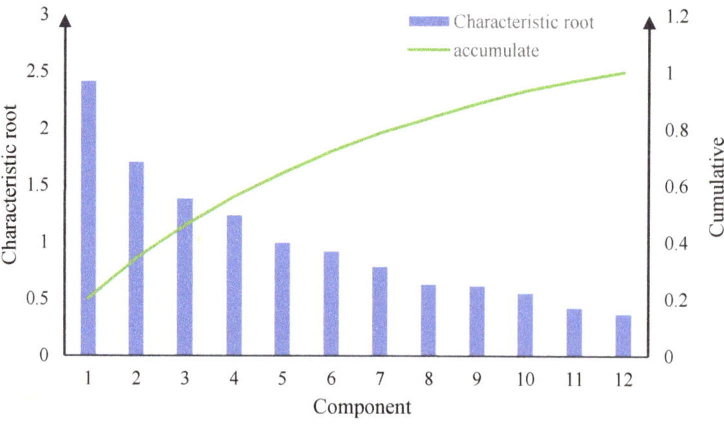

Figure 4. Explanation of variance of principal components.

Table 8. Results of PCA.

| | Factor Load Factor | | | |
	PC1	PC2	PC3	PC4
DO	−0.244	−0.149	0.231	0.493
PV	**0.644**	0.017	0.315	0.019
COD	**0.594**	0.329	−0.151	−0.205
NH_3-N	**0.563**	−0.052	0.484	−0.277
TP	**0.663**	0.234	−0.217	0.221
TN	0.135	−0.153	**0.746**	−0.009
Cu	−0.153	**0.572**	0.369	0.41
Zn	−0.319	**0.656**	0.371	−0.145
As	−0.217	0.238	−0.14	−0.324
Cd	0.13	**0.539**	−0.215	**0.554**
Cr	**0.787**	−0.113	−0.141	0.163
Pb	0.032	**0.612**	−0.086	−0.44
Characteristic root	2.412	1.704	1.381	1.233
Percentage variance	0.201	0.142	0.115	0.103
accumulate	0.201	0.343	0.458	0.561

The first principal component (PC1) pollutants were mainly Cr, TP, PV, COD, and NH_3-N, which accounted for 20.1% of the total variance. Table 7 shows that there were significant correlations between Cr and TP, PV, COD, and NH_3-N, all of which showed positive loadings indicating that the nutrient elements had the same source as Cr. Figure 4 shows that none of the five elements had significant spatial differences, indicating that these elements were mainly influenced by natural processes. Additionally, the increase in Cr content may be related to industrial activities and domestic waste emissions [51,52]. The pollutants in the second principal component (PC2) were mainly Zn, Cu, Pb, and Cd. The correlation coefficient between Zn and PC2 was high at 0.656, and there was also a significant correlation between Zn and Cu, Pb, and Cd, indicating that these heavy metals may have the same source and similar diffusion processes. The spatial distribution map of metal elements shows that, under the influence of the Gondwana mineralization belt, the mineral resources of Cu, Pb, and Zn are abundant [34]. The high Zn, Cu, and Pb contents can be attributed to geothermal activities and mineral extractions [53]. The pollutants in the third principal component (PC3) were mainly TN and NH_3-N, with a component variance contribution of 11.5%. The N and P in surface water are mainly from farmland water and municipal wastewater [54]. The NH_3-N and TN were highly correlated, reflecting the nutrient status of water bodies. The increase in NH_3-N and TN may be due to the rapid development of urbanization and agriculture, human production and life, and the application of pesticides and fertilizers [55]. PC3 represents the source of the agricultural activities. The fourth principal component (PC4) pollutant was mainly Cd, which is likely to be influenced by transportation or industrial activities. Fossil fuels and metallurgical industries release volatile Cd, which is then dissolved in surface water [56]. Table 3 shows no significant spatial difference for Cd, which is known to be relatively weakly influenced by anthropogenic activities.

According to the results of multivariate statistical analysis, PC1 pollutants (Cr, TP, PV, COD, and NH_3-N) are mainly influenced by natural processes, PC2 pollutants mainly arise from mineral extractions, PC3 (TN and NH_3-N) mainly arise from production activities and pesticides and fertilizers, and PC4 (Cd) is mainly influenced by transportation or industry. Therefore, the elemental indicators in Tibetan surface water are influenced by natural factors, and human activities aggravate the spread of pollution.

4. Discussion

In this study, the non-parametric Kruskal–Wallis test, production of box line plots, and IDW method to map the spatial distribution of pollution elements were used to explore the spatial and temporal distribution characteristics of pollution elements. Temporally, the results showed that only the test for Pb had a p-value > 0.05 (0.2774), this indicates that there were no significant inter-seasonal fluctuations. Moreover, the rest polluting elements had significant seasonal differences under the influence of temperature and runoff, but the magnitude of the differences was very small. Spatially, the p-values of PV, TN, Cu, Zn, As, and Pb were <0.05, with significant results, indicating significant differences among the watersheds. Nutrient elements are mainly influenced by natural processes, partly under the influence of fertilizers and pesticides, and are mainly concentrated in the agricultural and animal husbandry development areas in Tibet. High heavy metal contents are mainly found in areas rich in mineral resources and in areas of frequent geological activities. Subsequently, the weights determined by the combination of entropy value and fuzzy evaluation methods were used to evaluate the water quality, and the results were compared with the water quality standards according to the Surface Water Environmental Quality Standard (GB3838-2002). We found that the contents of nutrient and heavy metal elements in the surface water of the Tibet Autonomous Region were low, and the water quality evaluation of 41 cross-section monitoring points in the study area were all Class I. At GJ monitoring point, As exceeded the standard more remarkably from lithology and geological activities. Finally, the potential sources of pollutants were analyzed based on correlation analysis and PCA, which explained 56.08% of the total variance of the five pollution principal components and identified four pollution principal components. PC1 is mainly influenced by natural processes, PC2 can be attributed to geothermal activities and mineral extractions, PC3 arises from human production activities and the application pesticides and fertilizers, and PC4 is influenced by industrialization or transportation.

In summary, the NH_3-N, PV, COD, and TN contents are mainly influenced by natural processes, partly under the influence of fertilizers, pesticides, and biological emissions. High contents are mainly concentrated in the agricultural and animal husbandry development areas in central and northern Tibet. However, TP may be related to the phosphorus content of the soil-forming parent material and urban wastewater discharge and has the greatest magnitude of inter-basin variation under the influence of the Bangongcuo-Nujiang collision zone. The area with high Zn and Pb contents was located close to the Gangdise metallogenic belt in Tibet, which is mainly affected by mineral development and geological activities. Cu and Cd are mainly attributed to the development of non-ferrous metal minerals and human activities.

Currently, the surface water bodies in the Tibet Autonomous Region are less affected by human activities, and the degree of organic matter and heavy metal pollution is weak and at a low-risk level. The spatial heterogeneity of heavy metals increased and was significantly affected by human activities, and the potential risks might exceed our expectations. This is evidenced by the increasing influence of human activities on water quality deterioration. In the future, considering Tibet's fragile ecological environment and unique geographical location, the security of water resources in Tibet will face an increasingly complex situation. At the same time, the pollution sources of nutrients and heavy metals will shift from being influenced by single factors to being influenced by multiple factors and from natural causes to human causes. This study analyzed the spatial and temporal distribution characteristics of six nutrient elements and six heavy metal elements in the Tibet Autonomous Region in 2021, and further health risk evaluation can be performed subsequently. At present, it is urgent to improve the local water resource management level, and focusing on water resource security is of great strategic importance.

5. Conclusions

Based on the water quality data of Tibetan rivers in China, this study analyzed the temporal and spatial changes in surface water nutrients and heavy metals, water quality status and pollutant sources, and drew the following four conclusions:

1. Temporally, all 12 polluting elements, except Pb, had significant seasonal variations, but the magnitude of the differences was very small;

2. Spatially, nutrient elements were relatively concentrated in the agricultural and pastoral development areas in central and northern Tibet. High heavy metal content was mainly found in areas with rich mineral resources, more developed cities, or areas undergoing frequent geological activities;

3. The water quality of 41 monitoring sections belonged to Class I water standard as per the entropy method–fuzzy evaluation method, and the local exceedance factors were mainly affected by agricultural production activities and geological and geothermal activities;

4. Determine the main sources of four pollutants: the first principal component was mainly influenced by natural processes, the second principal component was mainly influenced by mineral extraction, the third principal component was mainly influenced by production activities and pesticides and fertilizers, and the fourth principal component was mainly influenced by transportation or industrial activities.

The results of this study can provide a scientific basis for pollution prevention and control in the Tibet Autonomous Region, and contribute to further research on water ecology and the environment.

Author Contributions: Conceptualization, J.H.; Data curation, J.H.; Methodology, J.H.; Supervision, J.Z., Y.S. and X.C.; Visualization, J.H.; Writing—Original draft, J.H.; Writing—Review & editing, J.Z. and Y.S. All authors have read and agreed to the published version of the manuscript.

Funding: This research was funded by the National Key R&D Program of China (2017YFC0406004) and the NSFC (41271004).

Informed Consent Statement: Not applicable.

Data Availability Statement: Data inquiries can be directed to the author.

Acknowledgments: We are grateful to the teachers and students who participated in the field collection of data.

Conflicts of Interest: The authors declare no conflict of interest.

References

1. Chen, W. *Hydrologic Characteristics and Mathematical Simulation of the Qinghai-Tibet Plateau*; Science Press: Beijing, China, 2016.
2. Wang, J.; Liu, G.; Liu, H.; Lam, P.K.S. Multivariate Statistical Evaluation of Dissolved Trace Elements and a Water Quality Assessment in the Middle Reaches of Huaihe River, Anhui, China. *Sci. Total Environ.* **2017**, *583*, 421–443. [CrossRef] [PubMed]
3. He, J.; Shi, D.; Wang, B.B.; Feng, C.L.; Su, H.L.; Wang, Y.; Qing, N. Ecological risk and water quality standard evaluation of 10 typical heavy metals in eight river basins. *China Environ. Sci.* **2019**, *39*, 2970–2982.
4. Zhu, Q.Q.; Wang, Z.L. Distribution characteristics and sources of heavy metals in sediments of major river systems in china. *Earth Environ.* **2012**, *40*, 305–313.
5. He, X.L.; Wu, Y.H.; Zhou, J.; Bing, H.J. Chemical Characteristics of Surface Water and Assessment of Water Environment Quality in Gongga Mountain Area. *Environ. Sci.* **2016**, *37*, 3798–3805.
6. Chen, Z.Y.; Wang, S.; Zhao, Y.Y. Characteristics of Heavy Metals in Soils and River Sediments of the Dobuza Copper Deposit in Tibet and Their Environmental Significance. *J. Earth Sci. Environ.* **2020**, *42*, 376–393.
7. Yang, R.Q.; Yao, T.D.; Xu, B.Q.; Jiang, G.B.; Xin, X.D. Accumulation Features of Organochlorine Pesticides and Heavy Metals in Fish from High Mountain Lakes and Lhasa River in the Tibetan Plateau. *Environ. Int.* **2007**, *33*, 151–156. [CrossRef]
8. Bayram, A.; Önsoy, H.; Kömürcü, M.İ.; Tüfekçi, M. Reciprocal Influence of Kürtün Dam and Wastewaters from the Settlements on Water Quality in the Stream Harşit, NE Turkey. *Environ. Earth Sci.* **2014**, *72*, 2849–2860. [CrossRef]
9. Vörösmarty, C.J.; McIntyre, P.B.; Gessner, M.O.; Dudgeon, D.; Prusevich, A.; Green, P.; Glidden, S.; Bunn, S.E.; Sullivan, C.A.; Liermann, C.R.; et al. Global Threats to Human Water Security and River Biodiversity. *Nature* **2010**, *467*, 555–561. [CrossRef]
10. Yu, C.; Hua, K.; Huang, C.-S.; Jin, H.; Sun, Y.; Yu, Z. Spatiotemporal Characteristics of Arsenic and Lead with Seasonal Freeze-Thaw Cycles in the Source Area of the Yellow River Tibet Plateau, China. *J. Hydrol. Reg. Stud.* **2022**, *44*, 101210. [CrossRef]
11. Radfard, M.; Yunesian, M.; Nabizadeh, R.; Biglari, H.; Nazmara, S.; Hadi, M.; Yousefi, N.; Yousefi, M.; Abbasnia, A.; Mahvi, A.H. Drinking Water Quality and Arsenic Health Risk Assessment in Sistan and Baluchestan, Southeastern Province, Iran. *Hum. Ecol. Risk Assess. Int. J.* **2019**, *25*, 949–965. [CrossRef]
12. Zhang, J.W.; Yan, Y.N.; Zhao, Z.Q.; Li, X.D.; Guo, J.Y.; Hu, D.; Cui, L.F.; Meng, J.L.; Liu, C.Q. Spatial and Seasonal Variations of Dissolved Arsenic in the Yarlung Tsangpo River, Southern Tibetan Plateau. *Sci. Total Environ.* **2021**, *760*, 143416. [CrossRef] [PubMed]

13. Wang, C.; Zhou, H.; Kuang, X.; Hao, Y.; Shan, J.; Chen, J.; Li, L.; Feng, Y.; Zou, Y.; Zheng, Y. Water Quality and Health Risk Assessment of the Water Bodies in the Yamdrok-Tso Basin, Southern Tibetan Plateau. *J. Environ. Manag.* **2021**, *300*, 113740. [CrossRef] [PubMed]

14. Zhang, Z.; Li, Y.; Wang, X.; Li, H.; Zheng, F.; Liao, Y.; Tang, N.; Chen, G.; Yang, C. Assessment of River Health Based on a Novel Multidimensional Similarity Cloud Model in the Lhasa River, Qinghai-Tibet Plateau. *J. Hydrol.* **2021**, *603*, 127100. [CrossRef]

15. Wang, X.P.; Liu, X.J.; Wang, L.Q.; Yang, J.; Wan, X.M.; Liang, T. A Holistic Assessment of Spatiotemporal Variation, Driving Factors, and Risks Influencing River Water Quality in the Northeastern Qinghai-Tibet Plateau. *Sci. Total Environ.* **2022**, *851*, 157942. [CrossRef] [PubMed]

16. Mahammad, S.; Islam, A.; Shit, P.K. Geospatial Assessment of Groundwater Quality Using Entropy-Based Irrigation Water Quality Index and Heavy Metal Pollution Indices. *Environ. Sci. Pollut. Res.* **2022**, *1*, 1–24. [CrossRef] [PubMed]

17. Zou, Z.H.; Yun, Y.; Sun, J.N. Entropy Method for Determination of Weight of Evaluating Indicators in Fuzzy Synthetic Evaluation for Water Quality Assessment. *J. Environ. Sci.* **2006**, *18*, 1020–1023. [CrossRef]

18. Zhong, C.; Yang, Q.; Liang, J.; Ma, H. Fuzzy Comprehensive Evaluation with AHP and Entropy Methods and Health Risk Assessment of Groundwater in Yinchuan Basin, Northwest China. *Environ. Res.* **2022**, *204*, 111956. [CrossRef]

19. Environmental Monitoring of China. Available online: http://www.cnemc.cn/sssj/ (accessed on 12 November 2022).

20. Han, B.; Liu, H.; Wang, R. Urban Ecological Security Assessment for Cities in the Beijing–Tianjin–Hebei Metropolitan Region Based on Fuzzy and Entropy Methods. *Ecol. Model.* **2015**, *318*, 217–225. [CrossRef]

21. Chen, S.Z.; Wang, X.J.; Zhao, X.J. An Attribute Recognition Model Based on Entropy Weight for Evaluating the Quality of Groundwater Sources. *J. China Univ. Min. Technol.* **2008**, *18*, 72–75. [CrossRef]

22. Chen, M.X.; Lu, D.D.; Zhang, H. Comprehensive Measurement of China's Urbanization Level and Its Dynamic Factor Analysis. *J. Geogr.* **2009**, *64*, 387–398.

23. Zhang, X.; Wang, C.; Li, E.; Xu, C. Assessment Model of Ecoenvironmental Vulnerability Based on Improved Entropy Weight Method. *Sci. World J.* **2014**, *2014*, 797–814. [CrossRef] [PubMed]

24. Icaga, Y. Fuzzy Evaluation of Water Quality Classification. *Ecol. Indic.* **2007**, *7*, 710–718. [CrossRef]

25. Mohiuddin, K.M.; Otomo, K.; Ogawa, Y.; Shikazono, N. Seasonal and Spatial Distribution of Trace Elements in the Water and Sediments of the Tsurumi River in Japan. *Environ. Monit. Assess.* **2012**, *184*, 265–279. [CrossRef] [PubMed]

26. Zhou, G.R. *Dictionary of Applied Statistical Methods*; China Statistics Press: Beijing, China, 1993.

27. Mao, G.; Zhao, Y.; Zhang, F.; Liu, J.; Huang, X. Spatiotemporal Variability of Heavy Metals and Identification of Potential Source Tracers in the Surface Water of the Lhasa River Basin. *Environ. Sci. Pollut. Res.* **2019**, *26*, 7442–7452. [CrossRef]

28. Huang, X.; Sillanpää, M.; Duo, B.; Gjessing, E.T. Water Quality in the Tibetan Plateau: Metal Contents of Four Selected Rivers. *Environ. Pollut.* **2008**, *156*, 270–277. [CrossRef]

29. Warnken, K.W.; Santschi, P.H. Delivery of Trace Metals (Al, Fe, Mn, V, Co, Ni, Cu, Cd, Ag, Pb) from the Trinity River Watershed Towards the Ocean. *Estuaries Coasts* **2009**, *32*, 158–172. [CrossRef]

30. Tian, Y.; Zha, X.; Gao, X.; Yu, C. Geochemical Characteristics and Source Apportionment of Toxic Elements in the Tethys–Himalaya Tectonic Domain, Tibet, China. *Sci. Total Environ.* **2022**, *831*, 154863. [CrossRef]

31. Chu, X.; Wu, D.; Wang, H.; Zheng, F.; Huang, C.; Hu, L. Spatial Distribution Characteristics and Risk Assessment of Nutrient Elements and Heavy Metals in the Ganjiang River Basin. *Water* **2021**, *13*, 3367. [CrossRef]

32. Zhang, W.; Jin, X.; Liu, D.; Lang, C.; Shan, B. Temporal and Spatial Variation of Nitrogen and Phosphorus and Eutrophication Assessment for a Typical Arid River—Fuyang River in Northern China. *J. Environ. Sci.* **2017**, *55*, 41–48. [CrossRef]

33. Zhang, Y. Overview of Mineral Resources in Tibet. *Tibet. Sci. Technol.* **2005**, *6*, 33–34.

34. Du, H.L.; Wang, Y.; Wang, J.S.; Yao, Y.B.; Zhou, Y.; Liu, X.Y.; Lu, Y.L. Distribution characteristics and ecological risk assessment of heavy metals in soils of typical watershed of Qinghai-Tibet Plateau. *Environ. Sci.* **2021**, *42*, 4422–4431.

35. Decena, S.C.P.; Arguilles, M.S.; Robel, L.L. Assessing Heavy Metal Contamination in Surface Sediments in an Urban River in the Philippines. *Pol. J. Environ. Stud.* **2018**, *27*, 1983–1995. [CrossRef]

36. Akele, M.L.; Kelderman, P.; Koning, C.W.; Irvine, K. Trace Metal Distributions in the Sediments of the Little Akaki River, Addis Ababa, Ethiopia. *Environ. Monit. Assess.* **2016**, *188*, 389. [CrossRef] [PubMed]

37. Zhang, J.; Shu, L.C.; Zhang, C.; Li, X.; Ji, Y.F.; Ma, C.Q. Application of comprehensive index method based on entropy weight in groundwater quality evaluation. *Hydropower Energy Sci.* **2010**, *28*, 30–32.

38. Zou, Z.H.; Sun, J.N.; Ren, G. Entropy Weighting Method for Fuzzy Evaluation Factor and Its Application in Water Quality Assessment. *J. Environ. Sci.* **2005**, *2005*, 552–556.

39. *GB3838-2002*; Environmental Quality Standards For Surface Water. Ministry of Ecology and Environment of the People's Republic of China: Beijing, China, 2002.

40. Wang, X.M.; Li, W.Y.; Zhou, Y.; Jing, X.Y.; Chen, S. Evaluation of surface water quality based on Stochastic Forest and fuzzy comprehensive evaluation. *Water Supply Drain.* **2022**, *58*, 128–132.

41. Hou, Y.H.; Dawa, Z.M.; Chen, G.H. Current situation and application countermeasure of chemical fertilizer in Tibet. *Tibet Agric. Sci. Technol.* **2010**, *32*, 43–46.

42. Yao, Q.-Z.; Zhang, J.; Wu, Y.; Xiong, H. Hydrochemical Processes Controlling Arsenic and Selenium in the Changjiang River (Yangtze River) System. *Sci. Total Environ.* **2007**, *377*, 93–104. [CrossRef]

43. Bo, J.K.; Li, C.L.; Kang, S.C.; Wang, J.L.; Chen, P.F. Distribution Characteristics of Dissolved Elements and Water Quality Assessment in Three Typical Rivers of Yarlung Zangbo River Basin. *Environ. Chem.* **2014**, *33*, 2206–2207.

44. Li, C.; Kang, S.; Zhang, Q.; Wang, F. Rare Earth Elements in the Surface Sediments of the Yarlung Tsangbo (Upper Brahmaputra River) Sediments, Southern Tibetan Plateau. *Quat. Int.* **2009**, *208*, 151–157. [CrossRef]

45. Li, Z.; Liu, J.; Chen, H.; Li, Q.; Yu, C.; Huang, X.; Guo, H. Water Environment in the Tibetan Plateau: Heavy Metal Distribution Analysis of Surface Sediments in the Yarlung Tsangpo River Basin. *Environ. Geochem. Health* **2020**, *42*, 2451–2469. [CrossRef] [PubMed]

46. Liang, B.; Han, G.; Zeng, J.; Qu, R.; Liu, M.; Liu, J. Spatial Variation and Source of Dissolved Heavy Metals in the Lancangjiang River, Southwest China. *Int. J. Environ. Res. Public Health* **2020**, *17*, 732. [CrossRef]

47. Shiller, A.M.; Boyle, E. Dissolved Zinc in Rivers. *Nature* **1985**, *317*, 49–52. [CrossRef]

48. Zhang, L.; Wang, S.; Wu, Z. Coupling Effect of PH and Dissolved Oxygen in Water Column on Nitrogen Release at Water–Sediment Interface of Erhai Lake, China. *Estuar. Coast. Shelf Sci.* **2014**, *149*, 178–186. [CrossRef]

49. Vega, M.; Pardo, R.; Barrado, E.; Debán, L. Assessment of Seasonal and Polluting Effects on the Quality of River Water by Exploratory Data Analysis. *Water Res.* **1998**, *32*, 3581–3592. [CrossRef]

50. Kaiser, H.F. The Application of Electronic Computers to Factor Analysis. *Educ. Psychol. Meas.* **1960**, *20*, 141–151. [CrossRef]

51. Simeonov, V.; Stratis, J.A.; Samara, C.; Zachariadis, G.; Voutsa, D.; Anthemidis, A.; Sofoniou, M.; Kouimtzis, T. Assessment of the Surface Water Quality in Northern Greece. *Water Res.* **2003**, *37*, 4119–4124. [CrossRef]

52. Kahn, A.H.; Nolting, R.F.; Van der Gaast, S.J. *Trace Element Geochemistry at the Sediment-Water Interface in the North Sea and the Western Wadden Sea*; BEON Report, 0924-6576; Netherlands Institute for Sea Research: Den Burg, The Netherlands, 1992.

53. Duan, J.; Tang, J.; Lin, B. Zinc and Lead Isotope Signatures of the Zhaxikang PbZn Deposit, South Tibet: Implications for the Source of the Ore-Forming Metals. *Ore Geol. Rev.* **2016**, *78*, 58–68. [CrossRef]

54. Alexander, R.B.; Smith, R.A.; Schwarz, G.E.; Boyer, E.W.; Nolan, J.V.; Brakebill, J.W. Differences in Phosphorus and Nitrogen Delivery to The Gulf of Mexico from the Mississippi River Basin. *Environ. Sci. Technol.* **2008**, *42*, 822–830. [CrossRef]

55. Wang, X.; Lu, Y.; Han, J.; He, G.; Wang, T. Identification of Anthropogenic Influences on Water Quality of Rivers in Taihu Watershed. *J. Environ. Sci.* **2007**, *19*, 475–481. [CrossRef]

56. Zhang, H.; Wang, Z.; Zhang, Y.; Ding, M.; Li, L. Identification of Traffic-Related Metals and the Effects of Different Environments on Their Enrichment in Roadside Soils along the Qinghai–Tibet Highway. *Sci. Total Environ.* **2015**, *521–522*, 160–172. [CrossRef] [PubMed]

 *water*

Article

Tributary Loadings and Their Impacts on Water Quality of Lake Xingyun, a Plateau Lake in Southwest China

Liancong Luo [1,*], Hucai Zhang [1], Chunliang Luo [2], Chrisopher McBridge [3], Kohji Muraoka [3], Hong Zhou [4], Changding Hou [5], Fenglong Liu [6] and Huiyun Li [7,*]

[1] Institute for Ecological Research and Pollution Control of Plateau Lakes, School of Ecology and Environmental Science, Yunnan University, Kunming 650504, China; zhanghc@ynu.edu.cn

[2] Research Institute for Plateau Lakes, Dianchi Management Bureau, Kunming 650504, China; chunliangluo@126.com

[3] Environmental Research Institute, The University of Waikato, Hamilton 3240, New Zealand; christopher.mcbride@waikato.ac.nz (C.M.); kohji.muraoka@gmail.com (K.M.)

[4] Waikato Institute of Technology, Hamilton 3204, New Zealand; hong.zhou@wintec.ac.nz

[5] Yuxi Environmental Research Institute, Yuxi 653100, China; hcdhappy@126.com

[6] Financial Department, Hunan University of Humanities, Science and Technology, Loudi 417000, China; 13907384452@139.com

[7] Nanjing Institute of Geography and Limnology, Chinese Academy of Sciences, Nanjing 210008, China

* Correspondence: billluo@ynu.edu.cn (L.L.); hyli@niglas.ac.cn (H.L.); Tel.: +86-(0)-181-1290-7530 (L.L.); +86-(0)-159-0515-7570 (H.L.)

Citation: Luo, L.; Zhang, H.; Luo, C.; McBridge, C.; Muraoka, K.; Zhou, H.; Hou, C.; Liu, F.; Li, H. Tributary Loadings and Their Impacts on Water Quality of Lake Xingyun, a Plateau Lake in Southwest China. *Water* **2022**, *14*, 1281. https://doi.org/10.3390/w14081281

Academic Editor: Liudmila S. Shirokova

Received: 23 February 2022
Accepted: 11 April 2022
Published: 15 April 2022

Publisher's Note: MDPI stays neutral with regard to jurisdictional claims in published maps and institutional affiliations.

Abstract: Lake Xingyun is a hypertrophic shallow lake on the Yunnan Plateau of China. Its water quality (WQ) has degraded severely during the past three decades with catchment development. To better understand the external nutrient loading impacts on WQ, we measured nutrient concentrations in the main tributaries during January 2010–April 2018 and modelled the monthly volume of all the tributaries for the same period. The results show annual inputs of total nitrogen (TN) had higher variability than total phosphorus (TP). The multi-year average load was 183.8 t/year for TN and 23.3 t/year for TP during 2010–2017. The average TN and TP loads for 2010–2017 were 36.6% higher and 63.8% lower, respectively, compared with observations in 1999. The seasonal patterns of TN and TP external loading showed some similarity, with the highest loading during the wet season and the lowest during the dry season. Loads in spring, summer, autumn, winter, and the wet season (May–October) accounted for 14.2%, 48.8%, 30.3%, 6.7%, and 84.9% of the annual TN load and 14.1%, 49.8%, 28.1%, 8%, and 84.0% of the annual TP load during 2010–2017. In-lake TN and TP concentrations followed a pattern similar to the external loading. The poor correlation between in-lake nutrient concentrations and tributary nutrient inputs at monthly and annual time scales suggests both external loading and internal loading were contributing to the lake eutrophication. Although effective lake restoration will require reducing nutrient losses from catchment agriculture, there may be a need to address a reduction of internal loads through sediment dredging or capping, geochemical engineering, or other effective measures. In addition, the method of producing monthly tributary inflows based on rainfall data in this paper might be useful for estimating runoff at other lakes.

Keywords: external loading; internal loading; water quality; tributary; Lake Xingyun

1. Introduction

Eutrophication refers to the enrichment of aquatic systems by nutrients, usually nitrogen (N) and phosphorus (P) [1]. Both N and P are required to support aquatic plant growth and are the key limiting nutrients in most aquatic ecosystems [1]. Eutrophication is an economic, recreational, and aesthetic problem that affects many lakes worldwide [2], resulting in high primary productivity, impaired WQ, and often leading to harmful algal blooms (HABs) [1,3,4]. Lake eutrophication is mainly caused by anthropogenic processes through industrial sewage discharge, agricultural farming, air deposition, and soil erosion [5,6] as well as internal nutrient releases [7] or sediment resuspension in shallow waterbodies [8–12]. These processes may be further exacerbated by climate change, which is predicted to result in more rainstorms that increase the frequency and strength of stormwater inflows and nutrient inputs through surface water runoff [13,14].

The nutrient sources and processes leading to freshwater eutrophication have been well described [10–12,15,16]. Excessive nutrient inputs lead to the degradation of ecological function, which was reported in large lakes, e.g., Victoria [17,18], Peipsi [19], Taihu [20], Winnipeg [21], and Erie [22], as well as small ponds [23]. The nutrients entering lake water through tributaries could sink to the bottom and readily enter the trophogenic zone of shallow lakes through sediment resuspension and the periodic anoxia conditions promoted by calm, warm weather [24]. Due to the longevity of internal loading, eutrophic shallow lakes have more difficulty decreasing algal biomass and increasing transparency than deep water bodies [24], which maintain seasonal plaguey HABs in shallow lakes (e.g., Lake Taihu [25]).

The Yunnan Plateau (YP) is located in Southwestern China, with a subtropical–temperate climate [26] and distinct wet and dry seasons. The YP receives an annual precipitation of 916 mm distributed as 133, 522, 219, and 42 mm through spring, summer, autumn, and winter, respectively [27]. There are nine great plateau lakes (NGPLs) with water surface areas > 30 km^2 distributed through the northwest and middle of the YP. The NGPLs are tectonic in origin, formed from crustal movement from the Pliocene to Holocene periods [28,29]. Most of the NGPLs have experienced considerable WQ deterioration in recent decades. Lake Xingyun is one of the NGPLs and is located at the center of the YP at an elevation of 1740 m a.s.l. (Figure 1). More than 2 million people work in the local vegetable farming industry, creating potential point and non-point nutrient sources to waterways and drawing water from the lake for irrigation. The primary water sources for Lake Xingyun are riverine inputs and precipitation, mainly during the wet season. HABs were documented in the lake as early as August 1957 [30]. The lake eutrophication reportedly accelerated from 2002 to 2012 [31], and now it is hypertrophic, with an average TP of 0.36 mg/L, TN of 1.0 mg/L, and chlorophyll *a* (Chl-*a*) of 82.4 µg/L in 2017, although the local government has initiated many lake restoration projects since 2000, including riverine nutrient input reductions through sewage treatment plant construction, sewage water deposition in settling ponds before direct discharge into the lake, sediment dredging, water hyacinth planting and harvest, algae collection for treatment, fishery regulations, water flushing, and wetland construction [32,33].

The paleolimnology of Lake Xingyun has been a topic of several recent studies, with reports that include the soil erosion history caused by human activities [34], sediment organic content [35], and environmental change in relation to local human activities [36]. There have been also many papers documenting algae [37], microcystins [38], biological responses and WQ [38], and air deposition [39,40], but due in part to a lack of tributary discharge measurements, there has been only one paper addressing external loading as early as 1999 [41]. In this paper, our objective was to calculate the tributary nutrients entering the lake from 2010 to 2017 based on the WQ measurements and modelled inflows of the 12 main tributaries at monthly and annual time scales and to understand the external load variation from the 1990s to 2017. The calculated tributary nutrient loads for 2010–2017 represent a key reference for future lake environmental management, and the methodology described here should be applicable for other NGPLs.

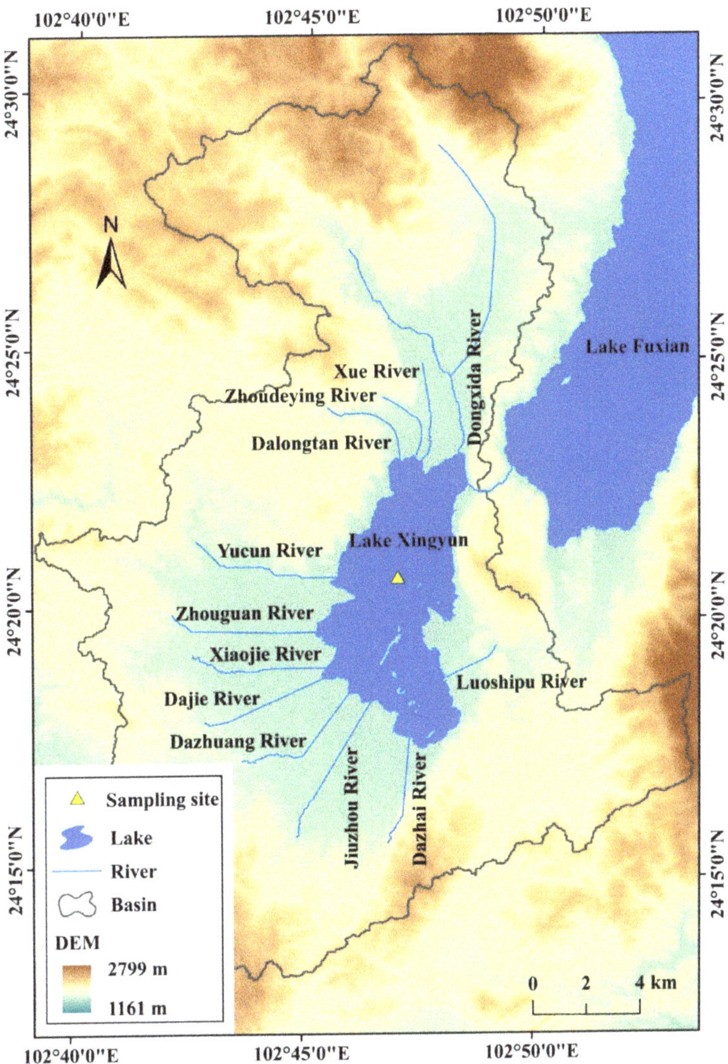

Figure 1. Map of Lake Xingyun catchment with the 12 main tributaries and the sampling site for lake water.

2. Materials and Methods

2.1. Study Site

Lake Xingyun is a shallow lake with a water volume of 2.1×10^8 m^3, and catchment area of 325 km^2. The average depth is 7.0 m, with a maximum depth of 10.8 m and a water surface area of 34.7 km^2 (Figure 1) [42]. There are 12 main shallow tributaries, including four rivers to the north (Figure 1, river numbers 1 to 4), two to the west (river numbers 5 to 6), four to the southwest (river numbers 7 to 10), one to the south (river number 11), and one river to the southeast (river number 12). All these sub-catchment proportions of water input, TP input, and TN input of total tributary inputs in 1999 are based on the only existing measurements of inflow volume (Table 1, [41]) and their proportions of paddy land, dry land, and the population of the catchment in 2015 [43]. So, the water and nutrients were mostly from the north (river numbers 1–4), southwest (river numbers 7–10), and east (river number 12) in 1999.

Table 1. The sub-catchment proportions of water input, TP input, and TN input of total tributary inputs in 1999 and their proportions of dry land, paddy land, and population of those of the whole catchment in 2015 (N—north, W—west, SW—southwest, S—south, E—east).

Sub-Catchment and River Numbers	1999 (%)			2015 (%)		
	Water	TP	TN	Dry Land	Paddy Land	Population
N (1–4)	30.8	40.9	27.7	34.6	6.8	33.2
W (5–6)	15.2	6.1	13.2	24.1	57.5	19.6
SW (7–10)	34.3	20.4	47.2	11.3	1.7	29.4
S (11)	2.4	1.1	1	7.2	24.8	11
E (12)	17.3	31.5	10.9	22.8	9.2	6.8

These rivers reportedly carried 41.0% of total water input, including surface and subsurface water inputs and rainfall, 34.3% of TP input, and 56.7% of TN input through surface water, subsurface water, dry deposition, and wet deposition to the lake in 1999 [41]. Before 2007, Lake Xingyun flowed to Lake Fuxian through the Ge River in the north, with an average water volume of 4.3644×10^7 m^3/year due to the 1 m higher elevation of the water surface of Lake Xingyun compared to Lake Fuxian, carrying significant nutrients to the other lake. To protect Lake Fuxian, the polluted water in Lake Xingyun has been diverted to the Jiuxi Wetland through a channel to the Dongfeng Reservoir for agricultural and industrial uses since May 2008. The water level at Lake Xingyun was then mainly regulated by the diversion channel for many years, but the diversion engineering is no longer working to guarantee water use for vegetable farming at the lake catchment.

The lake has experienced dramatic macrophyte degradation and significant phytoplankton biomass increase since the 1980s. The macrophyte coverage area percentage of the whole lake area was 22% in 1984, 2.2% in 2000, and 1.8% in 2008. The Chl-*a* concentration was 6.67 μg/L in 1984, 11.93 μg/L in 1993, and 18.18 μg/L in 1999. It dramatically increased during 2002–2009, with an average of 47.63 μg/L, and maintained this level during 2010–2017, with an average value of 61.16 μg/L [44]. At the lake catchment, the dominant land use types are agricultural land, accounting for 43.3%; forest, accounting for 23.4%; grass land, accounting for 15.4%; and construction land, accounting for 14.7% of the whole catchment area in 2016 [45].

2.2. Data Source

All the monthly WQ data for January 2010–April 2018 used here are from the Yuxi Environmental Monitoring Station (YEMS). In the lake, water samples were collected at surface, middle, and bottom layers of the sampling site (Figure 1) and were mixed for laboratory analysis at the middle of every month. For the 12 shallow tributaries, water samples were collected at the estuary. All the water samples were analyzed according to the standard methods issued by Ministry of Ecology and Environment of the People's Republic of China (MEEC) [46]. TN and TP were analyzed using combined persulphate digestion and spectrophotometry. The transparency was measured with a 20 cm diameter Secchi disc (SD) in situ. The Chl-*a* was determined spectrophotometrically after acetone (90%) extraction. The COD were analyzed by potassium permanganate oxidation methods (COD$_{Mn}$) with un-filtrated samples.

2.3. Reconstruction of Monthly Inflow Volumes

There have been very few inflow volume measurements recorded for Lake Xingyun. The only observations were conducted by YEMS in 1999 and reported by Wang [41]. The inflow volume at the main 12 tributaries was measured 3 times during January to April, 12 times during May to September, and 3 times from October to Decemberin 1999.

Mei [47] reported there was a good linear relationship between rainfall and inflow volume or runoff with a 24 h lag after the rainfall at three stations of the Lake Taihu catchment. This linear characteristic was also found in Xinjiang Province (correlation coefficient > 0.7) by Zhao [48], the Fujiang River of Sichuan Province at an annual time scale by Wang et al. [49], and the Jinghe River Basin [50]. Due to the extreme lack of inflow measurements at the 12 tributaries of Lake Xingyun, we hypothesize that there was also a linear relationship between the total inflow volume of the 12 tributaries and the precipitation at both monthly and annual time scales. We then calculated monthly inflow volumes for the 12 tributaries as follows:

(1) The ratio of inflow volume for each tributary to the total inflow volume in 1999:

$$R_i = Vol_i / TIV_{1999} \tag{1}$$

where Vol_i represents the inflow volume for the ith (i = 1, 2, 3, ... , 12) tributary from January to December in 1999, TIV_{1999} represents the total inflow volume of the 12 main tributaries in 1999, and R_i is the ratio of volume of the ith tributary in 1999 relative to TIV_{1999}.

(2) The annual total inflow volume estimation for 2010–2018:

$$TIV_j = TIV_{1999} \times \left(Rain_j / Rain_{1999} \right) \tag{2}$$

where TIV_j represents the total inflow volume of the 12 tributaries for Lake Xingyun in the jth year from 2010 to 2018 (a period when the river WQ was measured). TIV_{1999} represents the total inflow volume of the 12 tributaries in 1999. $Rain_{1999}$ and $Rain_j$ are the total annual rainfall in 1999 and in the jth year from 2010 to 2018, respectively.

(3) The annual total inflow volume for each tributary:

$$TIV_{ij} = TIV_j \times R_i \tag{3}$$

where TIV_{ij} represents the total inflow volume for the ith (i = 1, 2, 3, ... , 12) tributary in the jth (j = 2010, 2011, 2012 ... 2018) year.

(4) The monthly total inflow volume for every tributary:

$$TIV_{ijn} = TIV_{ij} \times \left(Rain_{jn} / Rain_j \right) \tag{4}$$

where TIV_{ijn} represents the total inflow volume for the ith tributary (i = 1, 2, 3, ... , 12) in the nth month (n = 1, 2, 3, ... , 12) of the jth year from 2010 to 2018. $Rain_{jn}$ and $Rain_j$ are the total rainfall in the nth month of the jth year and the total rainfall in the jth year.

2.4. Annual and Monthly Nutrient Loads for All the Tributaries

With the reconstructed monthly volume TIV_{ijn} for 2010 to 2018, the monthly nutrient load for each tributary was then calculated as:

$$Eload_{ijn} = TIV_{ijn} \times C_{ijn} \tag{5}$$

where $Eload_{ijn}$ represents the external nutrient load as either TN or TP (tons/month, abbreviated to t/mon) for the ith (i = 1, 2, 3, ... , 12) tributary in the nth (n = 1, 2, 3, ... , 12) month of the jth (j = 2010, 2011, 2012, ... , 2018) year from 2010 to 2018. C_{ijn} is the measured nutrient (TN or TP) concentration at the ith (i = 1, 2, 3, ... , 12) tributary estuary in the nth (n = 1, 2, 3, ... , 12) month of the jth (j = 2010, 2011, 2012, ... , 2018) year. The measurements in 2018 were only conducted during January to April, so these measurements were not included in the calculation of total annual external load.

2.5. Comprehensive Trophic Level Index (TLI)

The trophic state index (TSI) was developed by Carlson [51], reflecting lake trophic status. It was modified to the *TLI* by Wang et al. [52] and adapted to Chinese lakes. The aggregated *TLI* is calculated as the sum of individual *TLIs*, which are calculated as

$$TLI_{chl-a} = 10(2.5 + 1.086 \times lnChl - a) \tag{6}$$

$$TLI_{TP} = 10(9.436 + 1.624 \times lnTP) \tag{7}$$

$$TLI_{TN} = 10(5.453 + 1.694 \times lnTN) \tag{8}$$

$$TLI_{SDT} = 10(5.118 - 1.94 \times lnSDT) \tag{9}$$

$$TLI_{COD} = 10(0.109 + 2.661 \times lnCOD) \tag{10}$$

where the left side in the above equations is the sub-*TLI* for each parameter and the aggregated value is

$$TLI = \sum_{j=1}^{m} W_j \cdot TLI_j \tag{11}$$

where *TLI* represents the aggregated trophic level index for Lake Xingyun and W_j represents the weighting factor for the *j*th WQ parameter, including Chl-*a*, TN, TP, Secchi disk transparency (SDT), and chemical oxygen demand (COD) (*j* = 1, 2, 3, 4, 5). The weighting factor values are 0.2663 for Chl-*a*, 0.1879 for *TN*, 0.179 for *TP*, 0.1834 for SDT, and 0.1834 for COD based on investigations from 26 Chinese lakes by [53]. The *TLI* calculations were conducted by YEMS based on Equations (6)–(11). The *TLI* ranges and corresponding trophic status are shown in Table 2.

Table 2. Trophic Level Index (*TLI*) value ranges and corresponding trophic status for Chinese lakes.

TLI	*TLI* < 30	30 < *TLI* < 50	50 < *TLI* < 60	60 < *TLI* < 70	*TLI* > 70
Eutrophic status	Oligotrophic	Mesotrophic	Light-eutrophic	Middle-eutrophic	Hyper-eutrophic

2.6. Long-Term Lake WQ Data

To compare lake WQ after 2010 with that during 2000 to 2009 and before 2000, the lake WQ data before 2009 were retrieved from [54–56]. There were no TN and TP data for 1983–1987, no Chl-*a* data for 1983, 1985–1992, 1994–1998, and 2000–2001, and no *TLI* values for 1982–1995.

3. Results

3.1. Total Monthly and Yearly Inflows

The constructed yearly inflows for 2010–2017 and monthly inflows from January 2010 to April 2018 are shown in the Figure 2. It is obvious that the monthly inflows in the wet season (May–October) were much higher than those in the dry season (November–April), which is consistent with the rainfall distribution pattern through the year as well as the measurements in 1999 by Wang [41]. The highest monthly inflow was in Jul 2017, with a value of 8.918×10^6 m^3, while the lowest value was 0.018×10^6 m^3 in February 2018, i.e., the maximum monthly inflow was 495 times the minimum value. The annual inflow decreased from 2010 to 2011 but increased progressively from 2011 to 2017. The lowest annual inflow was 17.269×10^6 m^3 in 2011 and the highest was 31.09×10^6 m^3 in 2017, which is smaller than the annual inflow (37.995×10^6 m^3) measured in 1999 [41]. The estimated mean annual water inflow for 2010–2017 was 2.54×10^7 m^3, which is very close to the value (2.4×10^7 m^3) reported by Gao et al. [57], suggesting that the calculated inflows based on Equations (1)–(4) are reliable.

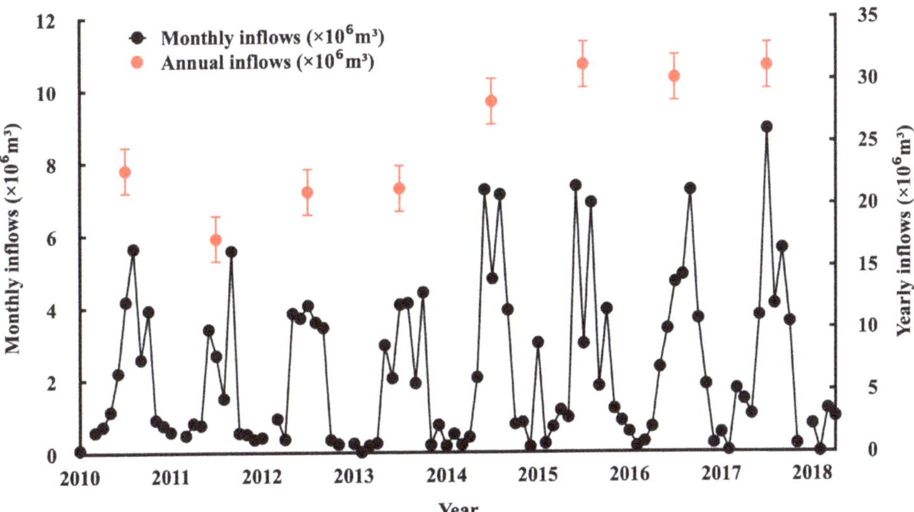

Figure 2. Calculated monthly inflows for January 2010–April 2018, represented by black solid circles and solid line (left *Y* axis), and yearly inflows ± standard error for 2010–2017, represented by red solid circles (right *Y* axis).

3.2. Monthly TN and TP Loads for 2010–2018

With the constructed inflow volumes and Equation (5), the monthly TN and TP loads from the 12 tributaries were calculated from January 2010 to April 2018. The monthly TN loads (TN_Eload, t/mon) and the lake TN concentrations (TN_Lake, mg/L) are shown in Figure 3. The corresponding monthly TP loads (TP_Eload, t/mon) and the lake TP concentrations (TP_Lake, mg/L) data are shown in Figure 4. The average values of TN_Lake, TP_Lake, TN_Eload, and TP_Eload were 1.87 mg/L, 0.36 mg/L, 15.8 t/mon, and 1.65 t/mon for Jan 2010–April 2018. The standard deviations (SD) for the time series of TN_Lake (SD_TN_Lake), TP_Lake (SD_TP_Lake), TN_Eload (SD_TN_Eload), and TP_Eload (SD_TP_Eload) were 0.58 mg/L, 0.16 mg/L, 17.86 t/mon, and 1.97 t/mon, respectively, demonstrating wide variations in the loads relative to the averages compared with comparable values for lake nutrient concentrations. There were obvious seasonal variations of TN_Lake, TP_Lake, TN_Eload, and TP_Eload, with peak values in summer and trough values in winter, suggesting that the rainfall in summer brought the lake abundant nutrients.

The Pearson correlation coefficient (*r*) between the two time series of lake TN and TN loads was 0.07 (*n* = 92). The maximum TN_Lake occurred in September 2010, with a value of 5.21 mg/L, and the minimum TN_Lake occurred in November 2017, with a value of 0.92 mg/L, while the TN_Eload in these two months was 17.9 t and 3.3 t. The maximum TN_Eload was 97.9 t/mon in Jul 2017, and the minimum TN_Eload was 0.15 t/mon in February 2018. The maximum values for both TN_Lake (September 2010) and TN_Eload (Jul 2017) were in the wet season, and the minimum values of TN_Lake (November 2017) and TN_Eload (February 2018) were both in the dry season.

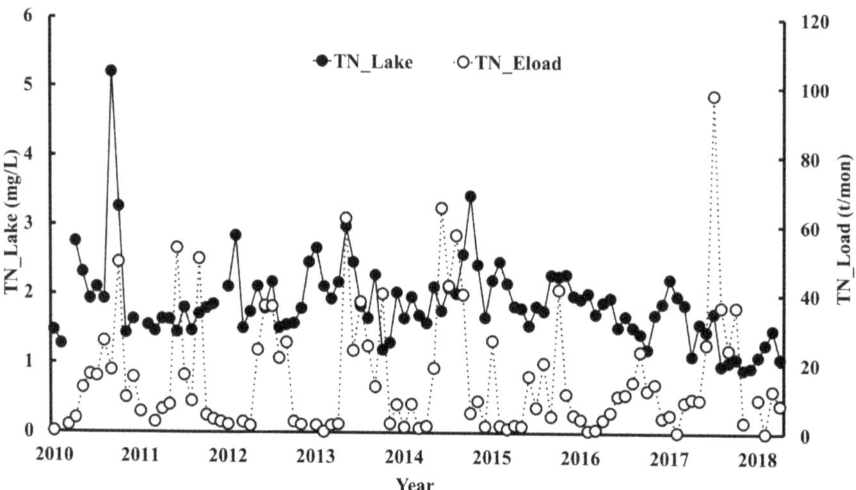

Figure 3. Monthly lake TN concentration (TN_Lake), represented by black solid circles and solid line (left Y axis), and total tributary TN inputs (TN_Eload) represented by black hollow circles and dash line (right Y axis), during January 2010–April 2018.

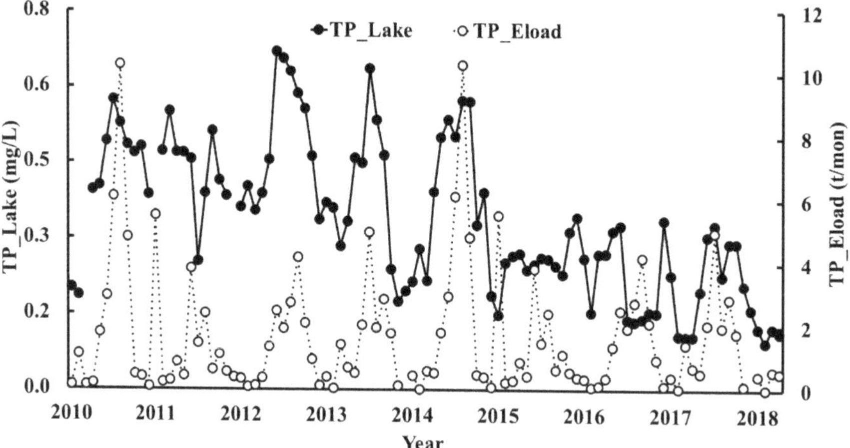

Figure 4. Monthly lake TP concentration (TP_Lake), represented by black solid circles and solid line (left Y axis), and total tributary TP inputs (TP_Eload), represented by black hollow circles and dash line (right Y axis), during January 2010–April 2018.

The r value between the two time series of lake TP and TP loads was 0.39 ($n = 95$), which was greater than r between TP_Eload and TP_Lake. The maximum TP_Lake concentration was 0.72 mg/L in June 2012, and the minimum value was 0.10 mg/L in February 2018, while the maximum TP_Eload was 10.3 t/mon in both August 2010 and August 2014, and the minimum TP_Eload was 0.0094 t/mon in both February 2014 and February 2018. Similar to *TN*, both the maximum TP_Eload (August 2010 and August 2014) and the maximum TP_Lake (June 2012) were observed in the wet season, and both the minimum TP_Eload (February 2014 and February 2018) and the minimum TP_Lake (February 2018) were observed in the dry season.

3.3. Seasonality of TN and TP Loading

The lake TN and TP concentrations and external loads for the 12 months (January–December) of each year are plotted in Figure 5. The average TN (Figure 5A) and TP (Figure 5B) concentrations in the wet season were 1.93 mg/L and 0.43 mg/L, respectively, and 1.81 mg/L and 0.3 mg/L in the dry season. Both the TN and TP concentrations were slightly greater in the wet season than in the dry season. The average TN load (Figure 5A) and TP load (Figure 5B) in the wet season were 26.3 t/mon and 3.2 t/mon, respectively, and 4.7 t/mon and 0.6 t/mon in the dry season. The highest TN and TP loads were 33.2 t in Jul and 4.4 t in September, respectively, while the highest lake TN and TP concentrations were 2.26 mg/L in September and 0.46 mg/L in June. The lowest monthly TN (1.3 t) and *TP* loads (0.2 t) were both in February, a month of relatively low rainfall. The lowest lake TN (1.61 mg/L) and TP (0.25 mg/L) concentrations occurred in August and January, respectively. Therefore, in winter and spring (i.e., the dry season), the average external loads and lake nutrient concentrations were lower compared with the wet season.

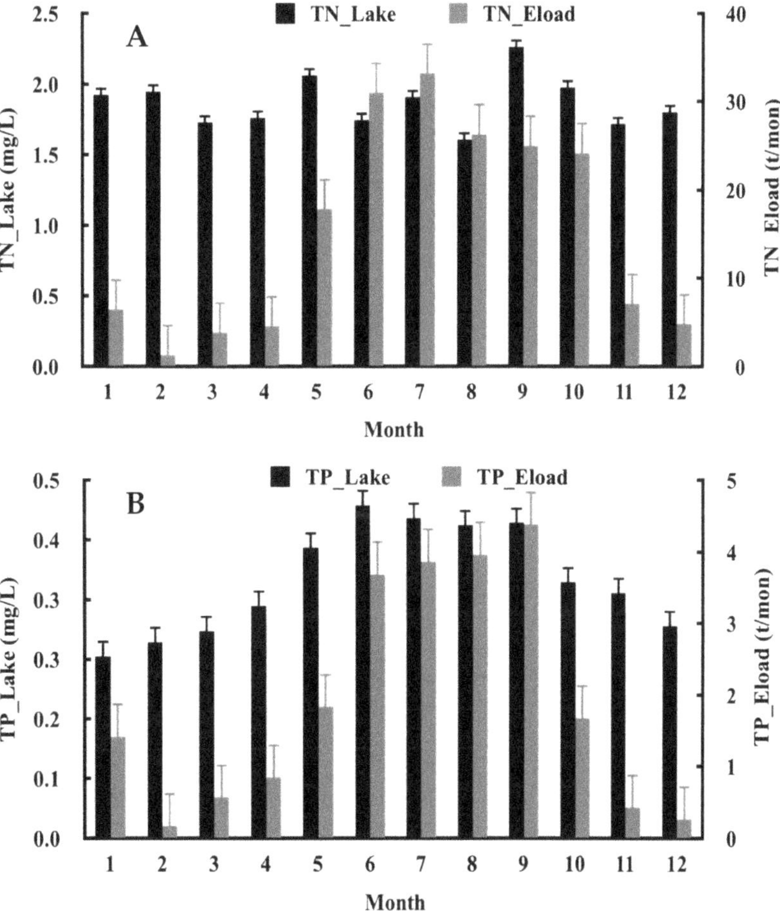

Figure 5. Mean monthly concentrations of lake TN (TN_Lake) with positive standard errors (black), and tributary loadings of TN (TN_Eload) with positive standard errors represented (grey) for January 2010–April 2018 (**A**). Mean monthly concentrations of lake TP (TP_Lake) with positive standard errors (black), and tributary loadings of TP (TP_Eload) with positive standard errors represented (grey) (**B**) for January 2010–April 2018.

The calculated total multi-year averaged external loads of TN and TP were 183.8 t/year and 23.3 t/year, respectively, during 2010–2017. The TN loads in spring, summer, autumn, and winter accounted for 14.2%, 48.8%, 30.3%, and 6.7%, respectively, of the total multi-year averaged TN load, and for TP loads the corresponding seasonal values were 14.1%, 49.8%, 28.1%, and 8% of the annual TP load (Table 3). In addition, the average load during the wet season accounted for 84.9% of the total TN load and 84.0% of the total TP load during the study period. The magnitude of the seasonal load was summer > autumn > spring > winter for both TN and TP.

Table 3. The proportions (percentage) of external loading in different seasons as a fraction of the sum of the 12 month loadings during January 2010–April 2018. Wet season is May–October and dry season is November–April for Yunnan Plateau.

	Spring (%)	Summer (%)	Autumn (%)	Winter (%)	Wet Season (%)	Dry Season (%)
TN	14.2	48.8	30.3	6.7	84.9	15.1
TP	14.1	49.8	28.1	8.0	84.0	16.0

3.4. Annual TN and TP Loads and Long–Term Variations in Lake WQ

The annual external loads of TN (TN_Eload) and TP (TP_Eload) are plotted in Figure 6. The highest and lowest TN_Eloads were 258.0 t/year and 106.1 t/year in 2017 and 2016, and for TP_Eload they were 31.0 t/year and 16.9 t/year in 2012 and 2016, respectively. The lowest external loads for both TP and TN were in 2016. The deviation coefficient (a coefficient with the standard deviation divided by the average value) for TN_Eloads and TP_Eloads were 0.28 and 0.24, respectively, showing TN_Eloads was slightly more variable than TP_Eloads.

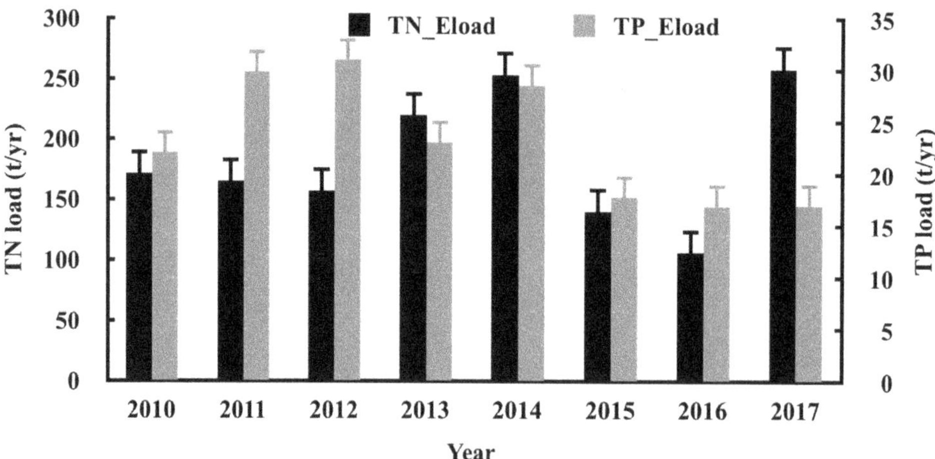

Figure 6. Annual tributary external loadings of TN (TN_Eload) with positive standard errors (black, left *Y* axis), and TP (TP_Eload) with positive standard errors (grey, right *Y* axis), in Lake Xingyun for 2010–2017.

The annual variations in TN, TP, Chl-*a*, and *TLI* during 1982–2017 are shown in Figure 7. TN and Chl-*a* showed greater relative interannual variability than TP and *TLI*. The highest observed TN, TP, Chl-*a*, and *TLI* were 2.3 mg/L, 0.52 mg/L, 109.5 μg/L, and 70.1 in 2010, 2012, 2013, and 2013, respectively, and the lowest values were 0.28 mg/L, 0.02 mg/L, 6.67 μg/L, and 47.1 in 1982, 1997, 1984, and 1996, respectively. There was a dramatic increase in TN, TP, Chl-*a*, and *TLI* after 2000 compared with the values from 1982 to 1999. The average TN, TP, Chl-*a*, and *TLI* for 1982–1999 (period 1), 2000–2009 (period 2), and 2010–2017 (period 3) are shown in Table 4. The average TN was > twice as high in period 2 compared to period 1 and increased again from period 2 (1.62 mg/L) to period 3 (1.89 mg/L). The relative increase in TP from period 1 to period 2 was much more dramatic than TN, and the increase in period 3 was > twice the increase in period 2. Chl-*a* and *TLI* increased considerably from period 1 to period 2 and increased again, but to a lesser extent, from period 2 to period 3. During period 3, the annual TN_Eload did not correspond closely with the lake TN concentration, but the highest TP_Eload (in 2012) corresponded with the highest TP_Lake concentration.

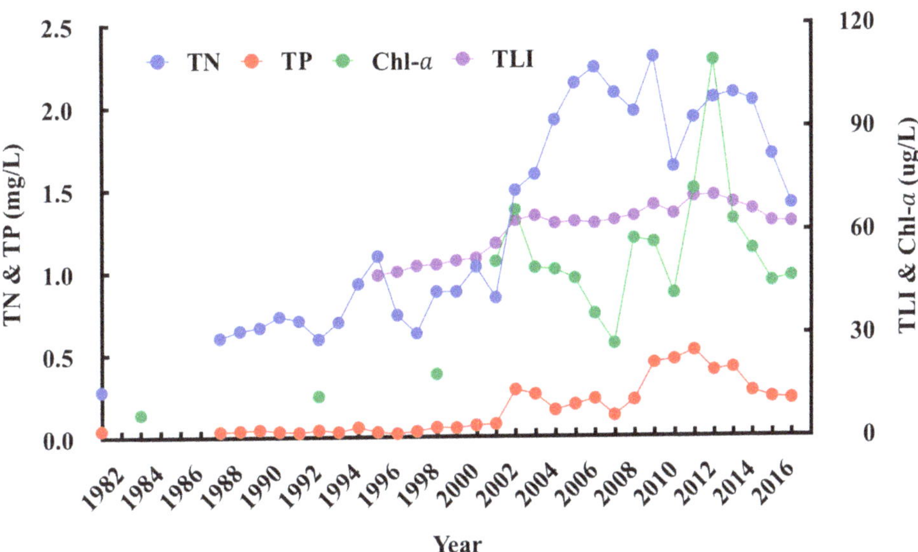

Figure 7. Annual concentrations of TN (blue), TP (red), Chl-*a* (green), and *TLI* (purple) at Lake Xingyun for 1982–2017 (no TN and TP data for 1983–1987, no Chl-*a* data for 1983, 1985–1992, 1994–1998, and 2000–2001, and no *TLI* data for 1982–1995).

Table 4. Average TN, TP, and Chl-*a* concentrations and *TLI* for the periods of 1982–1999, 2000–2009, and 2010–2017 at Lake Xingyun.

Year Period	1982–1999 (Period 1)	2000–2009 (Period 2)	2010–2017 (Period 3)
TN (mg/L)	0.71	1.62	1.89
TP (mg/L)	0.04	0.17	0.37
Chl-*a* (µg/L)	12.26	47.63	61.16
TLI	48.8	60.0	66.4

4. Discussion

Tributary nutrient loads of N and P are often the dominant driver of lake eutrophication. They increase in-lake concentrations, promote algal growth, and increase nutrient storage in sediments. The resuspension of sediment-associated nutrients in shallow waters or dissolved nutrients enhanced by periods of hypoxia/anoxia in the bottom waters of stratified water bodies can consequentially positively fuel internal nutrient loading and feedback to increase eutrophication. Sakamoto et al. [58] estimated that 25% of the total nitrogen inputs from the catchment of Lake Xingyun were stored in the bottom sediments. Based on an upper layer sediment of thickness of 27.7 cm, the sediment P storage was estimated to be 6967 t, and the N storage was estimated to be 8444 t [53]. The TP concentration in the bottom sediment ranged from 1682 to 6537 mg/kg (average 3059 mg/kg) based on observations in September 2010. This value is much higher than shallow Lake Taihu (mean depth 1.9 m), Lake Chao (mean depth 3 m), and Lake Dian (mean depth 4.4 m) [59] and likely reflects a greater relative deposition of autochthonous-generated organic material. The bottom sediments, particularly under hypoxic/anoxic conditions, therefore represent a potential major nutrient source that could stimulate eutrophication in the lake, although atmospheric deposition may be similarly important. The stores of nutrients in the bottom sediments have likely accelerated with agricultural and industrial development in the catchment [60], phosphorus mining [61], and livestock and poultry farming [62].

Wang [41] documented the external TN and TP loads through tributaries and wet and dry deposition based on measurements from 1999. He estimated TN and TP loads, including diffuse inputs at the shoreline and atmospheric deposition, of 236.7 t and 187.3 t in 1999, respectively. The TN and TP loadings from dry deposition and wet deposition were 29.3 t and 105.7 t in 1999 [41], accounting for 12.4% and 56.4% of the total TN and TP loads, respectively. The major TP contributor was air deposition resulting from phosphorus mining in 1999. The 12 main tributaries contributed 56.7% of the TN inputs and 34.3% of the TP inputs in 1999. Nowadays, all phosphorus mining factories have been closed or moved out of the lake catchment, leading to a remarkable decrease in both tributary and atmospheric TP loading. The wet and dry deposition estimates of TP loading dramatically decreased from 105.7 t in 1999 to 15.22 t in 2010 [61], likely reflecting the early stages of mine closure. The average TP loading through tributaries has decreased by 63.8% during 2010–2017 compared to that in 1999.

Despite P management efforts and mine closures, in-lake TP concentration have increased from 0.04 mg/L in period 1 (pre-2000) to 0.17 mg/L in period 2 (2000s) and 0.37 mg/L in period 3 (2010s), while the TP input through the main 12 tributaries in the three periods was gradually decreasing, with average annual loads of 61.2 t/year (Wang 2001), 46.6 t/year (Jin et al., 2010), and 23.3 t/year (this paper), respectively. The large increase in lake TP concentration from period 1 to period 3 implies that there might still be a large amounts of phosphorus entering the lake waterbody through sediment resuspension or, to some extent, through dry deposition, although four of five phosphorus mining factories have been closed by the local government, or that there might be significant phytoplankton P content since chl-*a* has dramatically increased from 12.26 µg/L in period 1 to 47.63 µg/L in period 2 and 61.16 µg/L in period 3. Therefore, the reason why the lake P has increased dramatically since 1980s need to be further studied for the mitigation of lake eutrophication and algal blooms.

The primary nitrogen source from the catchment is from vegetable farming [42]. The majority of TN reaches the lake via tributaries during the rainy season. We observed increased annual loading of TN, likely due to a progressive increase in agricultural development in the catchment over the past three decades [61]. Although there are now many pre-pools around the lake (~30 m to the lake shore) receiving water from all the tributaries, which are mostly used for vegetable farming, there is still significant N and P entering the lake water with overflow due to limited pre-pool storage during the rainy seasons. So, the buffering pre-pools cannot completely reduce the external loading from the catchment to be zero. Furthermore, they might make the lake water level decrease gradually. So, the effect of pre-pools around the lake shore on water quality improvement needs to be carefully and seriously evaluated in the future.

The increased TN loading and TP loading through tributary inputs and atmospheric deposition and internal loading, together with increases in in-lake TP have likely favored algal growth, as shown by a ~5-fold increase of in Chl-*a* concentrations for the period of 2010–2017 compared with pre-2000. The *TLI* values in the three periods also increased dramatically. Although most of the nutrients from tributaries are now entering the buffering pre-pools, there is inevitably some N and P entering the lake through overflow in the rainy seasons. The high N and P concentrations in surface sediment and the lake shallowness lead to considerable internal loadings with sediment resuspension during strong wind events. Therefore, to mitigate the eutrophication in Lake Xingyun, it is necessary to manage the reduction of internal TP and TN loads as well as reduced external loads. Directly taking the aggregated algae out of surface water regulated by the meteorological condition might be another optional method for mitigating eutrophication, as it is being carried out at this lake, Lake Dianchi, Lake Chao, and Lake Taihu with frequent and strong algal blooms.

The tributary loadings of TN and TP were poorly correlated with the in-lake TN and TP concentrations at both monthly and annual time scales, which implies internal loading is playing a vital role in lake eutrophication and that TN and TP concentrations in the bottom sediments have increased [54]. Cheng et al. [63] reported that the TN and TP concentrations in the surface sediment of Lake Xingyun were 5140 and 4790 mg/kg, respectively, which were both the highest among the NGPLs of the YP. During January 2010–April 2018, the monthly average TN of all the 12 tributaries was 1.44 mg/L, while the monthly average TN of lake water was 1.87 mg/L. The monthly average TP for inflows and lake water were 0.04 mg/L and 0.36 mg/L, respectively. So, the TN and TP concentrations in the lake were much higher than those of the inflow waters during the period, suggesting the nutrient-enriched surface sediment was an important N and P source that can enter the lake water through wind-induced sediment resuspension or release with hypoxia conditions at the sediment–water interface. The average water residence time was 7.6 years for 2010–2018. The long residence time has kept P and N from catchment cycling in the water–sediment and water–sediment–atmosphere (especially for N) systems, respectively.

The combination of persistent high nutrient loads from the catchment, while they have decreased slightly, and an apparent increase in internal loading based on ongoing increases in in-lake nutrient concentrations have caused frequent harmful algal blooms dominated by *Microcystis aeruginosa* since 2000 [64]. Concentrations of microcystins of up to 42.0 µg/L are observed during the warm season [65], which is much higher than the standard (1.0 µg/L) for drinking water safety suggested by the World Health Organization. Except for the internal load stored in the sediment originating from the lake catchment, cyanobacteria and water hyacinth in the lake water also contribute to internal loading through decomposition, with both vegetative types having increased in biomass in recent years [66].

It has been well documented that internal loading is a critical part of lake eutrophication [8,9,25]. For example, in Lake Pamvotis, internal P loading is expected to increase the response time of trophic status by a decade or longer after external reductions in the P load [67]. In shallow Lake Okeechobee, the average annual TP concentrations increased from 0.049 mg/L in 1973 to 0.098 mg/L in 1984 [68]. The increases in TP concentrations were not correlated with external phosphorus inputs but with changes in lake water levels. The resuspension of bottom sediments by wind action may also be a major factor influencing in-lake total phosphorus concentrations [69]. In large shallow Lake Taihu, nutrient loss from the water column and burial through sedimentation is hampered by frequent wind-induced sediment resuspension [25]. The internal loading induced by sediment resuspension in shallow waters and by hypoxia/anoxia in deep waters is a substantial barrier to reducing eutrophication. According to investigations into the WQ and sediment of 24 lakes in the YP in October 1994, Lake Xingyun's sediment had the highest TP concentration [70]. In this lake, eutrophication has reduced SDT and led to the near total elimination of submerged aquatic vegetation. Therefore, in managing eutrophication in severely degraded lakes, there is a need to reduce external nutrient loads [71,72] as well as internal loads. The management of internal loads can be difficult and expensive. Methods such as dredging [67,73,74] or geoengineering [75,76] have had variable levels of success.

Compared to 1999, with annual tributary TN loads of 134.2 t, the 2017 load and the average 2010–2017 load have increased by 92.2% and 37%, respectively, but the tributary phosphorus inputs have decreased by 73.6% in 2017 and 63.8% during 2010–2017 relative to a load of 64.3 t in 1999. Despite this sustained reduction of P loading, including from atmospheric deposition, in-lake TP has increased, apparently in response to increased internal loading from P-enriched sediment. *Microcystis aeruginosa* is commonly associated with high levels of internal P loading, and, as water transparency has decreased, it appears to have benefited from the lack of competition with submerged macrophytes, which have severely declined in recent years. The lake TN and TP concentrations have increased progressively since the 1980s, and effective lake restoration will require improved agricultural practices to reduce the nutrients, as well as sediment dredging, sediment capping, or geoengineering to reduce internal nutrient loadings at Lake Xingyun.

The estimation of total nutrient inputs to lakes from catchment likely looks very difficult if no monitored inflow volume of tributaries can be found. A hydrodynamic or catchment model [14] might be very useful to estimate inflow volume of all the tributaries for a lake. However, water levels and outflows are required by a hydrodynamic model, and measured daily inflows are required by a catchment model to conduct the estimation. At a monthly time scale, a statistical regression [47–50], genetic method [77], or wavelet method [78] to find the relationship between inflow and rain might be a good option to solve the problem. So, the process of producing monthly tributary inflows based on rainfall data in this paper might be a good example for estimating runoff at other lakes.

5. Conclusions

(1) During 2010–2017, the average yearly TN and TP loads from the main tributaries were 36.6% higher and 63.8% lower than the measurements in 1999.
(2) The TN and TP loads showed similar seasonality, with the highest loading during the wet season and the lowest during the dry season. The TN loading in the wet season accounted for 84.9% of the annual TN load, and the TP loading in the wet season accounted for 84.0% of the annual TP load during 2010–2017.
(3) In-lake TN and TP concentrations were poorly correlated with the corresponding external loadings from main tributaries, which suggests that the internal loading might significantly contribute to the lake eutrophication.
(4) A reduction in the internal loads might be seriously considered for the mitigation of lake eutrophication.

Author Contributions: Conceptualization, L.L., H.Z. (Hucai Zhang) and H.L.; methodology, L.L.; software, L.L.; validation, C.M., K.M. and H.Z. (Hong Zhou); formal analysis, L.L.; investigation, L.L., H.Z. (Hucai Zhang) and C.L.; resources, L.L. and H.Z. (Hucai Zhang); data curation, L.L., H.Z. (Hucai Zhang) and C.H.; writing—original draft preparation, L.L.; writing—review and editing, C.M. and H.Z. (Hong Zhou); visualization, F.L. and H.L.; supervision, L.L.; project administration, L.L., H.Z. (Hucai Zhang) and H.L.; funding acquisition, L.L. All authors have read and agreed to the published version of the manuscript.

Funding: This research was funded by Yunnan University (C176220100043), the Yunnan Provincial Department of Science and Technology (202001BB050078), the National Natural Science Foundation of China (41671205, 42171034), and the Yunnan Provincial Government Leading Scientist Program (2015HA024).

Institutional Review Board Statement: Not applicable.

Informed Consent Statement: Not applicable.

Data Availability Statement: The relevant data was available on the request of corresponding authors.

Acknowledgments: The authors would like very much to thank David Philip Hamilton from Griffith University of Australia for his valuable comments on the manuscript.

Conflicts of Interest: The authors declare no conflict of interest.

References

1. Conley, D.J.; Paerl, H.W.; Howarth, R.W.; Boesch, D.F.; Seitzinger, S.P.; Havens, K.E.; Lancelot, C.; Likens, G.E. Controlling eutrophication: Nitrogen and phosphorus. *Science* **2009**, *323*, 1014–1015. [CrossRef]
2. Greeson, P.E. Lake eutrophication—A natural process. *J. Am. Water Resour. Assoc.* **1969**, *5*, 15–30. [CrossRef]
3. Anderson, D.M.; Glibert, P.M.; Burkholder, J.M. Harmful algal blooms and eutrophication: Nutrient sources, composition, and consequences. *Estuaries* **2002**, *25*, 704–726. [CrossRef]
4. Smith, V.H.; Schindler, D.W. Eutrophication science: Where do we go from here. *Trends Ecol. Evol.* **2009**, *24*, 201–207. [CrossRef]

5. Smith, V.H. Eutrophication of freshwater and coastal marine ecosystems a global problem. *Environ. Sci. Pollut. Res.* **2003**, *10*, 126–139. [CrossRef]
6. Hamilton, D.P.; Salmaso, N.; Paerl, H.W. Mitigating harmful cyanobacterial blooms: Strategies for control of nitrogen and phosphorus loads. *Aquat. Ecol.* **2016**, *50*, 351–366. [CrossRef]
7. Mehner, T.; Diekmann, M.; Gonsiorczyk, T.; Kasprzak, P.; Koschel, R.; Krienitz, L.; Rumpf, M.; Schulz, M.; Wauer, G. Rapid recovery from eutrophication of a stratified lake by disruption of internal nutrient load. *Ecosystems* **2008**, *11*, 1142–1156. [CrossRef]
8. Søndergaard, M.; Kristensen, P.; Jeppesen, E. Phosphorus release from resuspended sediment in the shallow and wind– exposed Lake Arreso, Denmark. *Hydrobiologia* **1992**, *228*, 91–99. [CrossRef]
9. Søndergaard, M.; Jensen, J.P.; Jeppesen, E. Role of sediment and internal loading of phosphorus in shallow lakes. *Hydrobiologia* **2003**, *506–509*, 135–145. [CrossRef]
10. Matisoff, G.; Watson, S.B.; Guo, J.; Duewiger, A.; Steely, R. Sediment and nutrient distribution and resuspension in Lake Winnipeg. *Sci. Total Environ.* **2017**, *575*, 173–186. [CrossRef]
11. Paytan, A.; Roberts, K.; Watson, S.; Peek, S.; Chuang, P.; Defforey, D.; Kendall, C. Internal loading of phosphate in Lake Erie Central Basin. *Sci. Total Environ.* **2017**, *579*, 1356–1365. [CrossRef]
12. Wu, Z.; Liu, Y.; Liang, Z.; Wu, S.; Guo, H. Internal cycling, not external loading, decides the nutrient limitation in eutrophic lake: A dynamic model with temporal Bayesian hierarchical inference. *Water Res.* **2017**, *116*, 231–240. [CrossRef]
13. Whitehead, P.G.; Wilby, R.L.; Battarbee, R.W.; Kernan, M.; Wade, A.J. A review of the potential impacts of climate change on surface water quality. *Hydrol. Sci. J.* **2009**, *54*, 101–123. [CrossRef]
14. Zhang, Y.; Lan, J.; Li, H.; Liu, F.; Luo, L.; Wu, Z.; Yu, Z.; Liu, M. Estimation of external nutrient loadings from the main tributary (Xin'anjiang) into Lake Qiandao (2006–2016). *J. Lake Sci.* **2019**, *31*, 1534–1546. (In Chinese)
15. Sahoo, G.B.; Nover, D.M.; Reuter, J.E.; Heyvaert, A.C.; Riverson, J.; Schladow, S.G. Nutrient and particle load estimates to Lake Tahoe (CA-NV, USA) for total maximum daily load establishment. *Sci. Total Environ.* **2013**, *444*, 579–590. [CrossRef]
16. Robinson, C. Review on groundwater as a source of nutrients to the Great Lakes and their tributaries. *J. Great Lakes Res.* **2015**, *41*, 941–950. [CrossRef]
17. Verschuren, D.; Johnson, T.C.; Kling, H.J. History and timing of human impact on Lake Victoria, East Africa. *Proc. R. Soc. Lond. Ser. B Biol. Sci.* **2002**, *269*, 289–294. [CrossRef]
18. Sitoki, L.; Kurmayer, R.; Rott, E. Spatial variation of phytoplankton composition, biovolume, and resulting microcystin concentrations in the Nyanza Gulf (Lake Victoria, Kenya). *Hydrobiologia* **2012**, *691*, 109–122. [CrossRef]
19. Hensalu, A.; Alliksaar, T.; Leeben, A.; Nõges, T. Sediment diatom assemblages and composition of pore–water dissolved organic matter reflect recent eutrophication history of lake Peipsi (Estonia/Russia). *Hydrobiologia* **2007**, *584*, 133–143. [CrossRef]
20. Yang, M.; Yu, J.; Li, Z.; Guo, Z.; Burch, M.; Lin, T.F. Taihu Lake not to blame for Wuxi's woes. *Science* **2008**, *319*, 158. [CrossRef]
21. Schindler, D.W. The dilemma of controlling cultural eutrophication of lakes. *Proc. R. Soc. B Biol. Sci.* **2012**, *279*, 4322–4333. [CrossRef]
22. Michalak, A.M.; Anderson, E.J.; Beletsky, D.; Boland, S.; Bosch, N.S.; Bridgeman, T.B.; Chaffin, J.D.; Cho, K.; Confesor, R.; Daloğlu, I.; et al. Record-setting algal bloom in Lake Erie caused by agricultural and meteorological trends consistent with expected future conditions. *Proc. Natl. Acad. Sci. USA* **2013**, *110*, 6448–6452. [CrossRef]
23. Lewitus, A.J.; Schmidt, L.B.; Mason, L.J.; Kempton, J.W.; Wilde, S.B.; Wolny, J.L.; Williams, B.J.; Hayes, K.C.; Hymel, S.N.; Keppler, C.J.; et al. Harmful algal blooms in South Carolina residential and golf course ponds. *Popul. Environ.* **2003**, *24*, 387–413. [CrossRef]
24. Welch, E.B.; Cooke, G.D. Internal phosphorus loading in shallow lakes: Importance and control. *Lake Reserv. Manag.* **2005**, *21*, 209–217. [CrossRef]
25. Qin, B.; Paerl, H.W.; Brookes, J.D.; Liu, J.; Jeppesen, E.; Zhu, G.; Zhang, Y.; Xu, H.; Shi, K.; Deng, J. Why Lake Taihu continues to be plagued with cyanobacterial blooms through 10 years (2007–2017) efforts. *Chin. Sci. Bull.* **2019**, *64*, 354–356. [CrossRef]
26. Lou, Y.; Zhang, J.; Zhao, L.; Li, Y.; Yang, Y.; Xie, S. Aerobic and nitrite-dependent methane-oxidizing microorganisms in sediments of freshwater lakes on the Yunnan Plateau. *Appl. Microbiol. Biotechnol.* **2015**, *99*, 2371–2381. [CrossRef]
27. Zou, L.; Li, X.; Yang, K. The temporal and spatial distribution of precipitation over Kunming. *Yunnan Geogr. Environ. Res.* **2014**, *26*, 68–78. (In Chinese)
28. Yang, L. The Preliminary study on the original classification and distribution law of lakes in the Yunnan Plateau. *Trans. Oceanol. Limnol.* **1984**, *1*, 34–39. (In Chinese)
29. Zhu, H. Formation of fault lakes in Yunnan Province, and its sediment and evolution in lake cenozoic. *Oceanol. Limnol. Sin.* **1991**, *22*, 509–516. (In Chinese)
30. Ley, S.; Yu, M.; Li, G.; Zeng, J.; Chen, J.; Gao, B.; Huang, H. Limnological survey of the lakes of Yunnan Plateau. *Oceanol. Limnol. Sin.* **1963**, *5*, 87–114. (In Chinese)
31. Dong, Y. Research and development of algae in the nine plateau lakes in Yunnan. *Environ. Sci. Surv.* **2014**, *33*, 1–8. (In Chinese)
32. Zhang, X.; Liu, L.; Ma, S. Projects for comprehensive pollution treatment for the Xingyun Lake catchment. *J. Lake Sci.* **2001**, *13*, 175–179. (In Chinese)
33. Zhang, X.; Xu, H. Ecological preservation and pollution treatment of Xingyun Lake. *Environ. Sci. Surv.* **2014**, *33*, 20–24. (In Chinese)

34. Wu, D.; Zhou, A.; Liu, J.; Chen, X.; Wei, H.; Sun, H.; Yu, J.; Bloemendal, J.; Chen, F. Changing intensity of human activity over the last 2000 years recorded by the magnetic characteristics of sediments from Xingyun Lake, Yunnan, China. *J. Paleolimnol.* **2015**, *53*, 47–60. [CrossRef]

35. Zhou, X. Characterization and sources of sedimentary organic matter in Xingyun Lake, Jiangchuan, Yunnan, China. *Environ. Earth Sci.* **2016**, *75*, 1054. [CrossRef]

36. Zhang, H.; Li, S.; Feng, Q.; Zhang, S. Environmental change and human activities during the 20th century reconstructed from the sediment of Xingyun Lake, Yunnan Province, China. *Quat. Int.* **2010**, *212*, 14–20. [CrossRef]

37. Dong, Y.; Wu, X.; Sheng, S.; Yang, S. Paths for protection and treatment of the nine plateau lakes in Yunnan under ecological civilization development. *Ecol. Econ.* **2014**, *30*, 151–155. (In Chinese)

38. Li, R.; Dong, M.; Zhao, Y.; Zhang, L.; Cui, Q.; He, W. Assessment of water quality and identification of pollution sources of plateau lakes in Yunnan (China). *J. Environ. Qual.* **2007**, *36*, 291–297. [CrossRef]

39. Jin, S.; Jin, X.; Lu, Y.; Lu, H.; Yang, Y.; Li, Y. Research on amount of phosphorus entering into Xingyun Lake caused by phosphorite exploitation and processing in Xingyun Lake watershed. *Environ. Sci. Surv.* **2010**, *29*, 43–45. (In Chinese)

40. Jin, X.; Wang, L.; Qi, Y.; Liu, Y.; Lu, Y. Study on the influence of phosphorite exploration on the total phosphorus pollution in Xingyun Lake. *Environ. Sci. Surv.* **2011**, *30*, 78–80. (In Chinese)

41. Wang, J. In-lake pollution load of Xingyun Lake. *Yunnan Environ. Sci.* **2001**, *20*, 28–31. (In Chinese)

42. Zhang, W.; Ming, Q.; Shi, Z.; Chen, G.; Niu, J.; Lei, G.; Chang, F.; Zhang, H. Lake sediment records on climate change and human activities in the Xingyun Lake Catchment, SW China. *PLoS ONE* **2014**, *9*, e102167. [CrossRef]

43. Zheng, T.; Zhao, Z.; Zhao, X.; Gu, Z.; Pu, J.; Lu, F. Water quality change and humanities driving force in Lake Xingyun, Yunnan Province. *J. Lake Sci.* **2018**, *30*, 79–90. (In Chinese)

44. Wang, Y.; Wang, Z.; Wu, W.; Wang, K.; Liu, Y. Investigating and analyzing plankton and macrophytes in Xingyun Lake. In Proceedings of the 3rd Lake Forum, Wuhan, China, 24 October 2013. (In Chinese).

45. Zhang, X.; Shao, Y.; Chen, Z.; Zhang, F.; Wan, B.; Chen, G.; Nie, J. Response analysis between land use change and water quality in Xingyun Lake Basin. *J. Kunming Metall. Coll.* **2020**, *36*, 77–82. (In Chinese)

46. Ministry of Ecology and Environment of China (MEEC). *Methods of Analysis and Monitoring for Water and Wastewater*, 4th ed.; China Environmental Science Press: Beijing, China, 2002. (In Chinese)

47. Mei, Q.; Hu, Y.; Sun, H.; Ren, X. Relationship mining between rainfall and inflow volume. *China Econ. Informatiz.* **2008**, *26*, 44–45. (In Chinese)

48. Zhao, X. Analysis on correlation between annual runoff and precipitation in Xinjiang. *Water Resour. Dev. Manag.* **2016**, *1*, 73–75. (In Chinese)

49. Wang, Y.; Xu, H.; Cheng, B.; Huang, D.; Tan, Y.; Luo, Y. Impacts of precipitation change on the runoff change in the Fujiang River Basin during the period of 1951–2012. *Progress. Inquis. Mutat. Clim.* **2014**, *10*, 127–134. (In Chinese)

50. Guo, A.; Chang, J.; Wang, Y.; Li, Y. Variation characteristics of rainfall-runoff relationship and driving factors analysis in Jinghe river basin in nearly 50 years. *Trans. Chin. Soc. Agric Eng.* **2015**, *31*, 165–171. (In Chinese)

51. Carlson, R.E. A trophic state index for lakes. *Limnol. Oceanogr.* **1977**, *22*, 361–369. [CrossRef]

52. Wang, M.; Liu, X.; Zhang, J. Evaluation method and classification standard on lake eutrophication. *Environ. Monitor. China* **2002**, *18*, 47–49. (In Chinese)

53. Jin, X. *China Lake Environment*, 1st ed.; China Ocean Press: Beijing, China, 1995. (In Chinese)

54. Wang, H.; Tang, C. Advances on environmental research at Lake Xingyun. *Anhui Agric. Sci. Bull.* **2010**, *16*, 183–185. (In Chinese)

55. Liu, Y.; Chen, G.; Shi, H.; Chen, X.; Lu, H.; Duan, L.; Zhang, H.; Zhang, W. Response of a diatom community to human activities and climate changes in Xingyun Lake. *Acta Ecol. Sin.* **2016**, *36*, 3063–3073. (In Chinese)

56. Liu, Y.; Chen, G.; Hu, K.; Shi, H.; Huang, L.; Chen, X.; Lu, H.; Zhao, S.; Chen, L. Biological responses to recent eutrophication and hydrologic changes in Xingyun Lake, southwest China. *J. Paleolimnol.* **2017**, *57*, 343–360. [CrossRef]

57. Gao, C.; Yu, J.; Min, X.; Cheng, A.; Hong, R.; Zhang, L. Heavy metal concentrations in sediments from Xingyun lake, southwestern China: Implications for environmental changes and human activities. *Environ. Earth Sci.* **2018**, *77*, 666. [CrossRef]

58. Sakamoto, M.; Sugiyama, M.; Maruo, M.; Murase, J.; Song, X.; Zhang, Z. Distribution and dynamics of nitrogen and phosphorus in the Fuxian and Xingyun lake system in the Yunnan Plateau, China. *Yunnan Geogr. Environ. Res.* **2002**, *14*, 1–9.

59. Wu, X.; Qin, J.; Peng, T. Spatial distribution characteristics and different analysis of phosphorus in the overlying water and surface sediments of the Xingyun Lake. *J. Yuxi Normal. Univ.* **2013**, *29*, 12–17. (In Chinese)

60. Yi, D.; Gan, S.; Yu, Z. Relationship between water quality and catchment land use change at Lake Xingyun. *J. Zhejiang Agric. Sci.* **2019**, *60*, 459–461. (In Chinese)

61. Jin, X.; Lu, Y.; Jin, S.; Liu, H.; Li, Y. Research on amount of total phosphorus entering into Fuxian Lake and Xingyun Lake through dry ad wet deposition caused by surrounding phosphorus chemical factories. *Environ. Sci. Surv.* **2010**, *29*, 39–42. (In Chinese)

62. Wang, J.; Zhang, D.; Chen, Y. Livestock waste pollution loading and its environmental impact in Xingyun Lake watershed. *Shanghai Environ. Sci.* **2011**, *30*, 12–17. (In Chinese)

63. Cheng, N.; Liu, L.; Hou, Z.; Wu, J.; Wang, Q. Pollution characteristics and risk assessment of surface sediments in nine plateau lakes of Yunnan Province. *IOP Conf. Ser. Earth Environ. Sci.* **2020**, *467*, 012166. [CrossRef]

64. Liu, S.; Ji, Z.; Pu, F.; Liu, Y.; Zhou, S.; Zhai, J. On phytoplankton community composition structure and biological assessment of water trophic state in Xingyun Lake. *J. Saf. Environ.* **2019**, *19*, 1439–1447. (In Chinese)

65. Li, S.; Xie, P.; Xu, J.; Zhang, X.; Qin, J.; Zheng, L.; Liang, G. Factors shaping the pattern of seasonal variations of microcystins in Lake Xingyun, a subtropical plateau lake in China. *Bull. Environ. Contam. Toxicol.* **2007**, *78*, 226–230. [CrossRef] [PubMed]

66. Chen, P. Brief talk on protection and treatment of Xingyun Lake in Jiangchuan, Yuxi. *Environ. Sci. Surv.* **2018**, *37*, 53–57. (In Chinese)

67. Kagalou, I.; Papastergiadou, E.; Leonardos, I. Long term changes in the eutrophication process in a shallow Mediterranean lake ecosystem of W. Greece: Response after the reduction of external load. *J. Environ. Manag.* **2008**, *87*, 497–506. [CrossRef] [PubMed]

68. Canfield, D.E., Jr.; Hoyer, M.V. The eutrophication of Lake Okeechobee. *Lake Reserv. Manag.* **1988**, *4*, 91–99. [CrossRef]

69. Hansen, P.S.; Phlips, E.J.; Aldridge, F.J. The Effects of sediment resuspension on phosphorus available for algal growth in a shallow subtropical lake, Lake Okeechobee. *Lake Reserv. Manag.* **1997**, *13*, 154–159. [CrossRef]

70. Jiang, Z.; Wu, Y.; Song, X.; Whitmore, T.J.; Brenner, M.; Curtis, J.H.; Moore, A.M.; Engstrom, D.R. Water quality and sediment geochemistry in lakes of Yunnan Province, southern China. *Yunnan Geoglogy* **1997**, *16*, 115–128. (In Chinese)

71. Schindler, D.W.; Hecky, R.E.; Findlay, D.; Stainton, M.P.; Parker, B.R.; Paterson, M.J.; Beaty, K.G.; Lyng, M.; Kasian, S.E.M. Eutrophication of lakes cannot be controlled by reducing nitrogen input: Results of a 37-year whole-ecosystem experiment. *Proc. Natl. Acad. Sci. USA* **2008**, *105*, 11254–11258. [CrossRef]

72. Schindler, D.W.; Hecky, R.E.; McCullough, G.K. The rapid eutrophication of Lake Winnipeg: Greening under global change. *J. Great Lakes Res.* **2012**, *38* (Suppl. S3), 6–13. [CrossRef]

73. Brouwer, E.; Roelofs, J.G.M. Oligotrophication of acidified, nitrogen–saturated softwater lakes after dredging and controlled supply of alkaline water. *Archiv. Hydrobiol.* **2002**, *155*, 83–97. [CrossRef]

74. Smolders, A.J.P.; Lamers, L.P.M.; Lucassen, E.C.H.E.T.; Van der Velde, G.; Roelofs, J.G.M. Internal eutrophication: How it works and what to do about it—A review. *Chem. Ecol.* **2006**, *22*, 93–111. [CrossRef]

75. Mackay, E.B.; Maberly, S.C.; Pan, G.; Reitzel, K.; Bruere, A.; Corker, N.; Douglas, G.; Egemose, S.; Hamilton, D.; Hatton-Ellis, T.; et al. Geoengineering in lakes: Welcome attraction or fatal distraction? *Inland Waters* **2014**, *4*, 349–356. [CrossRef]

76. Lürling, M.; Mackay, E.B.; Peitzel, K.; Spears, B.M. Editorial—A critical perspective on geo-engineering for eutrophication management in lakes. *Water Res.* **2016**, *97*, 1–10. [CrossRef] [PubMed]

77. Rodriguez-Vazquez, K.; Arganis-Juarez, M.L.; Cruickshank-Villanueva, C.; Dominguez-Mora, R. Rainfall-runoff modelling using genetic programming. *J. Hydroinform.* **2012**, *14*, 108–121. [CrossRef]

78. Chou, C. Enhanced accuracy of rainfall-runoff modeling with wavelet transform. *J. Hydroinform.* **2013**, *15*, 392–404. [CrossRef]

MDPI

St. Alban-Anlage 66

4052 Basel

Switzerland

Tel. +41 61 683 77 34

Fax +41 61 302 89 18

www.mdpi.com

Water Editorial Office

E-mail: water@mdpi.com

www.mdpi.com/journal/water